Statistical Modelling of Occupant Behaviour

Do you have data on occupant behaviour, indoor environment or energy use in buildings? Are you interested in statistical analysis and modelling? Do you have a specific (research) question and dataset and would like to know how to answer the question with the data available?

Statistical Modelling of Occupant Behaviour covers a range of statistical methods and models used for modelling energy- and comfort-related occupant behaviour in buildings. It is a classical textbook on statistics, including many practical examples related to occupant behaviour that are either taken from real research problems or adapted from such.

The main focus is traditional statistical techniques based on the likelihood principle that can be applied to occupant behaviour modelling, including:

- General, generalised linear and survival models
- Mixed effect and hierarchical models
- Linear time series and Markov models
- Linear state space and hidden Markov models
- Illustration of all methods using occupant behaviour examples implemented in R

The built environment affects occupants who live and work in it, and occupants affect the built environment by adapting it to their needs – for example, by adapting their indoor environments by interacting with building components and systems. These adaptive behaviours account for great uncertainty in the prediction of building energy use and indoor environmental conditions. Occupant behaviour is complex and multi-disciplinary but can be successfully modelled using statistical approaches.

Statistical Modelling of Occupant Behaviour is written for researchers and advanced practitioners who work with real-world applications and modelling of occupant data. It describes the kinds of statistical models that may be used in various occupant behaviour modelling research. It gives a theoretical overview of these methods and then applies them to the study of occupant behaviour using readily replaceable examples in the R environment that are based on actual and experimental data.

Statistical Modelling of Occupant Behaviour

Jan Kloppenborg Møller, Marcel Schweiker,
Rune Korsholm Andersen, Burak Gunay,
Selin Yilmaz, Verena Marie Barthelmes
and Henrik Madsen

CRC Press
Taylor & Francis Group
Boca Raton London New York

CRC Press is an imprint of the
Taylor & Francis Group, an **informa** business
A CHAPMAN & HALL BOOK

First edition published 2024
by CRC Press
2385 Executive Center Drive, Suite 320, Boca Raton, FL 33431, U.S.A.

and by CRC Press
4 Park Square, Milton Park, Abingdon, Oxon, OX14 4RN

CRC Press is an imprint of Taylor & Francis Group, LLC

ISBN: 978-1-032-33460-8 (hbk)
ISBN: 978-1-032-69981-3 (pbk)
ISBN: 978-1-003-34081-2 (ebk)

DOI: 10.1201/9781003340812

Typeset in Computer Modern
by KnowledgeWorks Global Ltd.

Publisher's note: This book has been prepared from camera-ready copy provided by the authors.

Contents

Preface

This book is the result of collaboration between researchers with background in occupant behaviour and statistical modelling. The group was formed in connection with Annex 66 (http://annex66.org), on "Definition and Simulation of Occupant behaviour". Large parts of the book are adapted from Madsen and Thyregod (2011), which is written by one of the authors.

The overall idea of the book is to present a textbook covering a range of statistical methods and models used for modelling energy-related occupant behaviour in buildings. As such, it is a classical textbook on statistics, but combining the expertise from all authors, the book provides many practical examples related to occupant behaviour, that are either taken from real research problems or adapted from such.

The book is aimed at researchers and advanced practitioners working with practical applications and modelling of occupant data; hence, the book is driven by practical examples including R-code. The focus of the book is classical statistical methods, which have been used for statistical modelling of occupant behaviour or could/should be used for occupant modelling.

As mentioned, the examples are for the most part taken for practical research problems, and the theory is motivated by the structures that are needed in order to model the data. Hence, we believe that the book will be useful for solving practical problems within the domain of occupant modelling. The structure of the book is organised according to methods which relate clearly to structures in data that should be included in the modelling.

R-scripts and publicly available data used for this book will appear on;
`https://www2.imm.dtu.dk/~jkmo/`.

Author

Jan Kloppenborg Møller, PhD, Engineering, is an associate professor in stochastic dynamical systems at the technical university of Denmark (DTU). He received a master's of applied mathematics from DTU (2006), and a PhD in engineering from DTU (2011). His research is concentrated on modelling and forecasting of (continuous or discrete time) stochastic dynamical systems. He has been involved in projects and worked on many different systems such as ecosystems, urban drainage, waste water treatment, wind and solar power forecast and occupancy behaviour. In all cases, focus has been on models that have a clear interpretation of states and parameters.

Møller has taught six different courses in statistics on levels ranging from bachelor's to PhD level. This includes a redesign of the introductory statistics courses at DTU, and the redesign of a second-level course in statistics. His research concentrates on stochastic dynamical systems, using the appropriate tools and pushing advanced methods (like stochastic differential equation) towards application areas such as: wind power forecast, electricity load and occupant behaviour.

Marcel Schweiker is a full professor leading the Teaching and Research Area Healthy Living Spaces at the Institute for Occupational, Social and Environmental Medicine at University Hospital RWTH Aachen in Germany. He received a degree in Architecture from the University of Kassel in Germany, and spent four years at Tokyo City University in Japan, where he received a PhD in Environmental and Information Sciences in 2010. Until 2020, he was a researcher and lecturer at the Department of Architecture at the Karlsruhe Institute of Technology in Germany, and received from there the postdoctoral lecture qualification (habilitation) in 2017. In parallel, between 2019 and 2020, he had a temporary professorship for Energy and Technology at the Faculty of Architecture, TU Kaiserslautern in Germany.

Dr. Schweiker's research and teaching surround several disciplines including: architecture, engineering, medicine, and psychology. His core interests are in improving our understanding of indoor environmental conditions promoting well-being and health within the built environment. He was in charge of building the LOBSTER, a unique facility for experimental studies, and is a sub-task leader in two international working groups under the umbrella of the International Energy Agency – Energy in Buildings and Communities program. He is a member of DIN/ISO standardization committees on ergonomics of the physical environment, has published numerous research articles, and has acquired project funding from local sources, such as the Heidelberg Academy of Sciences and Humanities, national sources, such as the DFG, BMBF, BMWK, and international sources, such as the EU.

Rune Korsholm Andersen, PhD, is a senior researcher at the International Centre for Indoor Environment and Energy, Department of Environmental and Resource Engineering at the Technical University of Denmark (DTU). He received a Master of Science in Energy Engineering (2005) from DTU, and a PhD in indoor environment and occupant behaviour (2009) from DTU.

His research focuses on understanding, modelling and influencing the impact of occupants' behaviour on the indoor environment, and energy use in buildings. He uses long and short-term field measurements, questionnaire surveys, semi-structured interviews, building energy performance simulations and chamber studies with human subjects, combined with statistical analysis and modelling, to gain insight into the understanding of the underlying reasons for observed behaviour patterns. He has published numerous research articles, and acquired research funding from Danish ministries, state-financed research programs, private research foundations and EU funding sources.

Burak Gunay, PhD, PEng is an Associate Professor in the Department of Civil and Environmental Engineering at Carleton University, Ottawa, Canada. He holds a BASc degree in Civil Engineering from Bogazici University, Istanbul, Turkey (2009), an MASc (2011) and a PhD (2016) in Civil Engineering from Carleton University, Ottawa, Canada. He is the recipient of multiple prestigious awards including seven best paper awards and the Governor General's Gold Medal for his doctoral research. He is the principal investigator of the Datadriven Building Operation and Maintenance Laboratory. He has been affiliated with relevant professional organisations, and has held several leadership positions such as elected board member (2014–2024) and president (2022–2024) of the Canadian Chapter of the International Building Performance Simulation Association. He was a subtask co-leader at the International Energy Agency's Energy in Buildings and Communities Programme's Annex 79 (2019–2023).

His research examines methods to optimise the operation of buildings for comfort, energy use and demand flexibility. He uses operational data, gathered inside modern automation and control networks, to learn from the occupants' comfort, behaviour and presence patterns. Using operational data, he employs inverse modelling to characterise building equipment operation and the envelope performance. Aside from field-scale data analyses, his research commonly uses building performance simulation.

Selin Yilmaz is an assistant professor at the Institute of Geography and Sustainability at the University of Lausanne (UNIL). She is an experienced socio-technical researcher with extensive competencies in both theoretical and applied research in the field of energy, with a strong interest in environmental issues and sustainability transitions. She runs a research programme at University of Lausanne with inter- and trans-disciplinary approaches. She focuses actively on methods such as participatory methods, co-design, and common governance methods as part of Living Labs ecosystems, working to improve conceptually and empirically sustainability experimentation, with the goal of tackling social change for sustainability, and developing understanding of transition phenomena in various contexts and fields of study.

She conducts forward-looking research with in-depth analyses of institutions and governance, and practices changes in energy transition experiments to explore the possibility of how these processes and practices can be improved and promoted to accelerate just and inclusive energy transition towards sustainability. She zooms in and zooms out into micro and macro processes to arrive at more holistic research strategies that offer more comprehensive, valid explanations for energy transformations.

Verena Marie Barthelmes, PhD, is System Engineering Project Manager for building automation control in large scale projects at Siemens Schweiz AG, Switzerland. After receiving an MSc in Architecture, she pursued her PhD in Energetics at the Energy Department (DENERG) of the Polytechnic University of Turin, Italy, in close collaboration with the University of Cambridge, UK, and the Technical University of Denmark (DTU). Her doctoral studies revolved around occupant behaviour analysis and modelling, with specific focus on Bayesian network approaches, and survival analysis of national time-use survey data.

She was actively involved in European and national behavioural change programs, with the aim of defining ICT-based tools and feedback strategies towards empowering individuals and communities to adopt energy-friendly, healthy practices. From 2019 to 2022, she was a Postdoctoral Researcher at the Éole Polytechnique Fédérale de Lausanne (EPFL), Switzerland. Key focus was the analyses of cause-effect relationships between the human–building interaction, combined indoor environmental factors, and energy use in buildings based on a wide range of objective and subjective surveying methods. She is actively involved in national and international research projects surrounding the topics of human–building interaction and building automation control.

Henrik Madsen, in 1986, received a PhD in Statistics at the Technical University of Denmark (DTU). He was appointed Ass. Prof. of Statistics in 1986, Assoc. Prof. in 1989 and Professor of Mathematical Statistics with a special focus on Stochastic Dynamical Systems in 1999. In 2017, he was appointed Professor II at the Norwegian University of Science and Technology (NTNU) in Trondheim. His main research interest is analysis and modelling of stochastic dynamics systems. This includes signal processing, time-series analysis, identification, estimation, grey-box modelling, prediction, optimisation and control. The applications are mostly related to Energy Systems, Informatics, Environmental Systems, Bioinformatics, Biostatistics, Process Modelling, Finance and Indoor Climate. He has received several awards. In June 2016, he was appointed Knight of the Order of Dannebrog by Her Majesty the Queen of Denmark, and in June 2017, was appointed Doctor HC at Lund University.

He has authored or co-authored approximately 650 papers and 12 books. The most recent books are: *Time Series Analysis* (2008), *General and Generalised Linear Models* (2011), *Integrating Renewables in Electricity Markets* (2013) and *Statistics for Finance* (2015).

1

Introduction

The purpose of this chapter is two–fold; 1) to motivate the work on occupant models, and set the general context of the state of occupant modelling in historical context and 2) to briefly introduce the concrete content of the book and how to use it from the perspective of data and the perspective of models covered in the various chapters of the book.

1.1 General motivation

In general, the built environment affects people who live and work in it, and people affect the built environment by adapting it to their needs. Hence, understanding of how people interact with the environment and how the environment affects people is important. Such understanding come from collected data, either in controlled experimental settings or by observations made directly in the building that are operational and in use. In particular, the last category is emerging with the price of sensors decreasing. We start by a short historical summary of occupant modelling.

1.1.1 History and trends

Occupants adapt their indoor environments by interacting with building components and systems such as windows, thermostats, electric lighting, and window shades, or adapt to the indoor environment through actions such as adjusting clothing insulation levels and drinking hot/cold beverages (Haldi and Robinson, 2008; Nicol, 2001). All these actions, broadly classified as adaptive behaviours, directly or indirectly affect a building's energy use. Other than adaptive behaviours, task-driven behaviours such as plug-in equipment/ appliance use also affects energy performance. Finally, the presence of the occupants affects both the energy performance and the indoor environmental quality of indoor spaces. As expected, occupant behaviour patterns are far from being deterministic and exhibit significant variation from one occupant to another. An added layer of uncertainty emerges from the randomness in occupancy – and the obvious dependency of occupant behaviours on their presence. Although Mahdavi et al. (2021) found that direct evidence is lacking, it is widely recognised that occupancy and occupant behaviour account for great uncertainty in building energy use (Gram-Hanssen, 2010; Haldi and Robinson, 2011; Calì et al., 2016). This highlights the necessity to develop statistical models for occupant behaviour and occupancy, which enable design decisions informed by accurate building energy models.

Earliest models for occupant behaviour and occupancy date back to the late 1970s. Hunt (1979) developed a stochastic Markov probit regression model to predict occupants' light switch behaviour. This was followed by only a few others until the 2000s: Rea (1984) and Newsham (1994) on window shade use, and Love (1998) and Maniccia

DOI: 10.1201/9781003340812-1

et al. (1999) on light use. Starting with the 2000s, occupant modelling has expanded into the thermal domain, and we began seeing models for occupancy (Page et al., 2008; Wang et al., 2005), window use (Yun and Steemers, 2008; Rijal et al., 2009; Andersen, 2009), personal heater and fan use (Nicol, 2001), air-conditioning use (Schweiker and Shukuya, 2009), and clothing levels (Haldi and Robinson, 2008) as well as new models for lighting and shade use (Reinhart, 2004; Mahdavi et al., 2008). When we come to the 2010s, there is a thriving research community on occupant modelling, collaborating primarily within IEA EBC Annexes 53 (Yoshino et al., 2017), 66 (Yan et al., 2017), and 79 (O'Brien et al., 2020), and ASHRAE Technical Committee 7.10 to address research challenges in harmony. Currently, a large group of researchers are active in occupant modelling. This trend continues with the availability of low-cost occupant sensing and data archiving solutions for building automation, controls, and monitoring, advances in programming environments for statistical analysis and modelling, and accumulation of this expertise in the occupant modelling community. A surge of research on the topic has occurred over the past four decades. A few fundamental research inquiries in occupant modelling are: Which model formalisms should we be using for occupant behaviour and occupancy modelling? A wide range of approaches has been studied so far and deemed appropriate for certain types of behaviour. Some examples are discrete-time or -event Markov logistic regression models (Haldi and Robinson, 2010; Rijal et al., 2008), Bernoulli random processes as logistic regression models (Haldi and Robinson, 2008), Markov Chains (Wolf et al., 2019c), Inhomogeneous Markov Chains (Andersen et al., 2014), Hidden Markov Models (Liisberg et al., 2016), indirect methods based on state-space models (Wolf et al., 2019a; Wolf et al., 2019b), survival models (Haldi and Robinson, 2009), Bayesian inference (Lee et al., 2019), and machine learning (Kim et al., 2018; Markovic et al., 2018).

How should inter-occupant diversity be represented in occupant models? For example, (O'Brien et al., 2017) and (Haldi et al., 2017) examined different mixed-effects models to this end.

How should predictors be selected in occupant modelling? While the early focus is on univariate or multivariate modelling with primary factors affecting a single domain of indoor environmental quality (thermal, visual, and acoustic comfort, and indoor air quality), a recent focus has been on capturing factors overarching across multiple domains (Schweiker et al., 2020).

1.1.2 Current motivations

As mentioned, occupant behaviour is now widely recognised as a major contributing factor to uncertainty of building performance, predicting in building energy use and consumption, and indoor environments. Therefore, modelling the occupant behaviour, their interaction with the environment and building components is a key issue for building design optimisation, energy diagnosis, performance evaluation, and building energy simulation. Occupant behaviour is complex and multi-disciplinary, but can be successfully modelled using a stochastic approach. In contrast to other scientific disciplines, the research on occupant behaviour requires models with both explanatory and predictive power, which represents a particular challenge. Current motivations for statistical approaches for modelling occupant behaviour include:

- Identify and understand the probabilistic and complex relationships between a wide range of driving factors and human perceptions and (adaptive or non-adaptive) actions within the built environment

- Identify occupant movement and presence, and occupant actions between spaces both within the built environment including residential and commercial sectors to predict indoor environment conditions (e.g. indoor environment quality) and energy use

- Integrate human-related factors in building energy simulation programs to reduce the gap between real and predicted building energy performance
- To use the computational intelligence (e.g. machine learning) methods in building energy performance that would complement knowledge-driven models describing physical behaviour
- Gain deeper insights into human-building interaction and move forwards to an effective building operation and design.

1.1.3 Scope of the book

Despite significant efforts and advancements in literature and a further increasing set of mathematical approaches to capture the human dimension in buildings, the development and practical application of occupant behaviour models in building simulation, design and operation is somewhat exceptional. Even amongst researchers, the replication of modelling approaches and/or an accurate comparison of findings remain challenging since details on statistical computation and coding (e.g.in R) in combination with underlying open data sources are frequently not available or insufficiently reported. For researchers and practitioners without a strong statistical or computational background, the ad-hoc development of statistical models based on own collected data and/or the replication of existing models can result in a time–consuming process that especially in building practice might allegedly not constitute the priority. Examples of statistical modelling approaches directly applied to the topic of occupant behaviour – accompanied with statistical computing solutions and underlying example datasets – might facilitate the choice of the statistical approach according to desired outcomes and available datasets, as well as its practical development in a computational tool.

This book first provides an overview of commonly used statistical approaches in the field of occupant behaviour research. It furthermore describes what types of statistical models can be applied to different types of occupant behaviour–related modelling studies. It provides the theoretical description of these approaches which are accompanied with easily replaceable examples (based on real-world and experimental data) in the R environment directly applied to the topic of occupant behaviour.

Next to all readers interested in the topic, this book is particularly intended to a specific audience that includes:

- Researchers and "advanced" practitioners researching, studying occupant behaviour modelling in built environment and energy use.
- Civil, mechanical and building engineering to model occupant behaviour with regard to better building design and operation.

This book focuses on statistical models designed for inference about the relationships between variables based on mathematical equations. It largely excludes issues around computer intelligence and currently developed methods such as machine learning which deserve their own book. All models and examples are executed in R. Other languages, such as Python or Matlab, could be used to produce similar models, but an extra programming effort is required when choosing another language.

We believe that for an efficient reading of this book, familiarity with basic notions of statistics and basic knowledge of the R environment are beneficial. There are many online sources introducing R based on individual level, for example if one is familiar with another programming language, but also introductory books on statistics, for example, Brockhoff et al. (2015), based on R may be consulted.

We assume the following basic knowledge:

- Basic notion of statistical concepts is beneficial.
- Basic notion of R environment is beneficial.
- Basic knowledge of linear algebra (like matrix multiplication, and matrix inverse).

The basic knowledge from statistics and R could be obtained from an R-based introductory course in statistics (like e.g. https://02402.compute.dtu.dk/), but can of course also be obtained separately, for example; https://www.r-project.org/ contains tutorials on the R-programming language. In some of the more advanced parts of the book some basic knowledge of programming will also be an advantage. The basic knowledge from linear algebra should be covered by a basic math course for engineers.

1.2 How to use this book?

Having this book in your hands might have several reasons. You may be interested in learning the statistics required for occupant behavioural modelling from scratch or you have a specific (research) question and dataset and would like to know how to answer the question with the data available.

The first group of readers is more than welcome continuing reading from here on towards the last line of the book. For the latter group of readers, Figure 1.1 is intended to guide you through a set of questions in order to understand which statistical method is most suitable for your dataset.

Figure 1.1 visualises the series of questions that the reader should answer in order to identify the appropriate chapter. The listing below elaborates shortly on each of the questions and how to identify the answer:

1. **Can you assume Gaussian data?** In some cases data are clearly not Gaussian and we cannot transform data to be approximately Gaussian (e.g. binary outcomes). In other cases, e.g. positive and skewed data, we may be able to transform data to be approximate Gaussian (Chapter 3), or we may use e.g. the gamma distribution (Chapter 4). Readers who are unsure about distribution types are invited to read the first part of the book (Chapters 2–4, and Section 1.3 of the present chapter).

2. **Is data a time–series?** If there is a timestamp associated with each measurement and an expected correlation between time–points, then time series analysis should be applied.

3. **Do you assume random effects?** This is mostly a question of how to interpret the model and the features you are interested in extracting from the model, e.g. you might have multiple observations from a number of subjects; if you are interested in the particular subjects then the general linear model (Chapter 3) is the right choice. If you are more interested in the average behaviour, then mixed effect models (Chapter 5) is the right choice.

4. **Are there unobserved states?** Do you have an underlying state that determines the observed time series, e.g. you observe window interactions that may be related to temperature or number of occupants present in the room (that is not observed), or the other way around you observe temperature and want to determine the state of the window.

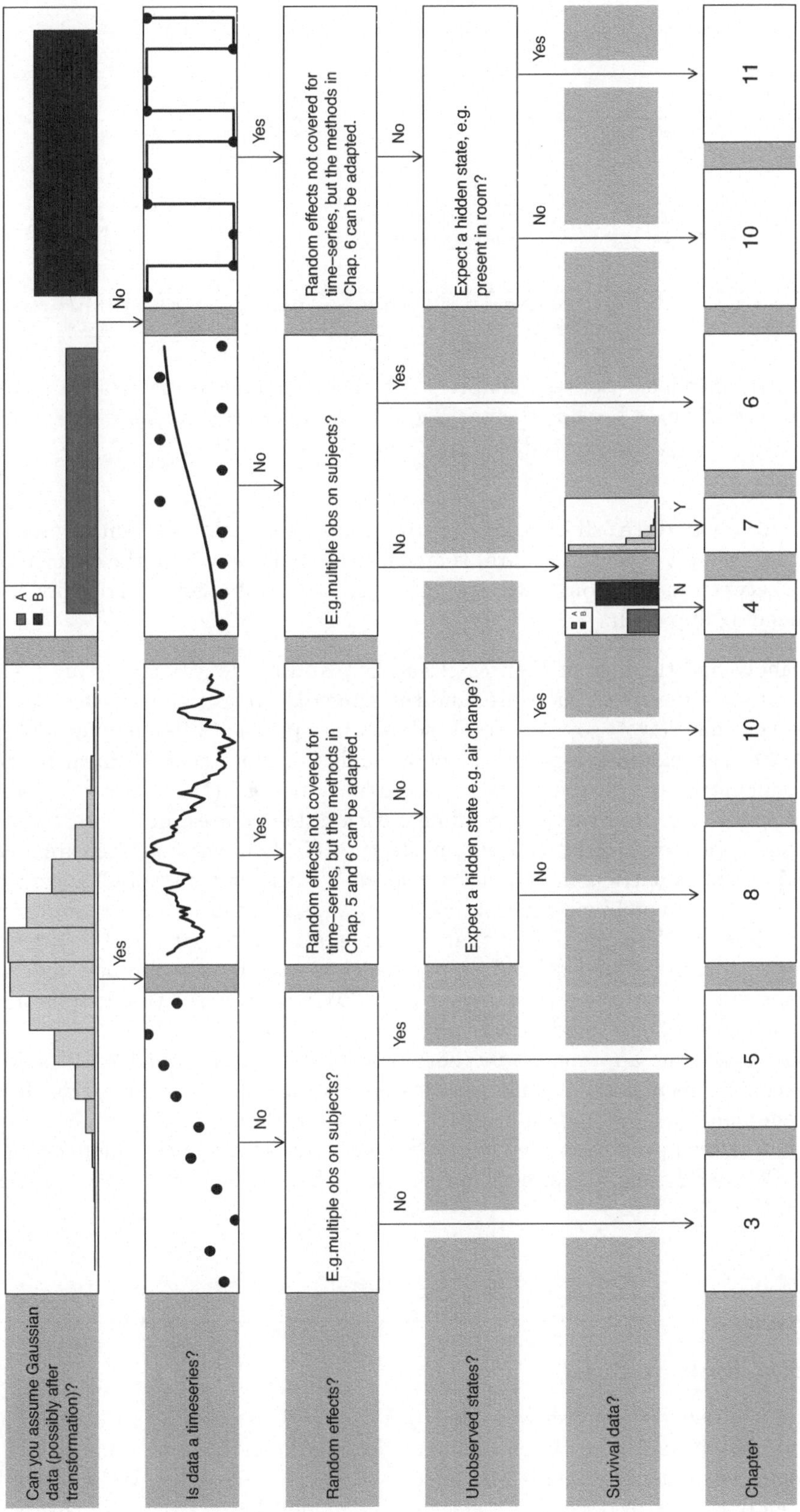

Figure 1.1: Guide to determine appropriate model based on data to be analysed.

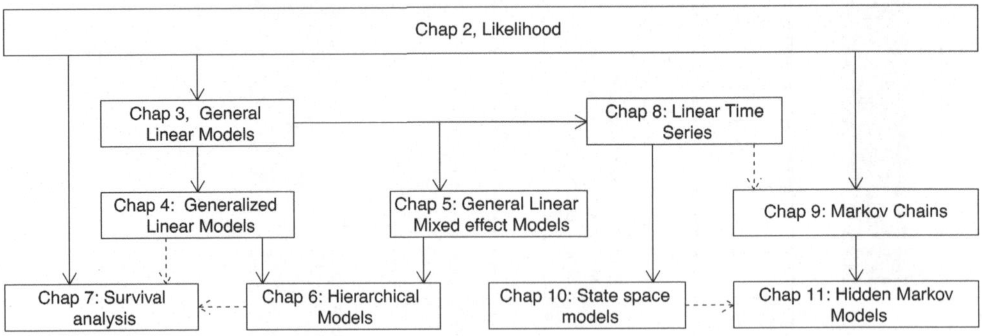

Figure 1.2: Interdependence between chapters. The figure should be understood as Chapter 2 is needed before Chapter 3, while there is not a direct link between Chapters 8 and 9, but the more general concepts are introduced in Chapter 8.

5. **Is data survival data?** If the data are records of the time between events and, in particular, if there are censored data, then survival analysis will be the right approach; e.g. time between interactions with a window then survival analysis models will be the right modelling approach.

The list above and the flow of Figure 1.1 can help you to identify a starting point, but modelling is often an iterative procedure, and the same data might be modelled in different ways. Hence you may have to revisit the fundamental approach after having looked at a simple approach. For example, in depth residual analysis from a general linear model may reveal that some structure in the covariance matrix is needed (i.e. Chapter 5). You also notice "black spots", e.g. survival data combined with hidden states, in Figure 1.1, another example is time–series combined with random effect; this is, however, to some extent covered in Chapters 5 and 6. In particular, the method covered in the first part of Chapter 6 can be extended to cover most combinations from Figure 1.1, i.e. most types of data and structure that one could imagine. You may notice that the explanatory or input variables are not part of the figure; this is intentional as it does not really matter for the methodology though these are sometimes given different names e.g. ANOVA, and ANCOVA (which are both included in the general linear model).

Once you realised the appropriate method for your dataset, Figure 1.2 will guide your reading. As can be seen from the figure the overall principle in the book is likelihood estimation, and hence readers unfamiliar with the concept should read Chapter 2 first, but while there is a natural progression in the book some chapters are fairly independent of the rest (except Chapter 2), e.g. Chapter 7, on the other hand Chapter 6 draws on concepts introduced in Chapters 3–5.

1.3 A first look at data

In general we will denote response variables by Y or Y_i. The use of capital letter stresses that these are random variables. When fitting models we do not have random variables, but rather outcomes of random variables, the latter will be denoted by y_i. We will denote the collection of all outcomes by $\boldsymbol{Y}$ or $\boldsymbol{y}$. Essentially any model that is treated in this book can be written in the generalised way

$$\boldsymbol{Y} \sim D(\boldsymbol{\mu}, \boldsymbol{\Sigma}), \tag{1.1}$$

where D is short for some distribution, $\boldsymbol{\mu}$ is the mean value (that often depends on some input variables) and $\boldsymbol{\Sigma}$ is related to the interdependence structure of the data (e.g. a variance-covariance structure).

The above implies that the modeller's task is to choose

- a distribution function (D)
- a mean ($\boldsymbol{\mu}$), this might be a function of some predictors
- an interdependence structure

With the distribution, the mean and the interdependence structure in place, parameters can be estimated and the model may be applied for the required purpose e.g. simulation, prediction, control, identifying or testing effects of interventions etc.

For the first part, we should identify the type of data/outcomes we are modelling. There are two overall types of data; these are continuous and discrete data, further for each type there are subgroups, as listed below.

1.3.1 Continuous data/distributions

Continuous data is data that originates from the set of real numbers or subsets of the real numbers. Formally, continuous data are uncountable, so we can have any outcome in some interval of the real line. Examples are:

i) *Continuous data* (e.g. $y_1 = 2.2143$, $y_2 = -0.453$, $y_3 = 1.2153$, $\ldots$, $y_n = 0.8$). Examples of distributions are

   - Normal (Gaussian). Used, e.g. for air temperatures in degrees Celsius.
   - Student-t distribution (heavy tailed distribution).

ii) *Continuous positive data* (e.g. $y_1 = 0.0238$, $y_2 = 1.0322$, $y_3 = 0.0012$, $\ldots$, $y_n = 0.8993$). Examples of distributions are

   - Log-normally distributed, often used for concentrations.
   - Gamma distribution, as the log-normal but different variance scaling.
   - Exponential distribution, often used for waiting times.
   - Weibull distribution, often used for waiting times. The Weibull is a generalisation of the exponential distribution.

iii) *Continuous data from an interval*, e.g. (0,1). An example is

   - Beta distribution, often used in combination with the binomial distribution to model probabilities.

The list of examples above does neither encompass all distributions functions that could be relevant nor are the categories completely defined by the way data is collected. An example could be the height of adult males. This will clearly be positive but often one would use the normal distribution to model this because the variation is symmetric around an overall mean.

1.3.2 Count and discrete data/distributions

Count data and discrete data originate from data that have a discrete or countable number of outcomes, i.e. all natural numbers or a subset of the natural numbers. Examples are

i) *Count data* without a natural upper bound (e.g. $y_1 = 57, y_2 = 67, y_3 = 54, \ldots, y_n = 59$). Examples of distributions are

- Poisson distributed, used, e.g. for number of accidents, number of interactions with a window in a fixed time interval.
- Negative binomial, as above but with more variability than implied by the Poisson distribution.

ii) *Binary (or quantal) data* (e.g. $y_1 = 0,\ y_2 = 0,\ y_3 = 1,\ \ldots,\ y_n = 0$), or proportion of counts (e.g. $y_1 = 15/297,\ y_2 = 17/242,\ y_3 = 2/312,\ \ldots,\ y_n = 144/285$). Examples of distributions are

- Binomial distribution, number of heads in ten tosses with a coin, have there been an interaction with the window within the last 10 minutes?
- Beta-binomial, as above but with a higher variability than implied by the binomial distribution.

iii) *Categorical data*, used when there is no numerical interpretation of the outcomes; these can be subdivided into

- Nominal, there is no natural ordering of outcomes (e.g. "red", "blue", "green").
- Ordinal, there is a natural ordering of outcomes (e.g. "Very unsatisfied", "Unsatisfied", "Neutral", "Satisfied", "Very satisfied").

In both cases, the multinomial distribution would be the first choice.

The data and the story of how data is obtained should guide us towards not only the choice of distribution but also quantitative methods should be taken into account when choosing the appropriate distribution. Both will be discussed throughout this book.

1.4 Introduction to the chapters: A first view on the models

The overall principle of the book is the likelihood principle and parameter estimation in all other models covered in the book can be seen as special cases of likelihood estimation. As outlined in the previous section the type of data influences the choice of distribution, and in addition the structure of the data also influences the structure of the model. The choice of methods, distributions and structures covered in this book is reflected in the chapters that are introduced below.

1.4.1 Likelihood-based estimation (Chapter 2)

The likelihood principle applies to general models where we assume some distribution for the observed data. An example on blind use is given in Figure 1.3 (see also Example 2.2); for this type of data, a first choice would be the exponential distribution, hence

$$Y_i \sim Exp(\theta). \tag{1.2}$$

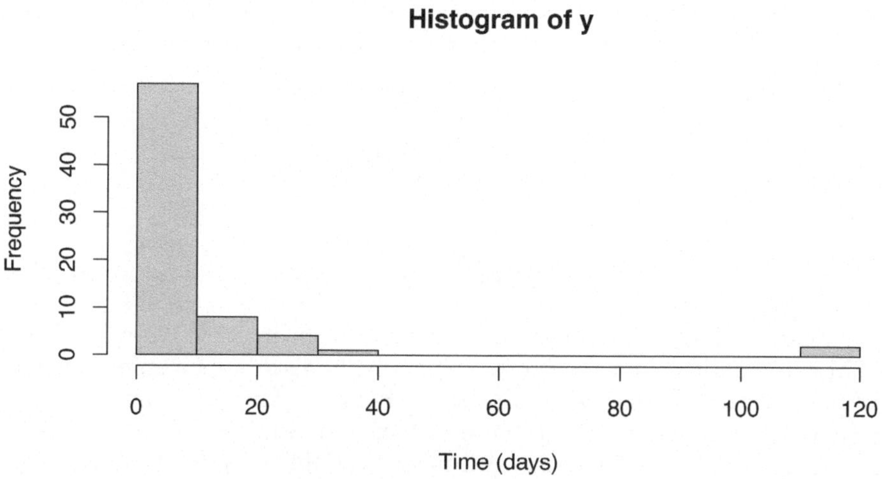

Figure 1.3: Time between interaction with blinds

The distribution assumption given in (1.2) also defines the likelihood, which is a function of the parameters in the model, in this case θ. The likelihood principle tells us that we should find the values of the parameter θ that maximise the joint density given the available data

$$L(\theta; \boldsymbol{y}) = f_Y(\boldsymbol{y} : \theta), \tag{1.3}$$

for independent observations, the joint density is the product of the individual densities

$$L(\theta) = \prod_{i=1}^{N} f_Y(y_i; \theta). \tag{1.4}$$

In addition to giving the point estimate of the parameter of interest, the likelihood function ($L(\theta)$) also contains information on the uncertainty of the parameters. Chapter 2 treats the estimation of parameters and uncertainty evaluation, e.g. standard errors and confidence intervals.

The evaluation of uncertainty might be done by considering the quadratic approximation of the log-likelihood, resulting in the Wald test and confidence interval, or by considering the entire log-likelihood function, resulting in the likelihood ratio test and the profile-likelihood confidence intervals. Both methods are treated in Chapter 2, as well as method for comparing non-nested models (based on so-called information criteria).

All other chapters in this book might be seen as special cases of applying the likelihood principle to specific distribution assumptions and structures of the data.

1.4.2 General linear models (Chapter 3)

In Chapter 3, general linear models (also called classical GLM) are considered. The expected value of the response variable Y can be represented as a linear combination of explanatory variables $(x_1, \ldots, x_k)$.

$$E[Y] = \beta_1 x_1 + \cdots + \beta_k x_k. \tag{1.5}$$

In practice, a number of observations $\mathbf{Y} = (Y_1, \ldots, Y_n)$, along with a vector of explanatory variables $\mathbf{x}^T = (x_{i1}, \ldots, x_{ik})$ are obtained. In order to estimate the unknown parameter vector $\boldsymbol{\beta} = (\beta_1, \ldots, \beta_k)$, the model can be formulated for all n observations.

$$\begin{pmatrix} E[Y_1] \\ \vdots \\ E[Y_n] \end{pmatrix} = \begin{pmatrix} \mathbf{x}_1^T \\ \vdots \\ \mathbf{x}_n^T \end{pmatrix} \boldsymbol{\beta}, \tag{1.6}$$

or

$$E[\mathbf{Y}] = \mathbf{X}\boldsymbol{\beta}, \tag{1.7}$$

where $\mathbf{X}$, the matrix of known coefficients, is called design matrix. We will see that the classical GLM is the natural approach to model normally distributed response variables.

One example of such models is illustrated in Figure 1.4 where a linear relationship between clothing insulation level, temperature and sex is shown together with the estimated model. This simple model contains both a continuous variable (temperature) and a discrete variable (sex). The model in the top left panel is given by

$$clo_i = \beta_0 + temp_i\beta_1 + \gamma(sex_i) + \epsilon_i; \quad \epsilon_i \sim N(0, \sigma^2) \text{ and iid}, \tag{1.8}$$

where "iid" is short for Independent Identically Distributed (see Chapter 2 for a formal definition).

The other panels include more complicated models such as transformations of the dependent variable (e.g. Box-Cox), transformation of the regressors (e.g. polynomials), and/or weighted analysis (e.g. $\sigma_i^2 = \sigma^2(sex_i)$). The figure in addition includes both confidence and prediction interval.

As stated above, all chapters are based on the likelihood principle. In the special case of the general linear model, exact results for likelihood ratio tests are available through the Student-t and the F-distribution. The latter is often referred to as Analysis of Variance or Analysis of Covariance (ANOVA/ANCOVA).

In Chapter 3, we will discuss model building, estimation, residual analysis, prediction, and transformation towards normality (e.g. the Box-Cox transformation). In addition, we will discuss several limitations of the GLM, in particular structures in the data pointing towards the mixed effect type of model discussed in Chapter 5.

1.4.3 Generalized linear models (Chapter 4)

Generalized linear models are an extension of the concept of general linear models. In this class of models, we relax the assumption of normally distributed response variables. Instead, the class of response variables is generalized to the exponential dispersion family, which includes a wide range of distributions including Gaussian, Poisson, binomial and gamma distributions.

The response variable Y can be considered as one of the three components of a generalized linear model. The other two are a linear combination of explanatory variables $\mathbf{X}\boldsymbol{\beta}$ (linear predictor or systematic component) and a link function $g(\cdot)$ which connects the response variable and the linear predictor.

Through transformation by the link function

$$g(E[\mathbf{Y}]) = \mathbf{X}\boldsymbol{\beta}, \tag{1.9}$$

the linear predictor is used as the basic component of the modelling and hence we need to choose a proper link function and the proper linear predictors.

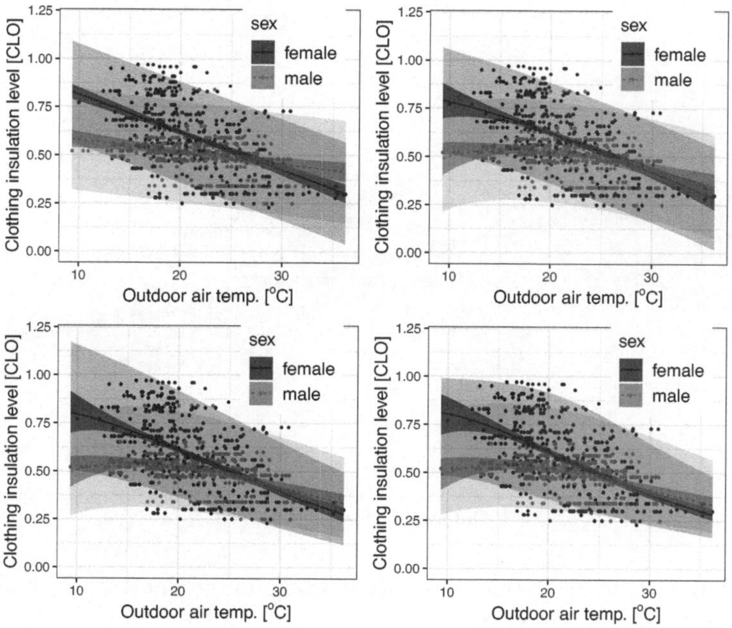

Figure 1.4: Relation between clothing insulation level and temperature for males and females (see also Example 3.3 in Chapter 3).

A special case of the generalized linear models is logistic regression, which corresponds to a binomial distributed response variable. This model finds many applications in occupant behaviour modelling, as it can be used to handle binary response data such as; presence, lighting or window opening/closing data. An example of the link function for binomial response is the logit–link

$$\log\left(\frac{E[\mathbf{Y}]}{1-E[\mathbf{Y}]}\right) = \mathbf{X}\boldsymbol{\beta}. \tag{1.10}$$

In the chapter we will discuss examples using:

- Number of changes in clothing insulation levels during a day (*Poisson*). Example of data: 1,4,3,6,2,..
- Window openings (*binomial*). Example of data: 1,0,0,1,0,1 or 41, 23, 12, 6 out of 98, 35, 18, 25.
- Thermal sensation vote (*multinomial*). Example of data: "satisfied", "very satisfied", "dis satisfied", "neutral",...
- Clothing insulation level (*gamma*). Example of data: 0.18, 0.95, 1.8, 0.75,...

Figure 1.5 features the gamma regression model for clothing insulation level, the particular model is given by

$$\log(\mu_i) = \beta_0(sex_i) + temp_i\beta_1 + \log(temp_i)\beta_2(sex_i) \tag{1.11}$$

$$clo_i \sim Gamma\left(\mu_i, \frac{1}{\lambda}\right), \tag{1.12}$$

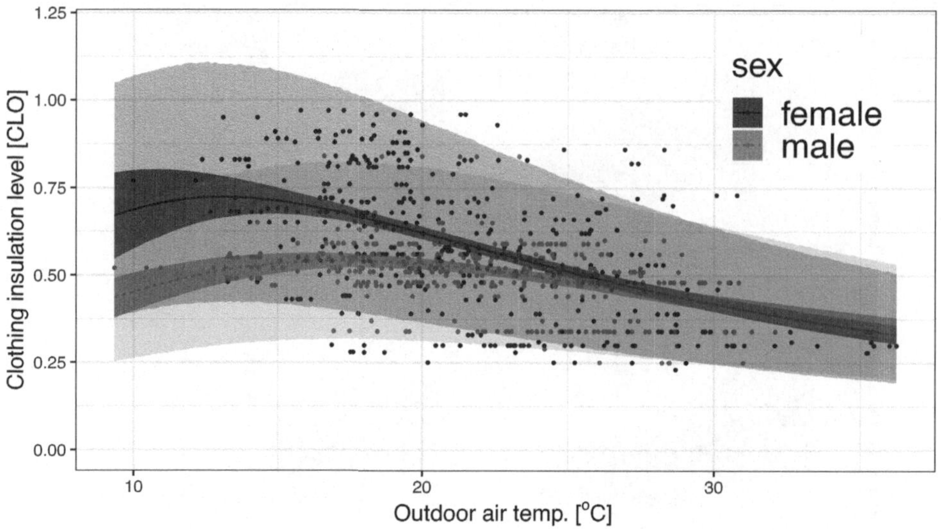

Figure 1.5: Relation between clothing insulation level and temperature for males and females (see also Example 4.9 in Chapter 4).

where the term $\frac{1}{\lambda}$ is related to the variance. As we saw above for the general linear model the figure shows both confidence and prediction intervals.

In Chapter 4, we discuss mostly the same points as for the general linear model. As most results are approximate, some of the specific results are somewhat different and the resulting distributions of the test statistics are approximate and are based on χ^2-distribution, normal, student-t, and F (depending on the specific situation).

1.4.4 Linear mixed effects models (Chapter 5)

In Chapter 5 mixed effects models, which are a natural extension of the general linear models, are considered. This class of models can hierarchically separate different sources of variation for normally distributed response variables. Let us consider heating set-point preferences. Different people certainly have different preferences for indoor temperatures, but also one person may have different preferences on different occasions, varying from one day to another day or along with the outdoor temperature. Therefore, we are dealing with two kinds of variations of different nature. One way to model this is a one-way random effects model.

$$Y_{i,j} = \mu + U_i + \varepsilon_{i,j}, \tag{1.13}$$

with $\varepsilon_{i,j} \sim N(0, \sigma^2)$ and $U_i \sim N(0, \sigma_u^2)$. The random variable U_i determines the deviation from the overall population mean μ for person i, whereas $\varepsilon_{i,j}$ models the variation of this person among different repetitions j. Hence, with mixed effects models, these two kinds of randomness are modelled separately. Note that for observed $U_i = u_i$, the model becomes an ordinary Gaussian model with mean $\mu + u_i$.

More generally, the class of models considered can be written as

$$\boldsymbol{Y} = \boldsymbol{X\beta} + \boldsymbol{ZU} + \boldsymbol{\varepsilon}, \tag{1.14}$$

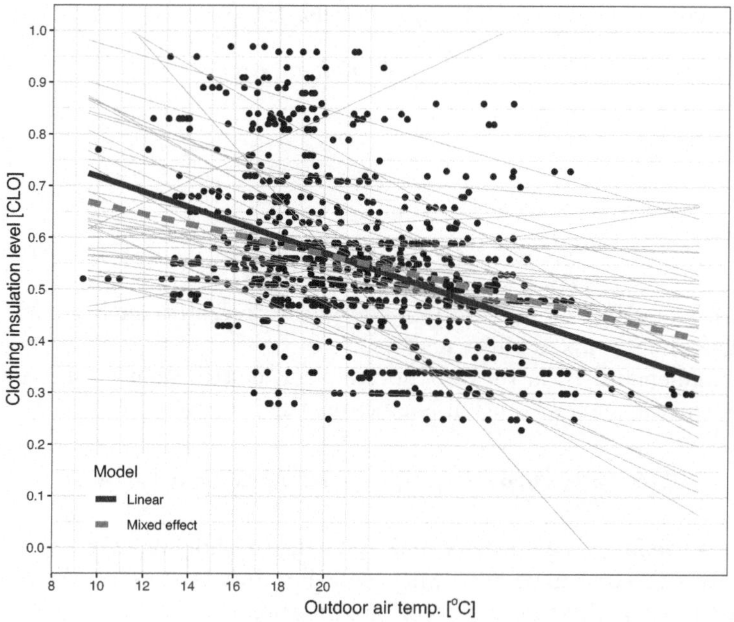

Figure 1.6: Relation between clothing insulation level and temperature, emphasising the individual models (see also Example 5.12 in Chapter 5).

where $\boldsymbol{U}_i \sim N(\boldsymbol{0}, \boldsymbol{\Psi})$ and $\boldsymbol{\varepsilon} \sim N(\boldsymbol{0}, \boldsymbol{\Sigma})$. Figure 1.6 shows the result for a simple mixed effect model of the type

$$clo_{i,j} = \beta_0 + temp_{i,j}\beta_1 + u_{0,i} + temp_{i,j}u_{1,i} + \varepsilon_{i,j}, \tag{1.15}$$

where i refers to the subject and j refers to measurements on the subject and $\boldsymbol{\varepsilon} \sim N(\boldsymbol{0}, \sigma^2\boldsymbol{I})$.

In addition to the estimation, testing and predictions using mixed effect model, we will emphasise structures in data, e.g. observations taken as small time-series (this is the case for the clothing insulation data). Such structures will lead to structure in the $\boldsymbol{\Sigma}$-matrix and we will emphasise how this can be implemented in the mixed effect structure. E.g. the structures present in the clothing insulation data is

- Between subject variation
- Between day variation (subject have observation from multiple days)
- Within day variation/correlation, observation from the same day are highly correlated.

The list above describes a hierarchy of variations and to get a good description of the data, all levels should be included in a reasonable way.

Even though the random effects are not parameters in the model, it is sometimes of interest to estimate them in order to produce figures like Figure 1.6. The chapter is concluded by discussing how those are estimated.

1.4.5 Hierarchical models (Chapter 6)

The chapter covers random effect models with arbitrary first (conditional on random effects) and second stage (random effects) models, hence e.g. the first stage models covered in

Chapter 4 and with arbitrary second stage models. An example of such a model is

$$Y_i|U_i \sim Binom(n_i, p_i) \tag{1.16}$$

with

$$p_i = \frac{e^{\boldsymbol{x}_i^T\boldsymbol{\beta}+U_i}}{1+e^{\boldsymbol{x}_i^T\boldsymbol{\beta}+U_i}} \tag{1.17a}$$

$$U_i \sim N(0,\sigma^2), \tag{1.17b}$$

hence adding a Gaussian random effect to the linear predictor, such models are in general called generalized linear mixed effect models.

The basic challenge here is to find the marginal distribution of the observations (the likelihood); in general, such a solution is given as the integral

$$L(\boldsymbol{\theta}) = f_{\boldsymbol{Y}}(\boldsymbol{y};\boldsymbol{\theta}) = \int f_{\boldsymbol{Y}|\boldsymbol{U}=\boldsymbol{u}}(\boldsymbol{y};\boldsymbol{X},\boldsymbol{u},\boldsymbol{\beta}) f_{\boldsymbol{U}}(\boldsymbol{u};\boldsymbol{\Phi})d\boldsymbol{u}, \tag{1.18}$$

where $f_{\boldsymbol{Y}|\boldsymbol{U}=\boldsymbol{u}}$ is the first stage density and $f_{\boldsymbol{U}}$ is the second stage density. For linear mixed effect models, this integral have an explicit solution, but in general, the solution needs to be approximated, e.g. using the Laplace approximation.

The method of the Laplace approximation can also be used for calculating the likelihood in more general setting using more general first and second stage models or non-linear models. An example of such general models is modelling of window opening, where the first stage model is a binomial with a non-linear model for the "linear predictor", including both additive effects and an estimate of timing of widow opening

$$\eta(\boldsymbol{x}, u) = \beta_0 + \beta_1 \cos\left(\frac{2\pi H}{24} - \beta_3 - u_2\right) + \beta_2 \log(CO_2) + u_1, \tag{1.19}$$

where $U_1 \sim N(0, \sigma_u^2)$ and U_2 follow a periodic distribution, exemplified by the VonMises distribution. An example of the output is shown in Figure 1.7.

Finally, the chapter treat models where the second stage model is chosen such that the integral, in (1.18), allow an explicit solution. An example of such an model is the Poisson-gamma model that can be defined by

$$Y|\gamma \sim Pois(\mu\gamma), \tag{1.20a}$$

$$\gamma \sim G(1, \phi), \tag{1.20b}$$

in which case the marginal is a negative binomial distribution.

The general development through the chapter focuses on choosing the right distribution for the right purpose while including random effects when appropriate. The generalized mixed effect model have the advantage that standard methods are implemented, e.g. through `glmmTMB`, while the other classes of models require more case–specific implementation by the user.

1.4.6 Survival models (Chapter 7)

Survival models are introduced as a class of models to predict the lifetime of an occupancy state (e.g., duration of uninterrupted presence before taking a break) or an operable building component's state (e.g., window state, light state). This class of models has been used in the occupant modelling literature primarily to capture non-adaptive actions, where indoor and outdoor environmental variables fail to explain the observed occupancy and occupant

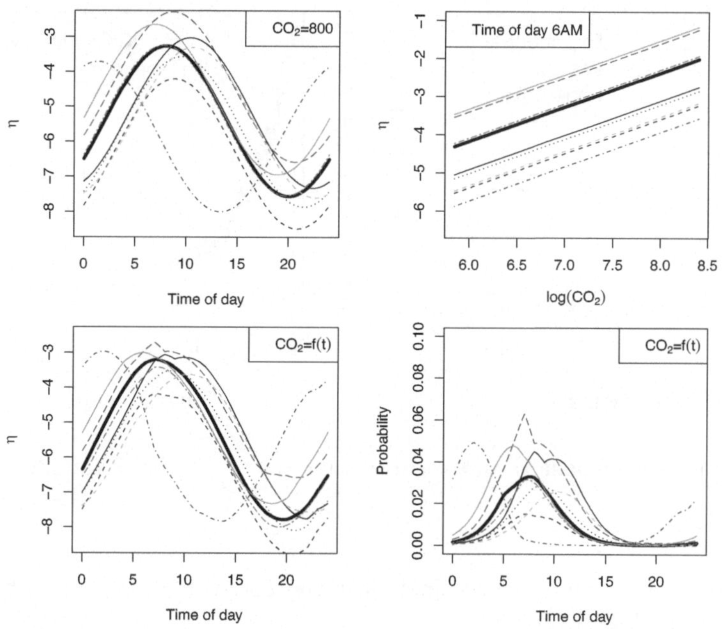

Figure 1.7: "Linear" predictors, and probability of window opening (see also Example 6.16 in Chapter 6).

behaviour patterns. Herein, the duration between observations of state change is continuous and positive.

The basic object that is modelled is the survival function

$$S(t) = P(T > t), \tag{1.21}$$

where T is the survival time. In this chapter, these models are first treated as non-parametric models where the survival function is estimated directly from data (the Kaplan-Meier estimate). In parameterising a survival model, a number of different parametric models can be used, all of which are expressed through the hazard function, that express the instantaneous risk of dying (change state). The resulting distribution or density is the distribution of the survival model; the most simple example is the exponential distribution while a more common and flexible example is the Weibull distribution

$$T_i \sim Weibull(\mu_i, \sigma), \tag{1.22}$$

where μ_i may include some regression parameters. An example of the hazard function, the Kaplan Meier estimate and parametric estimates of the survival function is given in Figure 1.8.

In many ways, the survival models can be considered in the same way as the generalized linear models in Chapter 4, and the chapter shows how random effects can be included in survival models (in line with generalized linear mixed effect models).

1.4.7 Linear time series models (Chapter 8)

Linear time series models should be used when data is expected to be correlated in time, e.g. temporal observations of CO_2 data.

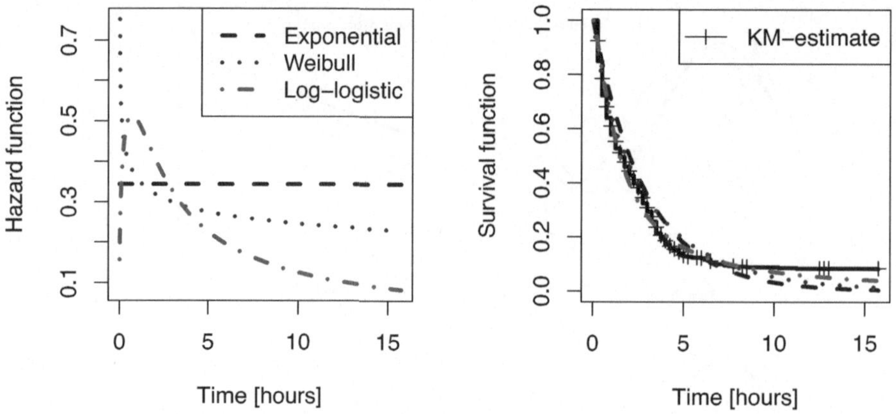

Figure 1.8: Examples of hazard and survival functions (see also Example 7.11 in Chapter 7).

In linear time series analysis, we include some correlation between observations, the simplest form is the AR(1) process

$$Y_t = \phi Y_{t-1} + \epsilon_t, \tag{1.23}$$

$\epsilon_t \sim N(0, \sigma^2)$ being iid random variables. The AR(1) process can be generalized to more general structures, and linear time series analysis is the study and identification of such processes. The process being Gaussian implies that the outcome should be continuous, e.g. time series of CO_2, which is used as the example in Chapter 8.

The chapter discusses a number of generalisations of the simple AR(1) model (1.23); such generalisations include the so-called ARMA(p,q) model

$$y_t = \phi_1 y_{t-1} + \phi_2 y_{t-2} + ... + \phi_p y_{t-p} + \epsilon_t + \theta_1 \epsilon_{t-1} + \theta_q \epsilon_{t-q}, \tag{1.24}$$

the formulation include observations close in time, but in many applications there will be a periodic pattern, e.g. diurnal; such patterns can be included using seasonal ARMA models, and finally we can of course also imagine some input, e.g. number of persons in a room, or window position. This can be included in the model using the so-called ARMAX models and ARIMAX models.

Identification relies on the observed autocorrelation and partial autocorrelation function, which can be compared to theoretical autocorrelation functions, and thereby identify possible model generalisations. The chapter covers theoretical autocorrelation functions and their empirical counterparts, and establishes proper ways of identifying model deficiencies and how to identify possible model extensions based on the residual analysis.

The chapter uses CO_2 concentration in a room to develop linear time series models based on residual analysis and available inputs as the driving example. The final model of the chapter is a ARX(2) model

$$y_t = \mu(\boldsymbol{x}_t) + \phi_1((y_{t-1} - \mu(\boldsymbol{x}_{t-1})) + \phi_2(y_{t-2} - \mu(\boldsymbol{x}_{t-2}) + \epsilon_t, \tag{1.25}$$

that includes the window position as an input for the model. Predictions, using the model, are shown in Figure 1.9. The figure illustrates the effect of a window opening (effect on mean value) and the effect of looking further into the future (increasing variance).

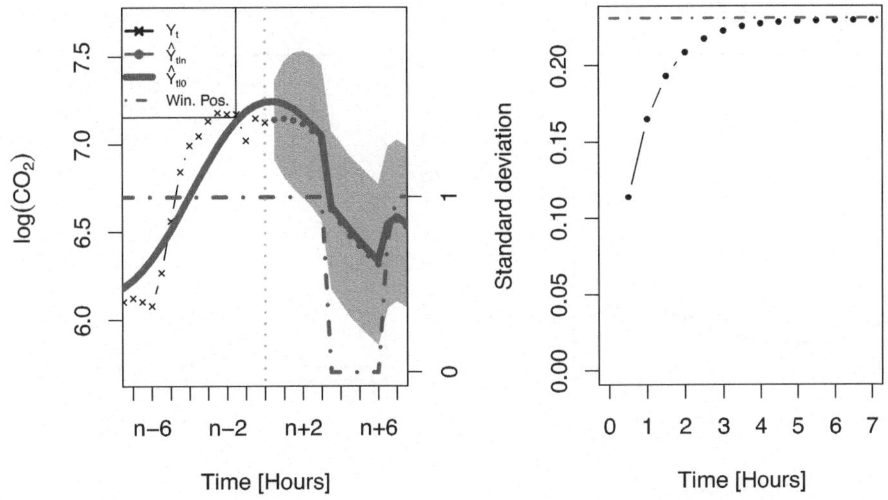

Figure 1.9: Out–of–sample predictions for CO_2 levels (see also Example 8.19 in Chapter 8).

1.4.8 Markov chain models (Chapter 9)

Markov chain models have a wide range of applications in occupant behaviour, for instance in presence models as well as in window opening and blind usage models. In general, time series whose outcome takes a finite number of values (which we refer to as the state), can be modelled using Markov chain models. The defining assumption of a Markov chain is that given knowledge about the current state, future states are independent of the past (referred to as the Markov property). In practice, this "memorylessness" is often a good compromise between realistic conditions and is a relatively convenient way of estimating future time steps.

Markov chain models include autocorrelation just as linear and Gaussian time series models, and actually, e.g. an AR(1) model is also a Markov chain. The specific meaning we use for Markov models in this book is time series models with a discrete outcome, e.g. number of people in a room (0,1,...), the state of a window (open or closed / 0 or 1). This implies that the Markov chain can take value $Y_t = i$, with $i \in \{1, \dots, m\}$

The basic object for modelling is the transition probability matrix, which defines the recursions from one time step to the next. If the probability distribution of the state at some time point (t) in the future is given by $\boldsymbol{u}_t$ with each element equal to the probability of observing $Y_t = i$. The transition probability matrix ($\boldsymbol{\Gamma}$) defines forward recursion of the probability vector $\boldsymbol{u}_t$, i.e.

$$\boldsymbol{u}_{t+1} = \boldsymbol{u}_t \boldsymbol{\Gamma}, \tag{1.26}$$

these forward recursions also define the likelihood function, and the parameters to be estimated are the elements of the transition probability matrix. In occupant models, it is natural to assume that the transition probabilities depends on some inputs, e.g. we might have higher probability of opening the window at a certain times of the day, or the probability is higher when temperature (inside or outside) is higher.

Figure 1.10 illustrates time–varying parameters in a Markov chain model and illustrates that, in the particular dwelling that is analysed, the probability of a window opening is much larger in the morning than the rest of the day (where it basically never happens).

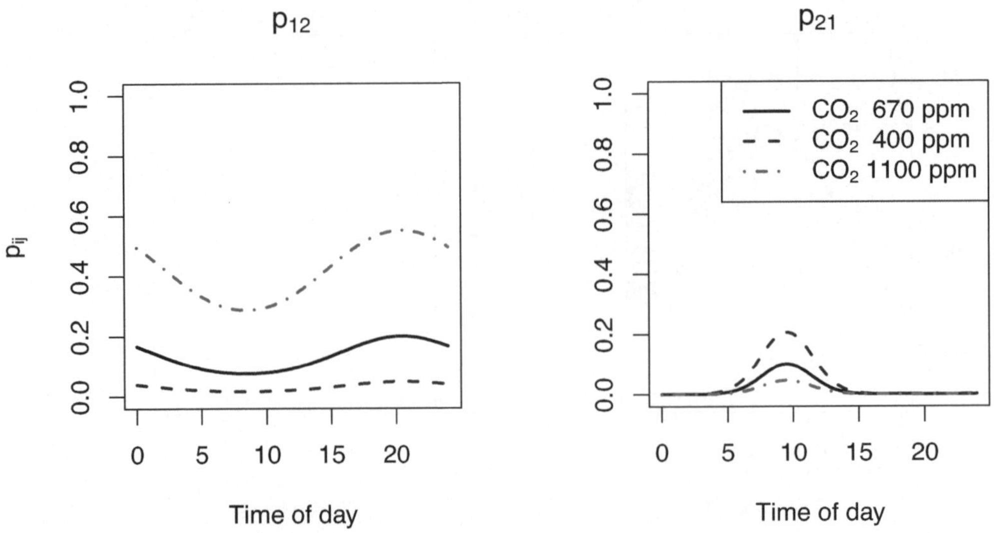

Figure 1.10: Transition probabilities from open to closed window (left) and from closed to open (right) as a function of time of day, for different indoor CO_2 concentrations (see also Example 9.15 in Chapter 9).

1.4.9 State space models (Chapter 10)

Indirect methods are methods where part of the variables in the model is not observed, e.g. we observe the temperature and CO_2 level in a room and from the measurements establish/estimate the number of person present in the room. This may include indirectly observed Markov processes or ARMA processes.

The state space models covered in this book are discrete time linear state space models; such models can be written as

$$\boldsymbol{x}_{t+1} = \boldsymbol{A}_t\boldsymbol{x}_t + \boldsymbol{B}_t\boldsymbol{u}_t + \boldsymbol{G}_t\boldsymbol{\epsilon}_t \tag{1.27}$$

$$\boldsymbol{y}_t = \boldsymbol{C}_t\boldsymbol{x}_t + \boldsymbol{D}_t\boldsymbol{u}_t + \boldsymbol{e}_t, \tag{1.28}$$

where $\boldsymbol{\epsilon}_t$ and $\boldsymbol{e}_t$ are Gaussian random variables, and $\boldsymbol{A}_t$-$\boldsymbol{G}_t$ may depend on time, but not the $\boldsymbol{x}$ of $\boldsymbol{y}$. The model can be seen as a generalisation of linear time series models and the state space models include time series models as a special example. This approach is useful for missing data, but the state space model does also include more general structures. An important part of the concept of state space models is that these include a hidden state ($\boldsymbol{x}_t$); this implies that part of the inference is to determine the conditional distribution of $\boldsymbol{x}_t$ given the observations up to some time (e.g. t). Such inference is done using the Linear Kalman filter, and from that the likelihood function can be derived.

Physical systems are often described as ordinary differential equations and such models can be approximated (including stochastic behaviour) using the state space approach. In such models, both the parameters driving the system and conditional distribution of the system may be important. As an example, the following system is considered

$$\begin{aligned} dC &= (-Q_r(t)C + E_r)dt + \sigma_1 dw_1 \\ dE_r &= \sigma_2 dw_2. \end{aligned} \tag{1.29}$$

This constitute a system of so-called stochastic differential equation (which is not covered in this book) that can be approximated by a linear state space model. In (1.29), C is the

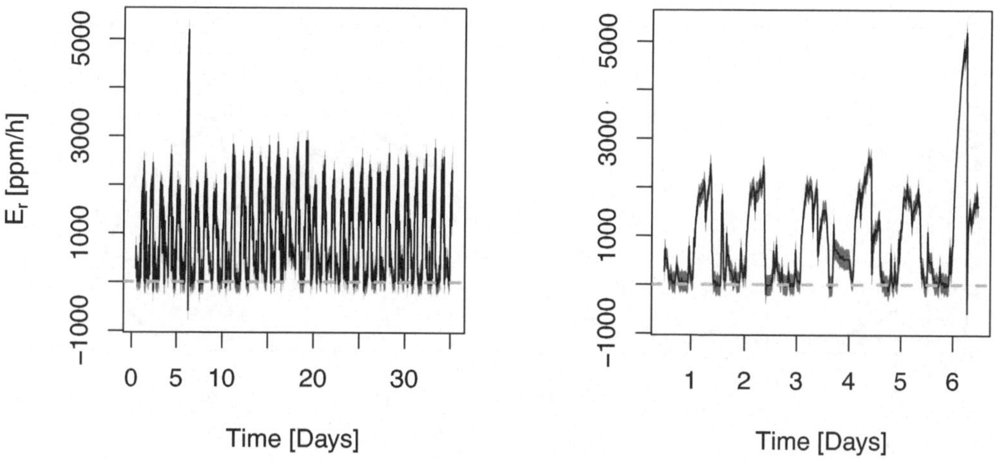

Figure 1.11: Estimated CO_2 production in a bedroom of 1 dwelling (see also Example 10.12 in Chapter 10).

concentration (above background) of CO_2 and E_r is the production of CO_2, which is an indication of activity in the room.

Figure 1.11 shows an example of inference of a hidden state (E_r) as a function of time, that is the production estimate is based on the observed air change ($Q(t)$) that differs depending on the position of the window (higher air change when window is open). For a specific problem, either the parameters or the state estimates are important (or both).

The Gaussian linear state space models can be generalised in a similar manner as the generalized linear model is an generalisation of the class of general linear models. This is also discussed using generalised state space models, and as an example we consider a time series of window openings. These have a direct link to generalised mixed effect models that are pointed out and discussed in the last part of the chapter.

1.4.10 Hidden Markov Models (Chapter 11)

Just like state space models can be seen as a generalisation of linear time series models, Hidden Markov Models can be seen as a generalisation of Markov Chain models where the underlying Markov process is not observed, the main extension is that the distribution of the observation now depend on the unobserved state, that hence have to be inferred from data. Formally we would write

$$p(Y_t = y_t | X_t = i) = p_i(y_t), \tag{1.30}$$

the problem is essentially the same as for the state space models, namely that the unobserved state have to be inferred from the observations and the model structure. The resulting likelihood function is formulated based on that inference. Further a number of different structure in the model are discussed throughout the chapter. This includes the distribution function ($p_i(\cdot)$) that can be formulated, for example using the distribution functions discussed in Chapter 4, structures related to the time series models, or/and the dimension of the observations.

The transition probability matrix, which relates to the Markov Chain, is of course also an important part of the model structure and different choices are also discussed in the chapter, for example extending the state space, considering higher order Markov models and time–varying transition probability matrices.

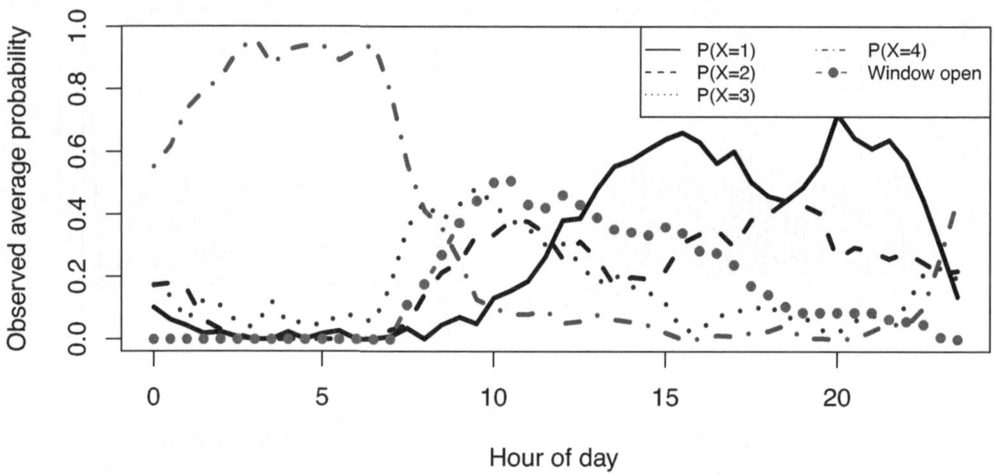

Figure 1.12: Examples of state inference in a Hidden Markov Model (see also Example 11.9 in Chapter 11).

Figure 1.12 shows an example of state inference for a model of CO_2 concentration using a Hidden Markov Model including time varying transition probabilities, and autoregressive structures. The 4 hidden states of the model all have meaningful interpretations, as

1. window closed and room empty
2. window open
3. transition state between low and high CO_2 concentration
4. high CO_2 concentration/activity and window closed.

Hence the model gives some nice insight into the dynamics of the room usage and interpretable parameters.

1.5 Use of R-software and R-packages

We use the statistical software R (R Core Team, 2022) for all analyses in the book, and in many cases also available contributed packages. While we aim at using well-established packages, packages will develop and may in the future include additional functionalities to include structures that are not handled today; further, there is of course always a risk that packages stop being maintained. In some chapters we rely entirely on contributed packages; in other chapters we have chosen to implement the needed algorithms along with theory, as we did not find that the available packages would support the development needed; finally in some chapters there is a mix between the two approaches.

When using R-functions that are part of the R-base we will not make additional citations, when contributing packages are used the first time we make the appropriate reference, but for an overview we also list these here (Chapter refer to the first time the package is used); `numDeriv` (Gilbert and Varadhan, 2019) (Chapter 2), `car` (Fox and Weisberg, 2019), `MASS` (Venables and Ripley, 2002) (Chapter 3), `ordinal` (Christensen, 2022) (Chapter 4), `lme4` (Bates et al., 2015), `glmmTMB` (Brooks et al., 2017), `mvtnorm` (Genz et al., 2021) (Chapter

5), TMB (Kristensen et al., 2016) (Chapter 6), survival (Therneau et al., 2022), tramME (Tamási and Hothorn, 2021) (Chapter 7), msm (Jackson, 2011), expm (Maechler et al., 2023) (Chapter 9), KFAS (Helske, 2017) (Chapter 10).

In addition a number of packages are used for graphics and data handling; these include ggplot2 (Wickham, 2016), reshape2 (Wickham, 2007), plyr (Wickham, 2011), data.table (Dowle and Srinivasan, 2020), grid (R Core Team, 2022), and gridExtra (Auguie, 2017).

Lastly this entire book is written based on the R-package knitr (Xie, 2014; Xie, 2015; Xie, 2020).

R-scripts and publicly available data used for this book will appear on https://www2.imm.dtu.dk/~jkmo/.

1.6 Structure of the book

The book is organised based on structures in data and the data collection; hence in Part I data are assumed independent, that is no correlation needs to be accounted for in data; in Part II, there is some correlation between data point, but the process is observed directly, and finally in Part III, the process that induces correlation is not directly observed. The individual chapters are summarised/introduced in Section 1.4, and the interdependence is given in Figure 1.2.

We hope this introduction gives you a feeling of the broad range of models that are covered in this book and an understanding of the methods needed for different data structures. Whether you will use the book for getting a broad understanding of the possibilities for modelling or for finding the specific method needed for you particular data, we do hope that you will find the book useful.

Part I

Time-Independent Observations

2

The likelihood principle, and statistical modelling

This chapter is a revision of a similar chapter in Madsen and Thyregod (2011), but tailored to occupant behaviour and the remaining chapters of this book. The chapter introduces the likelihood principle, which is an approach for supervised learning. In particular, it is a method for learning the parameters of a statistical model. The likelihood principle builds on a probability model for the outcome of interest. The likelihood is the underlying principle for all other chapters in this book. We introduce likelihood as a general tool for estimating parameters while special cases are treated in the later chapters.

Often we think of statistical models in terms of a mean value plus some noise,

$$Y_i = \mu_i + noise, \tag{2.1}$$

where μ_i is a function of environmental conditions that are either controlled or measured e.g. indoor temperature or sex. The noise introduces some uncertainty in our predictions, and describes the natural variation in data. This uncertainty also implies that there will be some parameter uncertainty. Often we use linear models for μ_i, resulting in the general linear model (when the noise is Gaussian) treated in the next chapter. The general linear model focus on the mean value structure, and basically imply that we should minimise the sum of squared error.

The simplest version of the above is

$$Y_i = \mu + \epsilon_i, \tag{2.2}$$

if $\epsilon_i \sim N(0, \sigma^2)$ and iid (independent and identically distributed, see Section 2.1 below), we could write

$$Y_i \sim N(\mu, \sigma^2). \tag{2.3}$$

The modellers task is to find estimates of μ and σ^2. Given this formulation (i.e. Gaussian observations), an obvious question for the modeller would be if the choice of distribution is the best choice or if other models should be considered. For this question, the answer is sometimes straightforward (e.g. flipping a coin results in the binomial distribution). In other cases, the modeller can come up with a number of candidate models that need to be compared. The next question is now how to estimate parameters in such models. The answer to the second question turns out to be that we should consider the so-called likelihood function

$$L(\theta; \boldsymbol{y}) = p_\theta(\boldsymbol{y}), \tag{2.4}$$

where $p_\theta(\cdot)$ is the probability mass function if the outcome is discrete (e.g. binomial or Poisson data) and the probability density function if the outcome is continuous (e.g. Gaussian or gamma data), and θ ("theta") is the parameters of the model, for example the success

DOI: 10.1201/9781003340812-2

probability (binomial), rate (Poisson), mean and variance (Gaussian) or shape and rate (gamma). The likelihood function can also be used for comparing different candidate distributions (i.e. help us answer the first question above). Hence to cut it short, the likelihood ($L(\theta; \boldsymbol{y})$) is a function of the parameters (θ) for fixed data ($\boldsymbol{y}$)) while the pdf (probability density function) is a function of the outcome ($\boldsymbol{y}$)) for fixed parameter values (θ). In order to illustrate this important principle, let us consider a simple example.

Example 2.1 (Tossing a coin) *Lets assume we toss a coin $n = 2$ times and observe one head and one tail, we further assign the value 1 to head and the value 0 to tail. The parameter of interest (θ) is the probability of getting head which is observed $y = 1$ time, and the probability model will in this case be the binomial*

$$p_\theta(y) = \binom{n}{y}\theta^y(1-\theta)^{n-1} = L(\theta; y, n). \tag{2.5}$$

In this particular example, we have $\binom{2}{1} = 2$ and we can calculated the likelihood of different values of the success probability, for example $L(0.5) = 0.5$, $L(0.25) = 0.375$ and $L(0.6) = 0.48$. As we see the highest likelihood is achieved with $\theta = 0.5$, but the values using θ equal 0.25 or 0.6 also hold information. In R the values can be calculated by;

```
dbinom(1, size = 2, prob = c(0.25, 0.5, 0.6))

## [1] 0.375 0.500 0.480
```

The likelihood function in (2.4) is the key inferential tool in this book. It is important to stress that the entire likelihood function holds information about the optimal parameter estimates and their uncertainty. Hence the entire function should be considered. Likelihood estimation builds on some parameterised distribution assumption, and we will therefore briefly discuss some common assumptions.

2.1 Independence

The most general statistical model for a set of observations $\boldsymbol{Y} = \{Y_1, Y_2, \ldots, Y_n\}$ is that the joint density (or joint probability mass) function is given by

$$f_Y(y_1, y_2, \ldots, y_n), \tag{2.6}$$

where $f_Y(y_1, y_2, \ldots, y_n)$ is a function from $\mathbb{R}^n$ into $\mathbb{R}_+$. It is often assumed that observations are *independent*; formally independence is defined by

$$f_Y(y_1, y_2, \ldots, y_n) = \prod_{i=1}^{n} f_{Y_1}(y_1) f_{Y_2}(y_2,) \cdots f_{Y_n}(y_n), \tag{2.7}$$

in some models (in particular the general linear model (GLM), see Chapter 3), we further assume the residuals are *independent and identically distributed* (iid), which is formally defined as

$$f_Y(y_1, y_2, \ldots, y_n) = \prod_{i=1}^{n} f_Y(y_1) f_Y(y_2,) \cdots f_Y(y_n) \tag{2.8}$$

for example if we consider the model

$$Y_i = \mu_i + \epsilon_i, \tag{2.9}$$

with $\epsilon_i \sim N(0, \sigma^2)$ and iid, μ_i depending on the observation, e.g. $\mu_i = \beta_0 + x_i\beta_1$. Then by definition, ϵ_i is iid while the observations (Y_i) are independent.

2.2 Likelihood estimation

More formally the likelihood function is defined by

Definition 2.1 (Likelihood function) *Given the parametric density* $f_Y(\boldsymbol{y}, \boldsymbol{\theta})$, $\boldsymbol{\theta} \in \Theta^k$, *for the observations* $\boldsymbol{y} = (y_1, y_2, \ldots, y_n)$ *the* likelihood function for $\boldsymbol{\theta}$ *is the function*

$$L(\boldsymbol{\theta}; \boldsymbol{y}) = c(y_1, y_2, \ldots, y_n), f_Y(y_1, y_2, \ldots, y_n; \boldsymbol{\theta}) \tag{2.10}$$

where $c(y_1, y_2, \ldots, y_n)$ *is a constant, not depending on* $\boldsymbol{\theta}$.

The definition above implies that the likelihood function is only meaningful up to a multiplicative constant (i.e. not depending on the parameters). It is usual practice to scale the likelihood such that it has unit maximum. That is we use

$$L_N(\boldsymbol{\theta}; \boldsymbol{y}) = \frac{L(\boldsymbol{\theta}; \boldsymbol{y})}{\max_\theta L(\boldsymbol{\theta}; \boldsymbol{y})}. \tag{2.11}$$

We will refer to $L_N(\boldsymbol{\theta}; \boldsymbol{y})$ as the normalised likelihood.

An important part of the modeller's task is to find the most appropriate statistical distribution $(f_Y(\boldsymbol{y}, \boldsymbol{\theta}))$.

Example 2.2 *The histogram below shows 72 times between opening and closing of blinds in office building in Ottawa Canada (see also examples in Chapter 7).*

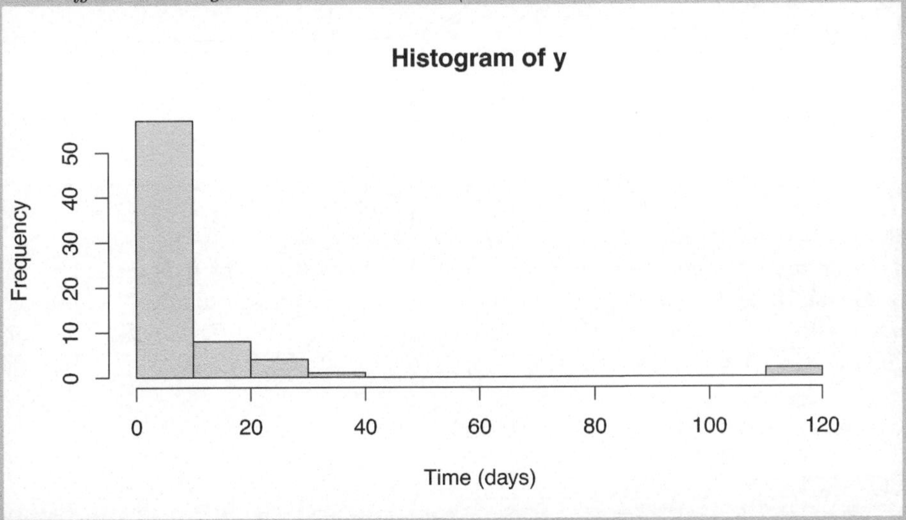

The mean and standard deviation for the waiting times (time between interactions with the blinds) are

```
c(mean(y),sd(y))

## [1]  9.05 19.90
```

Even though the large difference between average and standard deviation suggests that the exponential distribution is not the best choice, a first approach for modelling the time between opening and closing could be an exponential distribution. For example:

$$D_i \sim Exp(\theta), \tag{2.12}$$

or given in terms of the probability density function (with $\theta = E[Y_i]$*)*

$$f_D(y) = \frac{1}{\theta} e^{-y/\theta}. \tag{2.13}$$

Hence we can write the likelihood function as

$$L(\theta; \boldsymbol{y}) = \prod_{i=1}^{n} \theta^{-1} e^{-y_i/\theta} = \theta^{-n} e^{-\frac{1}{\theta}\sum_{i=1}^{n} y_i} = \theta^{-n} e^{-\frac{n}{\theta}\bar{y}}, \tag{2.14}$$

where $\bar{y}$ *is the average of the observed values. The figure below shows the normalised likelihood (i.e. maximum equal 1, see (2.11)) function.*

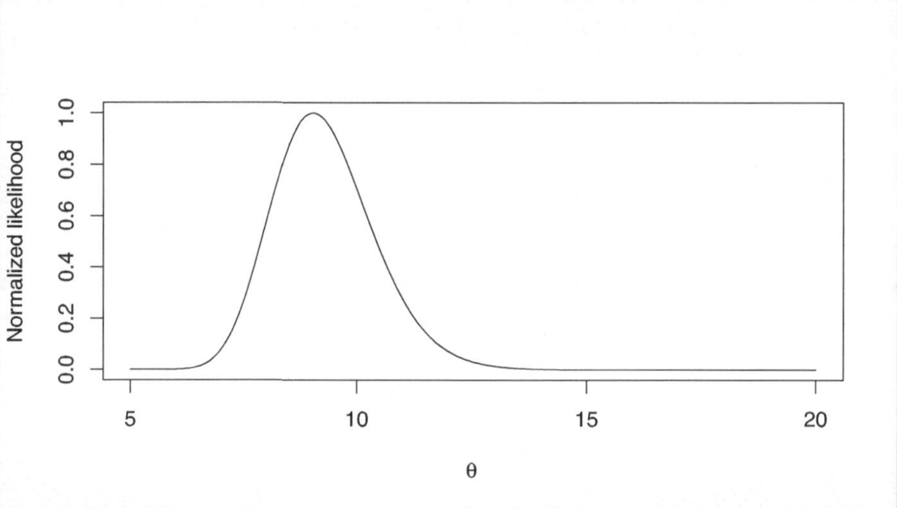

Hence we see that under the assumption of an exponential distribution the mean is around 9 (equal the average of the distribution, see also Example 2.3 below), but it seems clear that it could also be 10, but that it is not likely that it is 20.

In the above we have compared different parameter values in the case where we have chosen a specific distribution; if we would like to compare different distributions, we usually use the so-called log-likelihood ($\log(L(\theta))$*) at* $\theta = 9$*, this is;*

```
sum(dexp(y, rate = 1/9, log = TRUE))

## [1] -231
```

without getting into details we could have estimated a χ^2 distribution with θ degrees of freedom. The maximum is achieved with $\theta = 3.4$ degrees of freedom and the log-likelihood would be;

```
sum(dchisq(y,df=3.4,log=TRUE))

## [1] -358
```

which is a huge difference (we will address the details later), and we would conclude that choosing between the two we should use the exponential distribution.

In the example above, we saw that the likelihood function has a maximum around $\theta = 9$, and it seems intuitively clear that there is a range of values that we could accept. We will formalise this below. Note that in the example above we have not assessed the goodness of fit for the exponential model, and we will not treat this important part of modelling in this chapter, but discuss it in Chapter 7 for this specific family of models.

The example above indicates that we should optimise the likelihood, and formally this will give the Maximum Likelihood Estimate.

Definition 2.2 (Maximum Likelihood Estimate (MLE)) *Given the observations* $\boldsymbol{y} = (y_1, y_2, \ldots, y_n)$ *the* Maximum Likelihood Estimate (MLE) *is a function of the observations* $\widehat{\boldsymbol{\theta}}(\boldsymbol{y})$ *such that*

$$L(\widehat{\boldsymbol{\theta}}; \boldsymbol{y}) = \sup_{\boldsymbol{\theta} \in \Theta} L(\boldsymbol{\theta}; \boldsymbol{y}). \tag{2.15}$$

The function $\widehat{\boldsymbol{\theta}}(\boldsymbol{Y})$ *over the sample space of observations* $\boldsymbol{Y}$ *is called an* ML estimator.

Usually it is more convenient to use the log-likelihood function as the basis for estimation, the *log*-likelihood is denoted $l(\theta; \boldsymbol{y})$

$$l(\theta; \boldsymbol{y}) = \log L(\theta; \boldsymbol{y}). \tag{2.16}$$

Example 2.3 *Continuing the example above we have the* log-*likelihood*

$$l(\theta; \boldsymbol{y}) = -n \log(\theta) - \frac{n}{\theta} \bar{y}. \tag{2.17}$$

In order to find the maximum of this function, we will find the derivative of the log-*likelihood function*

$$l'(\theta; \boldsymbol{y}) = -\frac{n}{\theta} + \frac{n}{\theta^2} \bar{y} \tag{2.18}$$

setting the derivative equal zero and solving for θ gives $\hat{\theta} = \bar{y}$, that is for the exponential distribution, the maximum likelihood estimate of the expectation is equal the average of the observations. In this specific case we get the MLE and the log-likelihood at the MLE

```
n <- length(y)
theta.hat <- mean(y)
ll.m <- - n * log(theta.hat) - n / theta.hat * mean(y)
c(theta.hat, ll.m)

## [1]    9.05 -230.57
```

Hence the MLE is $\hat{\theta} = 9.05$ and $l(\hat{\theta}, \boldsymbol{y}) = -230.6$. As noted above, the parameter equals the average of the observations, while the value of the likelihood does not have a direct interpretation, but should be seen relative to other models (as we did above).

Even though we calculated the log-likelihood value above, this number does not give us any information in itself, but as we will see later, we can use it for comparing models.

Formally the MLE is found by solving the so-called estimating equations (setting the score function equal to zero).

Definition 2.3 (Score function) *Consider $\boldsymbol{\theta} = (\theta_1, \cdots, \theta_k)^T \in \Theta^k$. Assume that Θ^k is an open subspace of $\mathbb{R}^k$ and that the log-likelihood is continuously differentiable. Then consider the following first-order partial derivative of the log-likelihood function:*

$$\ell'_\theta(\boldsymbol{\theta}; \boldsymbol{y}) = \frac{\partial}{\partial \boldsymbol{\theta}} \ell(\boldsymbol{\theta}; \boldsymbol{y}) = \begin{pmatrix} \frac{\partial}{\partial \theta_1} \ell(\boldsymbol{\theta}; \boldsymbol{y}) \\ \vdots \\ \frac{\partial}{\partial \theta_k} \ell(\boldsymbol{\theta}; \boldsymbol{y}) \end{pmatrix}. \tag{2.19}$$

The function $\ell'_\theta(\boldsymbol{\theta}; \boldsymbol{y})$ is called the score function, *often written as $S(\boldsymbol{\theta}; \boldsymbol{y})$.*

Even though it is possible to solve $S(\boldsymbol{\theta}) = 0$ in simple examples, it is usually done by a numerical optimiser in real application; therefore we do not even have to calculate it. The score function can also be used to construct the so-called score-test, see for example Pawitan (2013).

So far we have discussed how to find the point estimate (the MLE), but we still need to quantify the uncertainty of the estimators. Before we go on to this part, we will shortly discuss some general statistical terms.

2.3 Properties of estimators

In the following we will think about our observations as random variables and consider properties of the implied estimators. It is important to note that an estimate is a number (a function of actual observations) while an estimator is a random variable (a function of random variables from the assumed distribution).

Two important properties of the estimators are unbiasedness and consistency.

Definition 2.4 (Unbiased estimator) *Any estimator $\widehat{\boldsymbol{\theta}} = \widehat{\boldsymbol{\theta}}(\boldsymbol{Y})$ is said to be* unbiased *if $E[\widehat{\boldsymbol{\theta}}] = \boldsymbol{\theta}$ for all $\boldsymbol{\theta} \in \Theta^k$.*

Definition 2.5 (Consistent estimator) *An estimator is* consistent *if the sequence $\boldsymbol{\theta}_n(\boldsymbol{Y})$ of estimators for the parameter $\boldsymbol{\theta}$ converges in probability to the true value of $\boldsymbol{\theta}$.*

The first definition tells us that we can expect (on average in many repetitions) to get the true value of the parameter. The second definition is quite technical (for more details and more precise definitions, see for example Lehmann and Casella (1998); p. 332), but we can think of this as the variance of the estimator going to zero as the number of observations increase.

Example 2.4 *If $Y_i \sim Exp(\theta)$, then we saw above that*

$$\hat{\theta} = \frac{1}{n}\sum_i Y_i. \tag{2.20}$$

In this case we also have $E[Y_i] = \theta$ and $V[Y_i] = \theta^2$, taking the expectation we get

$$E\left[\hat{\theta}\right] = \frac{1}{n}E\left[\sum_i Y_i\right] = \frac{1}{n}\sum_i \theta = \theta. \tag{2.21}$$

Hence the estimator is unbiased, and further we have

$$V\left[\hat{\theta}\right] = \frac{1}{n^2}V\left[\sum_i Y_i\right] = \frac{1}{n^2}\sum_i V\left[Y_i\right] = \frac{\theta^2}{n}. \tag{2.22}$$

Hence the variance goes to zero as we obtain more information (i.e. increase n).

Relating to Example 2.3 we observed $\hat{\theta} = 9.05$. If we made a new experiment under the same conditions we would not expect to get the exact same number. However, since the estimator is unbiased, we would expect that the average of many estimates would be close to the true value. The question of how close is answered by the variance of the estimator that gets smaller as we increase the sample size.

2.4 Likelihood ratio test and uncertainty evaluation

For general distribution assumptions, it is difficult to construct an exact test; a very prominent exception is the F-test in the analysis of variance set up (see Chapter 3). For general distribution assumptions, we have to rely on asymptotic results; this implies that in contrast to the F-test the distribution of the test statistics is unknown even when the distribution assumption holds. In such cases, we will need to rely on asymptotic results (implying that we have a large number of observations). We illustrate this with a small example:

Example 2.5 *Let's assume that we have observations from two different populations $x_1, ..., x_n$ and $y_1, ..., y_m$; now we want to compare the expected value in the two populations. In order to do so, we assume that the observations are realisations from different normal distributions; $X_i \sim N(\mu_1, \sigma^2)$, $Y_i \sim N(\mu_2, \sigma^2)$, these are the assumptions (normality and equal variance). Now, we want to test the hypothesis $H_0 : \mu_1 = \mu_2$, and calculate the test statistic*

$$t = \frac{\bar{x} - \bar{y}}{\sqrt{s_p^2\left(\frac{1}{n} + \frac{1}{m}\right)}} \tag{2.23}$$

where s_p^2 is the pooled variance. If the null hypothesis (and the assumptions) is true, then the test statistics (t) is a realization of a student t-distribution with $n + m - 2$ degrees of freedom. If however x_i and y_i do not follow normal distributions, then the distribution of t is not the t-distribution, and we will have to rely on asymptotic results.

For nested models, the results are given in terms of the likelihood ratio test.

Definition 2.6 (Likelihood ratio) *Consider the hypothesis $H_0\colon \boldsymbol{\theta} \in \Omega_0$ against the alternative $H_1\colon \boldsymbol{\theta} \in \Omega \setminus \Omega_0$ ($\Omega_0 \subseteq \Omega$), where $\dim(\Omega_0) = m$ and $\dim(\Omega) = k$.*

For given observations $y_1, y_2, \ldots, y_n$ the likelihood ratio *is defined as*

$$\lambda(\boldsymbol{y}) = \frac{\sup_{\boldsymbol{\theta}\in\Omega_0} L(\boldsymbol{\theta}; \boldsymbol{y})}{\sup_{\boldsymbol{\theta}\in\Omega} L(\boldsymbol{\theta}; \boldsymbol{y})}. \tag{2.24}$$

The above definition is quite technical, but basically it implies that the simple hypothesis can be obtained by a specific choice of parameters in the alternative hypothesis. A simple example is a mean value model with the null–hypothesis $H_0 : \mu_i = \beta_0$ and the alternative hypothesis $H_1 : \mu_i = \beta_0 + x_i\beta_1$.

The likelihood ratio test is given by the following theorem:

Theorem 2.1 (Wilk's Likelihood Ratio test) *Let $\lambda(\boldsymbol{y})$ be defined by (2.24); then under the null hypothesis H_0 (as in Definition 2.6), the random variable $-2\log\lambda(\boldsymbol{Y})$ converges in distribution to a χ^2-distributed random variable with $(k-m)$ degrees of freedom, that is,*

$$-2\log\lambda(\boldsymbol{Y}) \to \chi^2(k-m), \tag{2.25}$$

under H_0.

We illustrate the use of Theorem 2.1 by continuing the blind use example

Example 2.6 *Referring to the examples 2.2-2.4 above, we could test the (arbitrary) hypothesis $H_0 : \theta = 12$ against the two-sided alternative $H_1 : \theta \neq 12$; in order to do this, we need the likelihood under the null-hypothesis;*

```
theta0 <- 12
(l10 <- - n * log(theta0) - n / theta0 * mean(y))

## [1] -233
```

the p-value for the test is calculated using the χ^2-distribution with 1 degree of freedom (we fixed one parameter), as we look for evidence against the null–hypothesis we find the probability of observing something larger than the observed test statistics;

```
Q <- (ll.m - l10)
1 - pchisq(2 * Q, df = 1)

## [1] 0.022
```

Hence on the usual 5% significance level we will reject the null–hypothesis. In fact we will accept all hypotheses for which $\lambda(\boldsymbol{y}) > exp\left(\frac{1}{2}\chi^2_{1-\alpha}\right)$. The cut–off in the normalised likelihood can, in this example, be calculated as;

```
exp(-0.5*qchisq(0.95,df=1))

## [1] 0.147
```

We can give this a graphical interpretation:

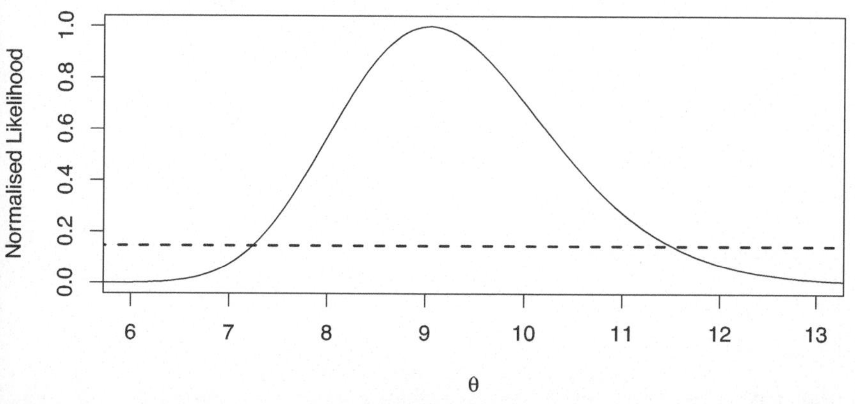

Hence, we can see that a 95% confidence interval is approximately from 7.25 to 11.5, i.e. we are 95% confident that the true waiting time is between 7.25 and 11.5, given that the assumptions are correct.

The analysis above does not imply that the exponential distribution is suitable for these data; we will discuss such goodness–of–fit approaches in the specialized chapters later on (a simple goodness–of–fit assesment is given in Figure 2.1 on page 42). Using the likelihood approach, we can, however, compare other assumptions for the waiting times, and another option is the Weibull distribution. The pdf of the Weibull distribution is given by

$$f(x) = \frac{\gamma}{\lambda}\left(\frac{x}{\lambda}\right)^{\gamma-1} e^{-(x/\lambda)^{\gamma}}. \tag{2.26}$$

For $\gamma = 1$, this is the exponential distribution, and hence the exponential and Weibull are nested. In the case of the Weibull distribution, the parameter is $\boldsymbol{\theta} = [\lambda, \gamma]$; the likelihood and log*-likelihood functions are given by*

$$\begin{aligned} L(\boldsymbol{\theta}, \boldsymbol{y}) &= \prod_{i=1}^{n} \frac{\gamma}{\lambda}\left(\frac{y_i}{\lambda}\right)^{\gamma-1} e^{-(y_i/\lambda)^{\gamma}} \\ l(\boldsymbol{\theta}, \boldsymbol{y}) &= n\log\left(\frac{\gamma}{\lambda}\right) - n(\gamma-1)\log(\lambda) + (\gamma-1)\sum_{i=1}^{n}\log(y_i) - \sum_{i=1}^{n}(y_i/\lambda)^{\gamma} \end{aligned} \tag{2.27}$$

rather than finding the score equations and solving for the parameters the MLE can be found numerically by;

```
## Negative log-likelihood function
nll.wei <- function(theta, y){
    - sum(dweibull(y, shape = theta[1], scale = theta[2],
                   log = TRUE))}

## Optimize parameters
fit.wei <- nlminb(c(1, 1), nll.wei, lower = 0.01, y = y)
## Parameter estimates
fit.wei$par

## [1] 0.6158871 5.7106386

## negative log-likelihood
fit.wei$objective

## [1] 209.195
```

Note that in the R-function for the negative log-likelihood function (`nll.wei`) it is important that θ is the first parameter; otherwise the optimisation in the next line will not work.

When reporting such numbers, we should give some measures of the uncertainty (e.g. confidence intervals). We will discuss how to construct these in the next section.

For comparing the exponential and the Weibull distribution, we can use the likelihood ratio test;

```
## test statistics
Q <- -fit.wei$objective-ll.m
## p-value
1 - pchisq(2 * Q, df = 1)

## [1] 6.225154e-11
```

Hence, we will reject the exponential distribution on any reasonable level.

In these examples, we have mostly looked at single parameter models. For multi-parameter models, the so-called information matrix plays an important role.

2.4.1 The information matrix

As we have discussed earlier in this chapter, the likelihood function contains all information about the parameters for a given choice of statistical model. Furthermore, the parameter that maximises the likelihood function is the maximum likelihood estimate and the derivative at the maximum (the score function) is zero. Therefore, we can use the curvature in order to summarise the function. This is called Fisher's information.

Definition 2.7 (Observed information) *The matrix*

$$j(\boldsymbol{\theta}; \boldsymbol{y}) = -\frac{\partial^2}{\partial \boldsymbol{\theta} \partial \boldsymbol{\theta}^T} \ell(\boldsymbol{\theta}; \boldsymbol{y}) \tag{2.28}$$

with the elements

$$j(\boldsymbol{\theta}; \boldsymbol{y})_{ij} = -\frac{\partial^2}{\partial \theta_i \partial \theta_j} \ell(\boldsymbol{\theta}; \boldsymbol{y}) \tag{2.29}$$

is called the observed information *corresponding to the observation* $\boldsymbol{y}$ *and the parameter* $\boldsymbol{\theta}$.

The observed information can be calculated directly from the likelihood either explicitly or by numerical method. The information matrix can also be viewed in terms of estimators, and we can obtain the expected information.

Definition 2.8 (Expected information) *The expectation of the observed information is*

$$\boldsymbol{i}(\boldsymbol{\theta}) = E[\boldsymbol{j}(\boldsymbol{\theta}; \boldsymbol{Y})], \tag{2.30}$$

where $\boldsymbol{j}(\boldsymbol{\theta}; \boldsymbol{Y})$ *is given by* (2.28)*, and the expectation is determined under the distribution corresponding to* $\boldsymbol{\theta}$. $\boldsymbol{i}(\boldsymbol{\theta})$ *is called the* expected information matrix *corresponding to the parameter* $\boldsymbol{\theta}$. *The expected information* (2.30) *is also known as the* Fisher information matrix.

Let us illustrate the use of information matrices with our exponential example:

Example 2.7 *From the score function for the exponential distribution, we can calculate the information as*

$$j(\theta; \boldsymbol{y}) = -\frac{n}{\theta^2} + 2\frac{n}{\theta^3}\bar{y}, \tag{2.31}$$

and the expected information becomes

$$i(\theta; \boldsymbol{Y}) = -\frac{n}{\theta^2} + 2\frac{n}{\theta^3}E[\bar{Y}] = \frac{n}{\theta^2}. \tag{2.32}$$

In this case, we have

$$j(\hat{\theta}; \boldsymbol{y}) = -\frac{n}{\hat{\theta}^2} + 2\frac{n}{\hat{\theta}^2} = i(\hat{\theta}; \boldsymbol{Y}). \tag{2.33}$$

Note that $\frac{1}{i(\theta, \boldsymbol{Y})} = V[\hat{\theta}]$ *(see Example 2.4), i.e. the variance is equal the inverse information.*

Referring to Example 2.3, would get the information

```
(j <- length(y) / mean(y) ^ 2)

## [1] 0.88
```

The fact that the expected and observed information, taken at $\theta = \hat{\theta}$*, are equal is a special feature of the so-called exponential family, but it does not hold in general.*

The importance of the information for uncertainty evaluation is given by the following theorem.

Theorem 2.2 (Distribution of the ML estimator) *We assume that* $\widehat{\boldsymbol{\theta}}$ *is consistent. Then, under some regularity conditions,*

$$\widehat{\boldsymbol{\theta}} - \boldsymbol{\theta} \to N(0, \boldsymbol{i}(\boldsymbol{\theta})^{-1}), \tag{2.34}$$

where $\boldsymbol{i}(\boldsymbol{\theta})$ *is the expected information or the information matrix (see* (2.30)*).*

Here, regularity conditions refer to properties of the likelihood function and are not a problem for the most common problems that we encounter.

Theorem 2.2 implies that the standard error of $\widehat{\theta}_i$ is given by

$$\widehat{\sigma}_{\widehat{\theta}_i} = \sqrt{V_{ii}[\widehat{\boldsymbol{\theta}}]}, \tag{2.35}$$

where $V_{ii}[\widehat{\boldsymbol{\theta}}]$ is the ith diagonal term of $\boldsymbol{j}^{-1}(\widehat{\boldsymbol{\theta}})$ (or $\boldsymbol{i}^{-1}(\widehat{\boldsymbol{\theta}})$). A test of an individual parameter $H_0 : \theta_i = \theta_{i,0}$ is given by the *Wald statistic*:

$$Z_i = \frac{\widehat{\theta}_i - \theta_{i,0}}{\widehat{\sigma}_{\widehat{\theta}_i}}, \tag{2.36}$$

which under H_0 is approximately $N(0,1)$-distributed. This also implies that we can construct confidence intervals based on the normal distribution.

Example 2.8 *In the Weibull model in Example 2.7 above we can find the observed information by (using the function* `hessian` *from the package* `numDeriv` *(Gilbert and Varadhan, 2019))*

```
## Information
## library(numDeriv)
H <- hessian(nll.wei, fit.wei$par, y = y)
```

The R-function `hessian` *find the numerical Hessian of a function in a given point (here the Hessian of the negative log-likelihood at the MLE). Further we can find 95% confidence intervals as;*

```
se <- sqrt(diag(solve(H)))
tab <- cbind(fit.wei$par, fit.wei$par + qnorm(0.025) * se,
             fit.wei$par + qnorm(0.975) * se)
rownames(tab) <- c("shape", "scale")
colnames(tab) <- c("theta", "lowerb", "upperb")
round(tab, digits=2)

##         theta lowerb upperb
## shape    0.62   0.51   0.72
## scale    5.71   3.44   7.98
```

Hence, we are 95% confident that $\gamma \in [0.51; 0.72]$ *and* $\lambda \in [3.44; 7.98]$.

Another way of obtaining confidence intervals is by the profile likelihood.

2.5 Profile likelihood

The Wald confidence interval given above assumes that the log-likelihood function can be approximated by a quadratic function (which will be the case in large samples). When this is not the case, we may use the so-called profile likelihood.

In single parameter cases, it is easy to check if the log-likelihood is well approximated by a quadratic function (by plotting). In the multiparameter case, we can simply assume that this is the case or we can check a few selected parameters (using profile likelihood as described below).

The profile likelihood is a function of the parameter(s) under investigation while the rest of the parameters are treated as nuisance parameters. Formally, the profile likelihood is given by:

Definition 2.9 (Profile likelihood) *Assume that the statistical model for the observations $Y_1, Y_2, \ldots, Y_n$ is given by the family of joint densities* (2.10)*, where $\boldsymbol{\theta} = (\boldsymbol{\tau}, \boldsymbol{\zeta})$ and $\boldsymbol{\tau}$ denoting the parameter of interest. Then the* profile likelihood function *for $\boldsymbol{\tau}$ is the function*

$$L_P(\boldsymbol{\tau}; \boldsymbol{y}) = \sup_{\zeta} L((\boldsymbol{\tau}, \boldsymbol{\zeta}); \boldsymbol{y}), \tag{2.37}$$

with the joint likelihood $L((\boldsymbol{\tau}, \boldsymbol{\zeta}); \boldsymbol{y})$ given by (2.10) *and where the maximisation is performed at a fixed value of $\boldsymbol{\tau}$.*

Based on the profile likelihood and using the result of the likelihood ratio test, we can also construct profile likelihood confidence intervals/regions by

$$\left\{\boldsymbol{\tau}; \frac{L_P(\boldsymbol{\tau}; \boldsymbol{y})}{L(\widehat{\boldsymbol{\theta}}, \boldsymbol{y})} > \exp(-\frac{1}{2}\chi^2_{1-\alpha}(p))\right\}. \tag{2.38}$$

Such regions can be visualised in up to two dimensions. For constructing likelihood-based confidence intervals/regions, there exist other methods than the profile likelihood (e.g. marginal and conditional likelihood, see e.g. Pawitan (2013)), which can be more accurate in some cases. The profile likelihood has the advantage of being very generic. The flip side of the coin is that we need an inner optimisation, which may be computationally expensive.

The steps needed in order to calculate the profile likelihood are summarised below:

1 Formulate a log-likelihood function $l((\tau, \zeta); \boldsymbol{y})$.

2 Formulate the profile log-likelihood function $l_P(\tau; \boldsymbol{y})$, with an inner optimisation over ζ.

3 Choose a fixed τ and calculate $l_P(\tau; \boldsymbol{y})$.

4 Repeat 3 for the values of τ of interest (usual a grid or sequence to define the function. $l_P(\tau; \boldsymbol{y})$)

The expensive part is that the optimisation in step 2 should be repeated a large number of times.

Example 2.9 *For the Weibull model above, we can calculate the profile likelihood of γ using the R-function;*

```
pll.wei.gam <- function(gamma,y){
    ## function for inner optiisation
    fun.tmp <- function(lambda,gamma,y){
        theta <- c(gamma, lambda)
        nll.wei(theta, y)
        }
    ## inner optimazation
    optimize(fun.tmp, c(0.001,100), gamma = gamma,
             y = y)$objective
}
```

Values of the profile log-likelihood can now be calculated by;

```
## Profile likelihood over gamma
gamma <- seq(0.4,0.9,length=100)
pll.gam <- sapply(gamma,pll.wei.gam,y=y)
```

The figure below shows the profile likelihood for each of the parameters; the rugs indicate the Wald confidence intervals.

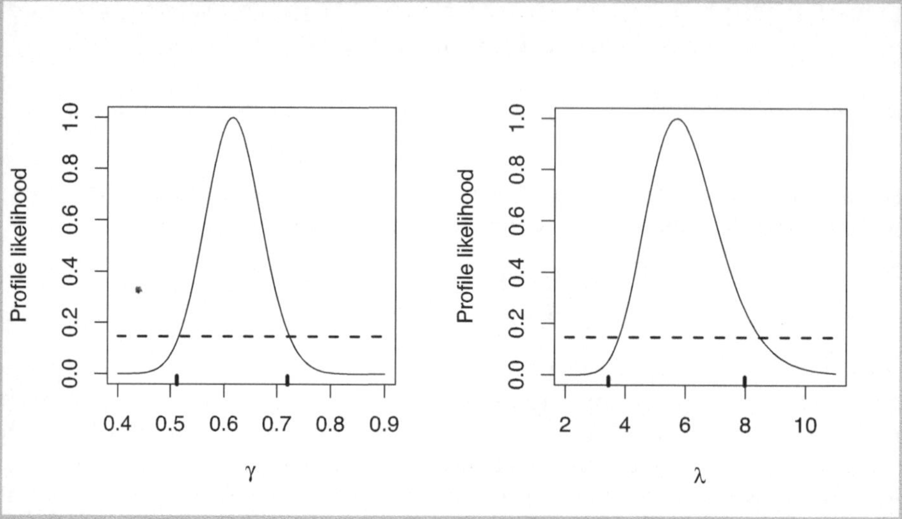

We see that in this case the profile likelihood and the Wald confidence intervals are quite similar though the confidence interval for λ is slightly shifted.

2.6 Parameter transformation

For numerical and practical reasons, parameters are sometimes optimised in a different space than the natural one. To give an example, we might estimate $\log(\theta)$ in the exponential distribution in order to have an unconstrained optimisation.

The likelihood function is invariant to transformation of the parameters (see e.g. Madsen and Thyregod (2011), Pawitan (2013)); therefore, the profile likelihood is invariant under parameter transformations. The Wald statistics relies on the second–order approximation of the likelihood, and it is therefore sensitive to deviations from the normal assumption. Recall that the (profile-) likelihood will be well approximated by the Gaussian if n is large (and hence the Wald statictics will work well in that case). The most common transformation is the log-transformation for positive parameters, but e.g. the *logit* transformation is usually used when estimating probabilities; further examples, related to the generalized linear model will be treated in Chapter 4. For small samples, there are also statistical arguments for transforming the parameter (see below).

The quadratic approximation (second–order Taylor expansion) of the likelihood function is

$$\ell(\theta) \approx \ell(\widehat{\theta}) + \ell'(\widehat{\theta})(\theta - \widehat{\theta}) - \frac{1}{2}j(\widehat{\theta})(\theta - \widehat{\theta})^2, \tag{2.39}$$

and hence

$$\log \frac{L(\theta)}{L(\widehat{\theta})} \approx -\frac{1}{2}j(\widehat{\theta})(\theta - \widehat{\theta})^2, \tag{2.40}$$

which implies that the likelihood ratio can be approximated by a normal distribution.

Example 2.10 *The plots below show the profile likelihood for λ and $\log(\lambda)$ in the Weibull model (lines); along with the quadratic approximation (dashed lines), we see that the quadratic approximation is much better for the transformed parameter than the parameter in the natural domain (the quadratic approximation is closer).*

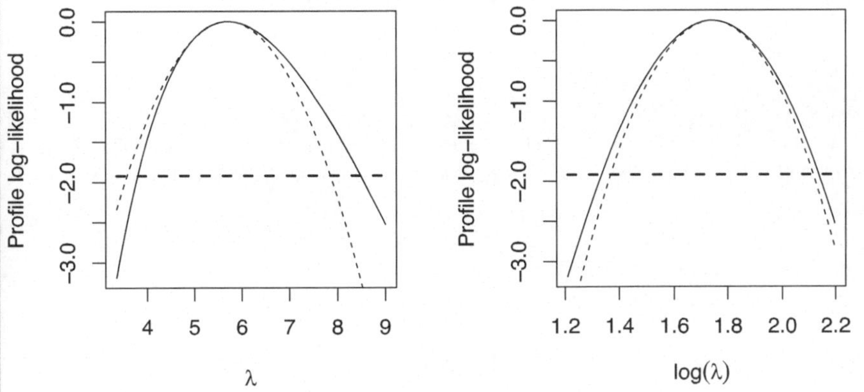

2.7 Data transformation

Sometimes it is beneficial to transform data in the hope transformed data will follow a known distribution better than the original data.

In principle and in practice, such data transformations can be based on diagnostic plots. While this approach is completely acceptable, it excludes quantitative comparison of the likelihood. The problem is that since we are actually changing the data, direct comparison of the likelihood is not possible.

In order to deal with this, we need the change of variable formula for random variables.

Theorem 2.3 (Change of variables) *Let X be a continuous random variable with pdf $f_X(x)$, and let Y be a random variable given by*

$$Y = g(X), \tag{2.41}$$

then the pdf of Y is

$$f_Y(y) = \frac{f_X(x)}{\left|\frac{dg}{dx}\right|}, \tag{2.42}$$

or

$$f_X(x) = f_Y(y)\left|\frac{dg}{dx}\right|. \tag{2.43}$$

For the likelihood, the change of variable formula implies that, if we assume that $g(Y)$ follows a specific distribution, then the likelihood on the original scale can be found as (subscript g refer to likelihood wrt. $g(\boldsymbol{Y})$)

$$L(\boldsymbol{\theta};\boldsymbol{y}) = L_g(\boldsymbol{\theta}; g(\boldsymbol{y})) \prod_i |g'(y_i)|, \tag{2.44}$$

and hence the log-likelihood is

$$l(\boldsymbol{\theta};\boldsymbol{y}) = l_g(\boldsymbol{\theta}; g(\boldsymbol{y})) + \sum_i \log |g'(y_i)|. \tag{2.45}$$

Example 2.11 *In modelling of waiting times, and in particular, in the so-called survival models, it is often assumed that the waiting times follow a* log-logistic distribution, *that is the logarithm of the waiting times follow a logistic distribution. We can find the MLE by;*

```
nll.loglogis <- function(theta,y){
  - sum(dlogis(y, location = theta[1], scale = theta[2],
               log=TRUE))}
fit.logis <- nlminb(c(0,1), nll.loglogis, y = log(y),
                    lower= c(-Inf, 0))
c(fit.logis$par,-fit.logis$objective)

## [1]    0.889    0.970 -140.457
```

The log-likelihood for the Weibull and exponential distribution is (see Examples 2.2 and 2.6)

```
c(-fit.wei$objective, ll.m)

## [1] -209 -231
```

The negative log-likelihood of the log-logistic model is not comparable to the other two models fitted for these data because we have transformed the data. The transformation of the data can be taken into account; in the case of the log-transformation, *we get*

$$l(\boldsymbol{\theta};\boldsymbol{y}) = l_{logis}(\boldsymbol{\theta}; \log(\boldsymbol{y})) + \sum_i \log(1/y_i) \tag{2.46}$$

Th actual number is;

```
(ll.logis <- - fit.logis$objective + sum(log(1/y)))

## [1] -206
```

hence a small improvement (from -209 to -206) of the likelihood, compared to the Weibull distribution. The Weibull and the log-logistic distribution are not nested and hence we cannot use the likelihood ratio test to compare the models.

The method above can be generalised to cases where the transformation depends on one or more parameters (e.g. the Box-Cox transformation, see Box and Cox (1964) and we will further elaborate in this matter in Chapter 3).

2.8 Comparing non-nested models

Whenever possible, we should base the model selection on tests and preferably exact tests. However, as we have seen in this chapter, the exact tests are not always available. In this case, we can use the likelihood ratio test for nested models. When models are not even nested (as the case of Weibull and log-logistic models treated above), we can use so-called information criteria. The most common criteria are Akaikes Information Criteria (AIC) and the Bayesian Information Criteria (BIC), defined by

$$\begin{aligned} AIC &= -2l(\boldsymbol{\theta}) + 2p \\ BIC &= -2l(\boldsymbol{\theta}) + \log(n)p, \end{aligned} \tag{2.47}$$

where p is the number of parameters used in the model and n is the number of observations. Both AIC and BIC penalise parameters and hence favour simple models if the improvement in the likelihood is small. For a large number of observation, BIC gives a larger penalty to increasing the number of parameters, hence favouring simpler models.

Example 2.12 *Since the number of parameters for both the Weibull and the log-logistic models are the same, we do not really have to calculate AIC/BIC to get the conclusion that we should use the log-logistic model. For completeness, these are calculated as shown below:*

```
AIC <- c(Wei = 2 * fit.wei$objective + 2 * 2,
         logis=-2 * ll.logis + 2 * 2)
BIC <- c(Wei = 2 * fit.wei$objective + log(length(y)) * 2,
         logis = -2 * ll.logis + log(length(y)) * 2)
cbind(AIC,BIC)

##       AIC BIC
## Wei   422 427
## logis 415 420
```

The conclusion is that we should go with the log-*logistic model. For completeness, we conclude this example with comparing the estimated distributions with the empirical cumulative density of the data (Figure 2.1).*

We see that the exponential distribution has a worse fit than the other two distributions. It is also difficult to see from the figure that the log-logistic is actually better than the Weibull.

We conclude this chapter with a small note on statistical modelling and a small guide of how to find the most appropriate chapter for the specific data at hand.

2.9 Statistical modelling

An integrated part of statistical modelling should be an appropriate choice of statistical distribution; this might include some choice of data transformation. In the next chapters, we will discuss different regression type of models. In such models, the appropriate choice

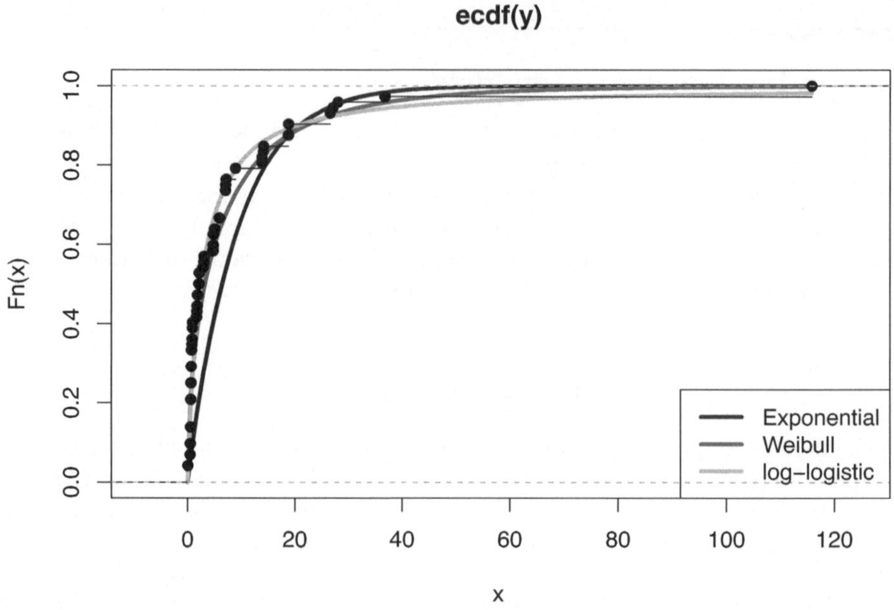

Figure 2.1: Comparing empirical distribution of the blind use data with the etimated distribution functions.

of regressors also becomes important. The question is how one should find the appropriate combination of regressors. While many options exist, the most common ones are forward (start from a simple model and add complexity) or backward selection (start from a complex model and remove terms), but these could also be combined into step-wise selection. In addition, structures in cross correlation matrices may also suggest paths for model development. Further methods related to auto- and cross–correlation functions will be treated in Chapters 8-11.

When the largest model is well–defined, backward selection is a good choice. However in many cases (e.g. polynomial regression and ARMAX models for time-series data), there does not exist a well defined largest model. In such cases forward selection is a good choice, but also computational reasons might lead to the choice of forward selection.

In data–rich situations, one might also use different kinds of cross validations, see for example Hastie et al. (2001). In this book, we do not consider this strategy, but rather focus on statistical test based on in-sample selection (i.e. all data are used for parameter estimation and model selection). The likelihood principle forms the foundation of all models discussed throughout the book, but the specific data of interest or the specific question to be answered will require different models, and each of the following chapters discusses a family of models covering specific situations. Figure 2.2 supplement Figure 1.1 (on page 5) and presents a series of questions one could ask in order to find the most appropriate chapter for further reading in specific situations.

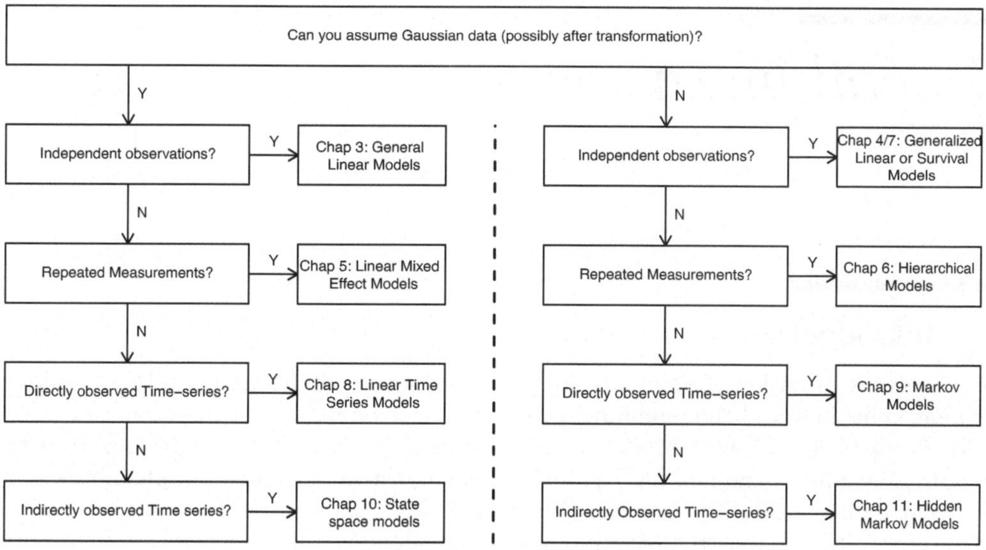

Figure 2.2: Guide to chapters depending on data structure and assumptions on data.

2.10 Concluding remarks

In this chapter, we have considered general likelihood estimation and uncertainty estimation based on properties of the log-likelihood, i.e. using the observed Fisher Information or profile likelihood. Even though we have considered specific examples, the discussion is kept general as likelihood estimation is the overarching principle throughout the book.

Estimation and testing in specific situations can in principle be based directly on the presented theory; however, in some cases the likelihood estimation simplifies with a closed form solution and in other cases a bit more structure than discussed in this chapter is needed in order to set up the likelihood. Finally, the important subject of residual analysis and is not covered at all in this chapter, as this is best covered along with the specific model structure and hence covered in subject–specific chapters.

3

General linear models

3.1 Introduction

This chapter is a revision of a similar chapter in Madsen and Thyregod (2011) but tailored to occupant behaviour and the remaining chapters of this book. One of the simplest regression models is the General Linear Model (GLM), where we assume a linear relationship between two or more variables and Normal residuals. As demonstrated in the example in this section, this could be a relationship between the clothing insulation level (as dependent variable) and the prevailing indoor temperature (as independent variable). There are several types of GLMs distinguished by the class of explanatory variables. The classical GLMs include

i) *regression analysis*
Explanatory variables: *Continuous* and/or *categorical.*

ii) *analysis of variance* ANOVA
Explanatory variables: *Ordinal* and/or *nominal* (i.e. *discrete*).

iii) *analysis of covariance* ANCOVA
Explanatory variables: *Nominal* or *continuous.*

Models in the class of general linear models can be written in the form

$$\boldsymbol{Y} = \boldsymbol{X}\boldsymbol{\beta} + \boldsymbol{\epsilon}; \quad \boldsymbol{\epsilon} \sim N(\boldsymbol{0}, \boldsymbol{\Sigma}). \tag{3.1}$$

If observations are independent, then $\boldsymbol{\Sigma} = \sigma^2 \boldsymbol{I}$, and in this case, the general linear model can be written as (iid is independent and identically distributed; see Section 2.1)

$$Y_i = \boldsymbol{x}_i^T \boldsymbol{\beta} + \epsilon_i; \qquad \epsilon_i \sim N(0, \sigma^2) \text{ and iid} \tag{3.2a}$$

$$= x_{1,i}\beta_1 + x_{2,i}\beta_2 + \cdots + x_{p,i}\beta_p + \epsilon_i; \qquad \epsilon_i \sim N(0, \sigma^2) \text{ and iid,} \tag{3.2b}$$

the formulation includes not only regression, ANOVA, and ANCOVA models but also non-parametric models can be given in terms of the general linear model.

3.2 The multivariate normal distribution

As we saw in Section 3.1, the error/residual structure in the general linear is given by the multivariate normal distribution, with the general formulation given in Definition 3.1.

Definition 3.1 (Multivariate normal distribution) *$\boldsymbol{Z}$ has a k-dimensional multivariate normal distribution if $\boldsymbol{Z}$ has the same distribution as $\boldsymbol{A}\boldsymbol{Y} + \boldsymbol{b}$ for some n, some $k \times n$ matrix $\boldsymbol{A}$, some k dimensional vector $\boldsymbol{b}$, and $\boldsymbol{Y} \sim N(\boldsymbol{0}, \boldsymbol{I})$. We indicate the multivariate normal distribution by writing $\boldsymbol{Z} \sim N(\boldsymbol{b}, \boldsymbol{A}\boldsymbol{A}^T)$.*

DOI: 10.1201/9781003340812-3

The general definition above includes cases where the variance-covariance matrix ($\boldsymbol{A}\boldsymbol{A}^T$) is singular (i.e. the inverse cannot be found; see the example below). When $\boldsymbol{A}\boldsymbol{A}^T$ is not singular, the density for the k-dimensional random vector $\boldsymbol{Z}$ with mean $\boldsymbol{\mu}$ and variance-covariance $\sigma^2\boldsymbol{\Sigma}$ is

$$f_Z(\boldsymbol{z}) = \frac{1}{(2\pi)^{k/2}\sigma^k\sqrt{\det\boldsymbol{\Sigma}}}\exp\left[-\frac{1}{2\sigma^2}(\boldsymbol{z}-\boldsymbol{\mu})^T\boldsymbol{\Sigma}^{-1}(\boldsymbol{z}-\boldsymbol{\mu})\right]. \tag{3.3}$$

Let us illustrate the points with a small example.

Example 3.1 *Assume that we have observations (or rather we plan to make the observations)* (Y_1, Y_2, Y_3)*; also assume* $Y_i \sim N(\mu, \sigma^2)$ *(and iid), the best estimator for* μ *is* $\hat{\mu} = \bar{Y}$*, or written as a general linear model*

$$\boldsymbol{Y} = \mathbf{1}\mu + \epsilon. \tag{3.4}$$

Inserting the estimator, we get

$$\hat{\boldsymbol{Y}} = \mathbf{1}\hat{\mu} = \boldsymbol{H}\boldsymbol{Y}, \tag{3.5}$$

with $\boldsymbol{H}_{ij} \in \mathbb{R}^{3\times 3}$ *and* $\boldsymbol{H}_{ij} = 1/3$*; hence, the variance-covariance matrix of* $\hat{\boldsymbol{Y}}$ *is*

$$V[\hat{\boldsymbol{Y}}] = \boldsymbol{H}V[\boldsymbol{Y}]\boldsymbol{H}^T = \sigma^2\boldsymbol{H}^2 = \sigma^2\boldsymbol{H}, \tag{3.6}$$

where the last equality holds because $\boldsymbol{H}$ *is a projection matrix (see Definition 3.8). Since the rank of* $\boldsymbol{H}$ *is one, we cannot give the formulation (3.3), but we still have* $\hat{\boldsymbol{Y}} \sim N(\mathbf{1}\mu, \boldsymbol{H})$*;* $\boldsymbol{H}$ *is an example of a singular matrix (rank equal 1).*

3.2.1 The normal density as a statistical model

In general, we can formulate a statistical model (with the Gaussian assumption) as

$$\boldsymbol{Y} \sim N_n(\boldsymbol{\mu}, \sigma^2\boldsymbol{\Sigma}), \tag{3.7}$$

and the modelling task is to formulate a model for $\boldsymbol{\mu}$. The variance covariance matrix is also called the *dispersion matrix*

$$D[\boldsymbol{Y}] = \sigma^2\boldsymbol{\Sigma}. \tag{3.8}$$

As stated earlier, we will often assume that $\boldsymbol{\Sigma}$ is a diagonal matrix or that the structure is known. The dispersion matrix also defines an inner product and a corresponding norm.

Definition 3.2 (Inner product and norm) *The bilinear form*

$$\delta_\Sigma(\boldsymbol{y}_1, \boldsymbol{y}_2) = \boldsymbol{y}_1^T\boldsymbol{\Sigma}^{-1}\boldsymbol{y}_2, \tag{3.9}$$

defines an inner product *in* $\mathbb{R}^n$. *Corresponding to this inner product, we can define* orthogonality,*which is obtained when the inner product is zero, and the* norm *is defined by* $\|\boldsymbol{y}\|_\Sigma = \sqrt{\delta_\Sigma(\boldsymbol{y}, \boldsymbol{y})}$.

The definition implies that distances should be seen relative to the distribution or correlation structure. As a small example, to illustrate this, consider the dispersion matrix

$$\begin{aligned}\boldsymbol{\Sigma} &= \begin{bmatrix} 1 & \rho \\ \rho & 1 \end{bmatrix} \\ \boldsymbol{\Sigma}^{-1} &= \frac{1}{1-\rho^2}\begin{bmatrix} 1 & -\rho \\ -\rho & 1 \end{bmatrix}.\end{aligned} \tag{3.10}$$

The norm of a vector is in this case equal

$$\|\boldsymbol{y}\|_{\Sigma}^2 = \frac{1}{1-\rho^2}[y_1 - \rho y_2, -\rho y_1 + y_2]\boldsymbol{y} = \frac{1}{1-\rho^2}(y_1^2 + y_2^2 - 2\rho y_1 y_2). \tag{3.11}$$

Hence, the squared norm (the distance between $(0,0)$ and $(1,1)$) is $\frac{2-2\rho}{1-\rho^2}$; for $\rho = 0$, we have $\|\boldsymbol{y}\|_{\Sigma}^2 = 2$; for $\rho = 0.5$ and $\rho = -0.5$, we have $\|\boldsymbol{y}\|_{\Sigma}^2 = 4/3$ and 4, respectively. So the distance should be seen through the glasses of the distribution.

Definition 3.3 (Deviance for normal distributed variables) *Let us introduce the notation*

$$D(\boldsymbol{y};\boldsymbol{\mu}) = \delta_{\Sigma}(\boldsymbol{y}-\boldsymbol{\mu}, \boldsymbol{y}-\boldsymbol{\mu}) = (\boldsymbol{y}-\boldsymbol{\mu})^T\boldsymbol{\Sigma}^{-1}(\boldsymbol{y}-\boldsymbol{\mu}) \tag{3.12}$$

to denote the quadratic norm of the vector $(\boldsymbol{y}-\boldsymbol{\mu})$ *corresponding to the inner product defined by* $\boldsymbol{\Sigma}^{-1}$.

With Definition 3.3, we can write the normal probability density function in terms of the deviance

$$f(\boldsymbol{y};\boldsymbol{\mu},\sigma^2) = \frac{1}{(\sqrt{2\pi})^n\sigma^n\sqrt{\det(\boldsymbol{\Sigma})}} \exp\left[-\frac{1}{2\sigma^2}D(\boldsymbol{y};\boldsymbol{\mu})\right]. \tag{3.13}$$

The distribution function directly gives the likelihood formulation; it is to be understood that $\boldsymbol{y} \in \mathbb{R}^n$, and that the expected values $\boldsymbol{\mu}$ are parameterised through a set of parameters and some known input (just as in the design matrix given in (3.1)). From (3.13), we obtain the likelihood function

$$L(\boldsymbol{\mu},\sigma^2;\boldsymbol{y}) = \frac{1}{(\sqrt{2\pi})^n\sigma^n\sqrt{\det(\boldsymbol{\Sigma})}} \exp\left[-\frac{1}{2\sigma^2}D(\boldsymbol{y};\boldsymbol{\mu})\right]. \tag{3.14}$$

The log-likelihood function is (apart from an additive constant)

$$\ell_{\mu,\sigma^2}(\boldsymbol{\mu},\sigma^2;\boldsymbol{y}) = -\frac{n}{2}\log(\sigma^2) - \frac{1}{2\sigma^2}(\boldsymbol{y}-\boldsymbol{\mu})^T\boldsymbol{\Sigma}^{-1}(\boldsymbol{y}-\boldsymbol{\mu}). \tag{3.15}$$

The score function wrt. $\boldsymbol{\mu}$ is thus

$$\frac{\partial}{\partial\boldsymbol{\mu}}\ell_{\mu,\sigma^2}(\boldsymbol{\mu},\sigma^2;\boldsymbol{y}) = \frac{1}{\sigma^2}\left[\boldsymbol{\Sigma}^{-1}\boldsymbol{y} - \boldsymbol{\Sigma}^{-1}\boldsymbol{\mu}\right] = \frac{1}{\sigma^2}\boldsymbol{\Sigma}^{-1}(\boldsymbol{y}-\boldsymbol{\mu}), \tag{3.16}$$

and the observed information (wrt. $\boldsymbol{\mu}$) is

$$j(\boldsymbol{\mu};\boldsymbol{y}) = \frac{1}{\sigma^2}\boldsymbol{\Sigma}^{-1}. \tag{3.17}$$

Note that the observed information is independent from the expected value while it depends on the variance-covariance matrix. It is sometimes useful to consider the full (or saturated) model.

Definition 3.4 (The full (or saturated) model) *In the full (or saturated) model,* $\boldsymbol{\mu}$ *varies freely in* $\mathbb{R}^n$, *and, hence, each observation fits the model perfectly, i.e.* $\widehat{\boldsymbol{\mu}} = \boldsymbol{y}$.

The saturated model does not play a very important role for the general linear model, because we usually estimate the dispersion parameter (σ^2), but we will later see that in the class of generalized linear model, we might use the saturated model to test for model sufficiency.

3.3 General linear models

In this chapter, we, as mentioned above, consider the statistical model given by

$$\boldsymbol{Y} \sim N_n(\boldsymbol{\mu}, \sigma^2\boldsymbol{\Sigma}) \text{ for } \boldsymbol{y} \in \mathbb{R}^n. \tag{3.18}$$

The formulation above include general mean–value and covariance structures, but is restricted to Gaussian models. This is an assumption that should be checked as a part of the modelling. The formulation of a distribution assumption (in this case the normal) is an integrated part of the modelling.

3.3.1 The general linear model

In the general linear model, we further make the restriction that the mean value is in a linear (strictly speaking affine, i.e. an additive constant) subspace of $\mathbb{R}^n$.

Definition 3.5 (The general linear model) *Assume that* $Y_1, Y_2, \ldots, Y_n$ *is normally distributed as described by* (3.18). *A* general linear model *for* $Y_1, Y_2, \ldots, Y_n$ *is a model of the form* (3.18), *where an affine hypothesis is formulated for* $\boldsymbol{\mu}$. *The hypothesis is of the form*

$$H_0 \; : \; \boldsymbol{\mu} - \boldsymbol{\mu}_0 \in \Omega_0 \; , \tag{3.19}$$

where Ω_0 *is a linear subspace of* $\mathbb{R}^n$ *of dimension* k *and* $\boldsymbol{\mu}_0$ *denotes a vector of* a priori known offset values.

The dimension of a general linear model is equal to the number of free parameters to be estimated, that is

Definition 3.6 (Dimension of general linear model) *The dimension of the subspace* Ω_0 *for the linear model* (3.19) *is the* dimension of the model.

The basic initial job of the modeller is to specify the design matrix as defined below.

Definition 3.7 (Design matrix for classical GLM) *Assume that the linear subspace* $\Omega_0 = span\{x_1, \ldots, x_k\}$, *i.e. the subspace is spanned by* k *vectors* ($k < n$).

Consider a general linear model (3.19) *where the hypothesis can be written as*

$$H_0 \colon \boldsymbol{\mu} - \boldsymbol{\mu}_0 = \boldsymbol{X}\boldsymbol{\beta} \quad \text{with } \boldsymbol{\beta} \in \mathbb{R}^k, \tag{3.20}$$

where $\boldsymbol{X}$ *has full rank. The* $n \times k$ *matrix* $\boldsymbol{X}$ *of known deterministic coefficients is called the* design matrix.

The ith *row of the design matrix is given by the* model vector

$$\boldsymbol{x}_i^T = \begin{bmatrix} x_{i1} & x_{i2} & \ldots & x_{ik} \end{bmatrix}. \tag{3.21}$$

When modelling, the modeller should define an initial design matrix that can then be developed using statistical methods either by removing non–significant terms or by adding more terms to a simple model. The set up is illustrated in the example below:

Example 3.2 (Clothing insulation level) *The level of clothing a person is wearing at office is one of the key factors influencing their level of comfort (Fanger, 1970). In addition, the level of comfort influences the need for cooling and/or heating.*

The following figure shows a histogram of clothing insulation levels worn by subjects of three experimental studies in the Laboratory of Occupant Behaviour, Satisfaction, Thermal comfort and Environmental Research (LOBSTER) (Schweiker and Wagner,

2015).

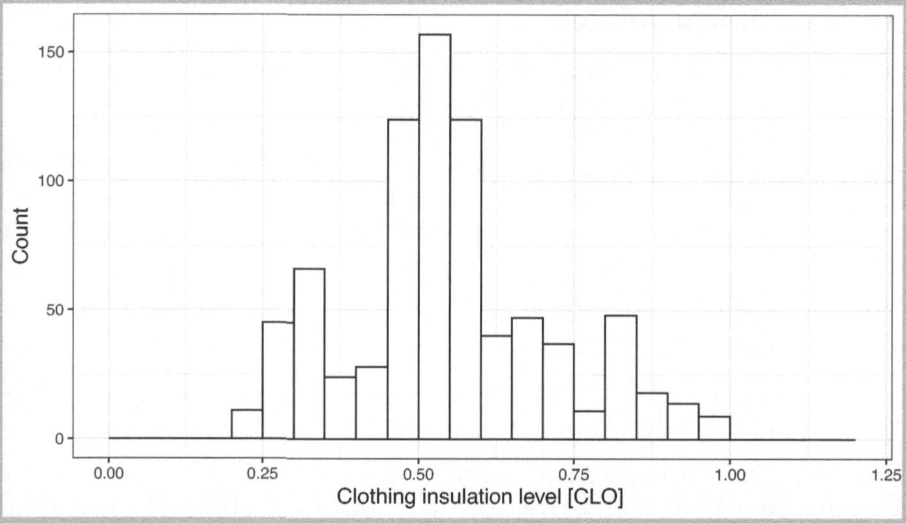

Studies exist, showing the influence of the outdoor air temperature on the level of clothing (De Carli et al., 2007; Schiavon and Lee, 2013; Schweiker and Wagner, 2015).

Based on these findings, it can be assumed that the clothing insulation level (`clo`*) is affected by the outdoor air temperature (*`tOut`*). Additionally, we can assume that there is an additional effect of the indoor conditions, represented in the following example with the variable of indoor operative temperature (*`tInOp`*). A simple scatterplot between* `clo` *and* `tOut`*, and* `tInOp` *shows that both temperatures have a negative impact on clothing insulation level (higher temperature give lower insulation level); also note that there is a positive correlation between indoor and outdoor temperature; we will discuss the implication of correlations between explanatory variables later (Section 3.7).*

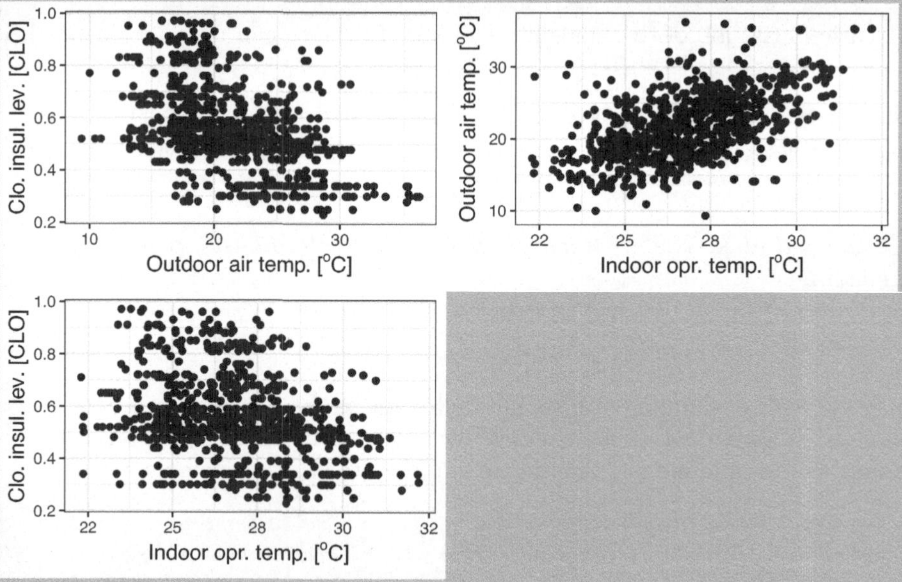

With `clo` *as outcome variable and* `tOut` *and* `tInOp` *as continuous explanatory vari-*

ables, an initial general linear model could be

$$\texttt{clo}_i = \beta_0 + \texttt{tOut}_i\beta_1 + \texttt{tInOp}_i\beta_2 + \epsilon_i, \tag{3.22}$$

or in a more general formulation

$$y_i = \beta_0 + x_{i1}\beta_1 + x_{i2}\beta_2 + \epsilon_i; \quad \epsilon_i \sim N(0, \sigma^2). \tag{3.23}$$

Also, we assume that ϵ_i is a white noise (i.e. independent) sequence (we will discuss this assumption further in Section 3.11).

In order to give the matrix-vector formulation, we can write regressors and parameters as

$$\begin{aligned} \boldsymbol{x}_i^T =& (1, x_{i1}, x_{i2}) \\ \boldsymbol{\beta} =& (\beta_0, \beta_1, \beta_2)^T, \end{aligned} \tag{3.24}$$

In matrix vector notation, we can write

$$\boldsymbol{Y} = \boldsymbol{X}\boldsymbol{\beta} + \boldsymbol{\epsilon}; \quad \boldsymbol{\epsilon} \sim N(\boldsymbol{0}, \sigma^2 \boldsymbol{I}), \tag{3.25}$$

where the first column in $\boldsymbol{X}$ is a vector of ones, for example

$$\boldsymbol{X} = \begin{bmatrix} 1 & x_{1,1} & x_{1,2} \\ 1 & x_{2,1} & x_{2,2} \\ \vdots & \vdots & \vdots \\ 1 & x_{n,1} & x_{n,2} \end{bmatrix}. \tag{3.26}$$

From the example above, please note that the (unconditional) clothing insulation level is clearly not Gaussian. The Gaussian assumption is, however, on conditional distribution (i.e. different expected values depending on the temperature). The Gaussian assumption is, therefore, best treated by examining the residuals; we will get back to this in Section 3.4.3 after we have discussed estimation of parameters.

3.4 Estimation of parameters

With the formulation given above, we need to find the parameters of the model. The parameters are $\boldsymbol{\beta}$ and σ^2, while $\boldsymbol{\Sigma}$ is assumed known (and often set to $\boldsymbol{I}$). In addition to the point estimate for the parameters, we will need to quantify the uncertainty of the parameters as well as mean value and uncertainty of predictions.

3.4.1 Estimation of mean value parameters

The mean value parameter is found by the so-called normal equation that can be derived from Definition 3.3 by differentiating the log-likelihood with respect to (wrt.) $\boldsymbol{\beta}$. The solution is given in the theorem below.

Theorem 3.1 (ML estimates of mean value parameters) *For a hypothesis of the form*

$$H_0\colon \boldsymbol{\mu}(\boldsymbol{\beta}) = \boldsymbol{X}\boldsymbol{\beta}, \tag{3.27}$$

the maximum likelihood estimate for $\boldsymbol{\beta}$ is found as a solution to the normal equation

$$\boldsymbol{X}^T\boldsymbol{\Sigma}^{-1}\boldsymbol{y} = \boldsymbol{X}^T\boldsymbol{\Sigma}^{-1}\boldsymbol{X}\widehat{\boldsymbol{\beta}}. \tag{3.28}$$

If $\boldsymbol{X}$ has full rank, the solution is uniquely given by

$$\widehat{\boldsymbol{\beta}} = (\boldsymbol{X}^T\boldsymbol{\Sigma}^{-1}\boldsymbol{X})^{-1}\boldsymbol{X}^T\boldsymbol{\Sigma}^{-1}\boldsymbol{y}. \tag{3.29}$$

Note that the estimates are independent of the variance (σ^2) in the system, but depend on the correlation structure ($\boldsymbol{\Sigma}$). Also, note that for $\boldsymbol{\Sigma} = \boldsymbol{I}$, we have

$$\widehat{\boldsymbol{\beta}} = (\boldsymbol{X}^T\boldsymbol{X})^{-1}\boldsymbol{X}^T\boldsymbol{y}. \tag{3.30}$$

In this case, it is clear that the estimates are independent from the residual variance. In the equations above, we have formulated the results in terms of observed values of the outcome ($\boldsymbol{y}$), this leads to the estimates. We could also formulate the parameters in terms of the random vector $\boldsymbol{Y}$, i.e.

$$\widehat{\boldsymbol{\beta}} = (\boldsymbol{X}^T\boldsymbol{\Sigma}^{-1}\boldsymbol{X})^{-1}\boldsymbol{X}^T\boldsymbol{\Sigma}^{-1}\boldsymbol{Y}. \tag{3.31}$$

In this formulation, $\widehat{\boldsymbol{\beta}}$ is an estimator and we can study the statistical properties of this estimator; the mean and variance of the estimator become (using $\boldsymbol{Y} = \boldsymbol{X}\boldsymbol{\beta} + \boldsymbol{\epsilon}$)

$$\begin{aligned} E[\widehat{\boldsymbol{\beta}}] =& E[(\boldsymbol{X}^T\boldsymbol{\Sigma}^{-1}\boldsymbol{X})^{-1}\boldsymbol{X}^T\boldsymbol{\Sigma}^{-1}\boldsymbol{Y}] \\ =&(\boldsymbol{X}^T\boldsymbol{\Sigma}^{-1}\boldsymbol{X})^{-1}\boldsymbol{X}^T\boldsymbol{\Sigma}^{-1}E[\boldsymbol{X}\boldsymbol{\beta} + \boldsymbol{\epsilon}] \\ =&(\boldsymbol{X}^T\boldsymbol{\Sigma}^{-1}\boldsymbol{X})^{-1}\boldsymbol{X}^T\boldsymbol{\Sigma}^{-1}\boldsymbol{X}\boldsymbol{\beta} \\ =&\boldsymbol{\beta}. \end{aligned} \tag{3.32}$$

Hence the estiomator is unbiassed. The variance is calculated by

$$\begin{aligned} V[\widehat{\boldsymbol{\beta}}] =& V[(\boldsymbol{X}^T\boldsymbol{\Sigma}^{-1}\boldsymbol{X})^{-1}\boldsymbol{X}^T\boldsymbol{\Sigma}^{-1}\boldsymbol{Y}] \\ =&(\boldsymbol{X}^T\boldsymbol{\Sigma}^{-1}\boldsymbol{X})^{-1}\boldsymbol{X}^T\boldsymbol{\Sigma}^{-1}V[\boldsymbol{\epsilon}]\left((\boldsymbol{X}^T\boldsymbol{\Sigma}^{-1}\boldsymbol{X})^{-1}\boldsymbol{X}^T\boldsymbol{\Sigma}^{-1}\right)^T \\ =&(\boldsymbol{X}^T\boldsymbol{\Sigma}^{-1}\boldsymbol{X})^{-1}\boldsymbol{X}^T\boldsymbol{\Sigma}^{-1}V[\boldsymbol{\epsilon}]\boldsymbol{\Sigma}^{-1}\boldsymbol{X}(\boldsymbol{X}^T\boldsymbol{\Sigma}^{-1}\boldsymbol{X})^{-1} \\ =&\sigma^2(\boldsymbol{X}^T\boldsymbol{\Sigma}^{-1}\boldsymbol{X})^{-1}\boldsymbol{X}^T\boldsymbol{\Sigma}^{-1}\boldsymbol{X}(\boldsymbol{X}^T\boldsymbol{\Sigma}^{-1}\boldsymbol{X})^{-1} \\ =&\sigma^2(\boldsymbol{X}^T\boldsymbol{\Sigma}^{-1}\boldsymbol{X})^{-1}. \end{aligned} \tag{3.33}$$

Since the estimator is a linear combination of Gaussian random variables, we have the following result on the distribution of the mean value estimator.

Theorem 3.2 (Properties of the ML estimator) *For the ML estimator* (3.29), *we have*

$$\widehat{\boldsymbol{\beta}} \sim N_k(\boldsymbol{\beta}, \sigma^2 \, (\boldsymbol{X}^T\boldsymbol{\Sigma}^{-1}\boldsymbol{X})^{-1}). \tag{3.34}$$

The distribution of the estimator is used for formulating so-called marginal test; for known variance (σ^2), we will have

$$\frac{\hat{\beta}_i - \beta_i}{se_i} \sim N(0,1), \tag{3.35}$$

where $se_i = \sigma\sqrt{(\boldsymbol{X}^T\boldsymbol{\Sigma}^{-1}\boldsymbol{X})^{-1})_{ii}}$. Of course, we usually do not have access to the true variance and we will have to rely on an estimate of the variance. The marginal maximum likelihood estimator (which is also a central estimator) for the variance is:

Theorem 3.3 (Estimation of the variance) *Under the hypothesis* (3.27), *the maximum marginal likelihood estimator for the variance* σ^2 *is*

$$\widehat{\sigma}^2 = \frac{D(\boldsymbol{y}; \boldsymbol{X}\widehat{\boldsymbol{\beta}})}{n-k} = \frac{(\boldsymbol{y} - \boldsymbol{X}\widehat{\boldsymbol{\beta}})^T\boldsymbol{\Sigma}^{-1}(\boldsymbol{y} - \boldsymbol{X}\widehat{\boldsymbol{\beta}})}{n-k}. \tag{3.36}$$

Further, $\widehat{\sigma}^2 \sim \sigma^2\chi^2_{n-k}/(n-k)$.

As can be seen from Theorem 3.3, the variance estimator is a random variable related to the χ^2 distribution. Inserting the estimator for the variance in (3.35), we get a marginal t-test.

Theorem 3.4 (Marginal t-test) *Under the null–hypothesis* $H_0 : \beta_i = \beta_{i,0}$ *the test statistic*

$$T = \frac{\hat{\beta}_i - \beta_{i,0}}{se_i}, \tag{3.37}$$

with $se_i = \hat{\sigma}\sqrt{(\boldsymbol{X}^T\boldsymbol{\Sigma}^{-1}\boldsymbol{X})^{-1})_{ii}}$ *follows a t-distribution with* $n-k$ *degrees of freedom* ($T \sim t(n-k)$). *Also* $1-\alpha$ *confidence intervals are given by*

$$\hat{\beta}_i \pm t_{1-\alpha/2}(n-k)se_i, \tag{3.38}$$

where $t_{1-\alpha/2}(n-k)$ *is short for the* $1-\alpha/2$ *quantile in a t-distribution with* $n-k$ *degrees of freedom.*

As can be seen from the theorem, we can test general hypothesis. However, we are usually interested in testing if a particular parameter is equal to zero. Also note that when we apply marginal tests, we should do so sequentially, i.e. removing one parameter at the time. Now that we have the tool to analyse the linear model, we can get back to our clothing example.

Example 3.3 (Clothing insulation level (cont.)) *Coming back to our multiple linear regression model with clothing insulation level,* `clo`, *as the response variable and outdoor air temperature,* `tOut`, *and indoor operative temperature,* `tInOp`, *as the explanatory variables, we can define and fit a linear model in the following way:*

```
## Define and fit linear model
modLMEx11 <- lm(clo ~ tOut * tInOp, data = dfLMEx1)
## Summary of linear model fit
summary(modLMEx11)

##
## Call:
## lm(formula = clo ~ tOut * tInOp, data = dfLMEx1)
##
## Residuals:
##     Min      1Q  Median      3Q     Max
## -0.3339 -0.0957 -0.0102  0.0725  0.4142
##
## Coefficients:
##              Estimate Std. Error t value Pr(>|t|)
## (Intercept)  1.337448   0.318936    4.19 0.000031 ***
## tOut        -0.028154   0.014167   -1.99    0.047 *
## tInOp       -0.017630   0.011988   -1.47    0.142
## tOut:tInOp   0.000504   0.000521    0.97    0.334
## ---
## Signif. codes:  0 '***' 0.001 '**' 0.01 '*' 0.05 '.' 0.1 ' ' 1
##
## Residual standard error: 0.14 on 799 degrees of freedom
## Multiple R-squared:  0.215,Adjusted R-squared:  0.212
## F-statistic: 73.1 on 3 and 799 DF,  p-value: <2e-16
```

From the R-output, we can see that the estimated standard deviation (residuals standard error) is $\hat{\sigma} = 0.14$*, and the standard error for the individual parameters is also given in the table. The last two columns give the t-test statistics and the associated p-value, we can see that both the interaction term and the indoor operative temperature are non-significant, hence we should remove the interaction term.*

The output also provides a summary of the residuals (min, 1Q, Median, 3Q, max), this can be used as a quick check of the residuals (i.e. skewnees), the numbers are the ones needed to construct a standard box-plot.

The parameters of the reduced model is given below (reduced R-output);

```
# Define and fit linear model
modLMEx12 <- lm(clo ~ tOut + tInOp, data = dfLMEx1)

## [...]
## Coefficients:
##             Estimate Std. Error t value Pr(>|t|)
## (Intercept)  1.03785    0.07606   13.65   <2e-16 ***
## tOut        -0.01451    0.00130  -11.14   <2e-16 ***
## tInOp       -0.00647    0.00327   -1.98    0.048 *
## ---
## Residual standard error: 0.14 on 800 degrees of freedom
## Multiple R-squared:  0.215,Adjusted R-squared:  0.213
## [...]
```

We see that both indoor and outdoor temperatures are now significant. The R-squared value signifies this for the dataset we used, the explained variance does not decrease much (from 0.2154 to 0.2145) by removing the interaction.

Another hypothesis based on field studies, e.g. Karjalainen (2012), suggests that there are differences between males and females with respect to the level of clothing worn. Therefore, we add the binary factor, sex of the person, to the model and refit (reduced R-output);

```
# define and fit linear model with one continuous and one
# binary variable and interaction term
modLMEx13 <- lm(clo ~ tOut * sex + tInOp * sex,
                data = dfLMEx1)

## [...]
## Coefficients:
##                Estimate Std. Error t value Pr(>|t|)
## (Intercept)     1.31524    0.10312   12.76  < 2e-16 ***
## tOut           -0.01786    0.00177  -10.11  < 2e-16 ***
## sexmale        -0.62354    0.14528   -4.29 0.000020 ***
## tInOp          -0.01268    0.00461   -2.75   0.0062 **
## tOut:sexmale    0.01070    0.00249    4.30 0.000019 ***
## sexmale:tInOp  0.01158    0.00621    1.86   0.0626 .
## ---
## [...]
```

Again, we see non-significant parameter estimates, namely interaction between sex and indoor operative temperature. We therefore remove the interaction;

```
# define and fit linear model with one continuous and one
# binary variable and interaction term
modLMEx14 <- lm(clo ~ tOut * sex + tInOp, data = dfLMEx1)

## [...]
## Coefficients:
##               Estimate Std. Error t value Pr(>|t|)
## (Intercept)    1.17812    0.07238   16.28  < 2e-16 ***
## tOut          -0.01949    0.00154  -12.69  < 2e-16 ***
## sexmale       -0.36665    0.04612   -7.95  6.4e-15 ***
## tInOp         -0.00628    0.00309   -2.03    0.043 *
## tOut:sexmale   0.01321    0.00210    6.30  4.8e-10 ***
## ---
## Signif. codes:  0 '***' 0.001 '**' 0.01 '*' 0.05 '.' 0.1 ' ' 1
##
## Residual standard error: 0.13 on 798 degrees of freedom
## Multiple R-squared:  0.314,Adjusted R-squared:  0.31
## [...]
```

In this case, we can observe that the coefficient for sex is significantly different from

0 and that adding sex increases the explained variance to about 30%. We also see that females are (slightly) more responsive to outdoor temperatures than males.

The output of the model can be visualised as follows (with the indoor temperature fixed at the average observed level): It shows that females are varying their level of clothing much stronger with varying outdoor conditions than males. The shaded areas represent 95% confidence intervals (the meaning of which will be discussed below).

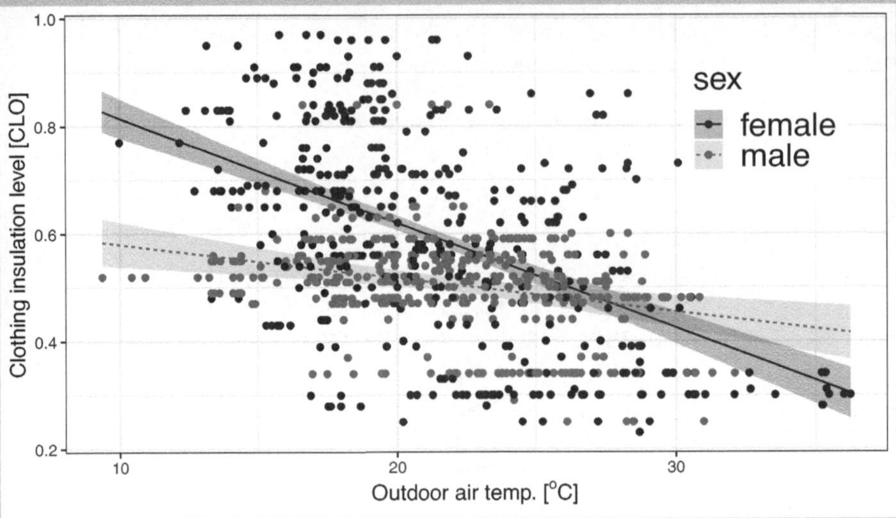

3.4.2 Fitted values

In order to keep the notation simple, we will assume that $\boldsymbol{\Sigma} = \boldsymbol{I}$ in this subsection. The statistical properties of the fitted values and hence also the residuals are best described through projections.

Definition 3.8 (Projection matrix) *A matrix $\boldsymbol{H}$ is a* projection matrix *if and only if (a) $\boldsymbol{H}^T = \boldsymbol{H}$ and (b) $\boldsymbol{H}^2 = \boldsymbol{H}$, i.e. the matrix is* idempotent.

We can write the general linear model in terms of the estimators of the parameters,

$$\begin{aligned}\hat{\boldsymbol{Y}} =& \boldsymbol{X}\hat{\boldsymbol{\beta}} \\ =& \boldsymbol{X}(\boldsymbol{X}^T\boldsymbol{X})^{-1}\boldsymbol{X}^T\boldsymbol{Y} \\ =& \boldsymbol{H}\boldsymbol{Y}.\end{aligned} \tag{3.39}$$

It is a simple exercise to show that $E[\hat{\boldsymbol{Y}}] = \boldsymbol{X}\boldsymbol{\beta}$, and it can also be shown that $\boldsymbol{H}$ is an projection matrix, and hence we have

$$\begin{aligned}V[\widehat{\boldsymbol{Y}}] =& \boldsymbol{H}V[\boldsymbol{Y}]\boldsymbol{H}^T \\ =& \sigma^2\boldsymbol{H}\boldsymbol{H} = \sigma^2\boldsymbol{H}.\end{aligned} \tag{3.40}$$

In addition, we can write the observed residuals as

$$\boldsymbol{r} = \boldsymbol{y} - \boldsymbol{X}\widehat{\boldsymbol{\beta}} = (\boldsymbol{I} - \boldsymbol{H})\boldsymbol{y}. \tag{3.41}$$

When $\boldsymbol{H}$ is a projection matrix, then $\boldsymbol{I} - \boldsymbol{H}$ is also a projection matrix

$$\begin{aligned}(\boldsymbol{I} - \boldsymbol{H})^T =& \boldsymbol{I}^T - \boldsymbol{H}^T = \boldsymbol{I} - \boldsymbol{H} \\ (\boldsymbol{I} - \boldsymbol{H})^2 =& \boldsymbol{I} + \boldsymbol{H}^2 - 2\boldsymbol{H} = \boldsymbol{I} - \boldsymbol{H}.\end{aligned} \tag{3.42}$$

Hence the variance of the observed residuals (when viewed as random variables) is

$$V[\boldsymbol{r}] = V[(\boldsymbol{I}-\boldsymbol{H})\boldsymbol{Y}] = \sigma^2(\boldsymbol{I}-\boldsymbol{H}). \tag{3.43}$$

This implies that the variance of the residuals depend on the design–matrix, and in the residual analysis, we should take this into account; if σ^2 is known, then

$$\frac{r_i}{\sigma\sqrt{1-h_{ii}}} \sim N(0,1), \tag{3.44}$$

but as usually we do not have direct access to the variance and hence the observed residuals are usually normalised in one of the two following ways:

Definition 3.9 (Standardized residual) *The standardized residual is given by*

$$r_i^{rs} = \frac{r_i}{\sqrt{\widehat{\sigma}^2(1-h_{ii})}}. \tag{3.45}$$

Definition 3.10 (Studentized residual) *The studentized residual is given by*

$$r_i^{rt} = \frac{r_i}{\sqrt{\widehat{\sigma}^2_{(i)}(1-h_{ii})}}, \tag{3.46}$$

where $\widehat{\sigma}^2_{(i)}$ denotes the estimate of σ^2 determined by deleting the ith observation.

The standardized residuals are simply constructed by replacing σ with $\hat{\sigma}$ in (3.44). In this construction, the denominator and enumerator are not independent resulting in a complicated distribution for the residuals. This problem is dealt with in the studentized residuals by excluding the ith observation in the estimation of the variance, hence making the enumerator and the denominator independent.

The diagonal terms of $\boldsymbol{H}$ are called leverage.

Definition 3.11 (Leverage) *The ith diagonal element in the hat–matrix ($\boldsymbol{H}$) is called the* leverage *of the ith observation.*

Leverage is a measure of how unusual a point is in terms of the explanatory variables. An unusual point might also be very influential, and hence we should keep an eye on (and identify) those point in the estimation.

3.4.3 Model diagnostics

In order for the inference made in the estimation to be reliable the model assumptions should hold, and hence we need to check these. The simplest way to do this is by examining diagnostic plots. The assumptions can be boiled down to $\boldsymbol{\epsilon} \sim N(\boldsymbol{0}, \sigma^2\boldsymbol{I})$ or that the residuals are iid and normal. This implies that we should check that

1: There are no systematic effects in the mean (plot residuals vs. fitted).

2: There are no systematic effect in the variance (plot absolute residuals vs. fitted).

3: The residuals are normal (normal quantile quantile plot).

In addition, usually the fitted model is checked for influential observations (combination of size of residual and leverage). If there is a natural ordering in data (e.g. time series), the residual autocorrelation (see Section 3.11 and Chapter 8) should also be checked.

Example 3.4 (Clothing insulation level (cont.)) *The figure below shows diagnostic plots for the model fitted in Example 3.3.*

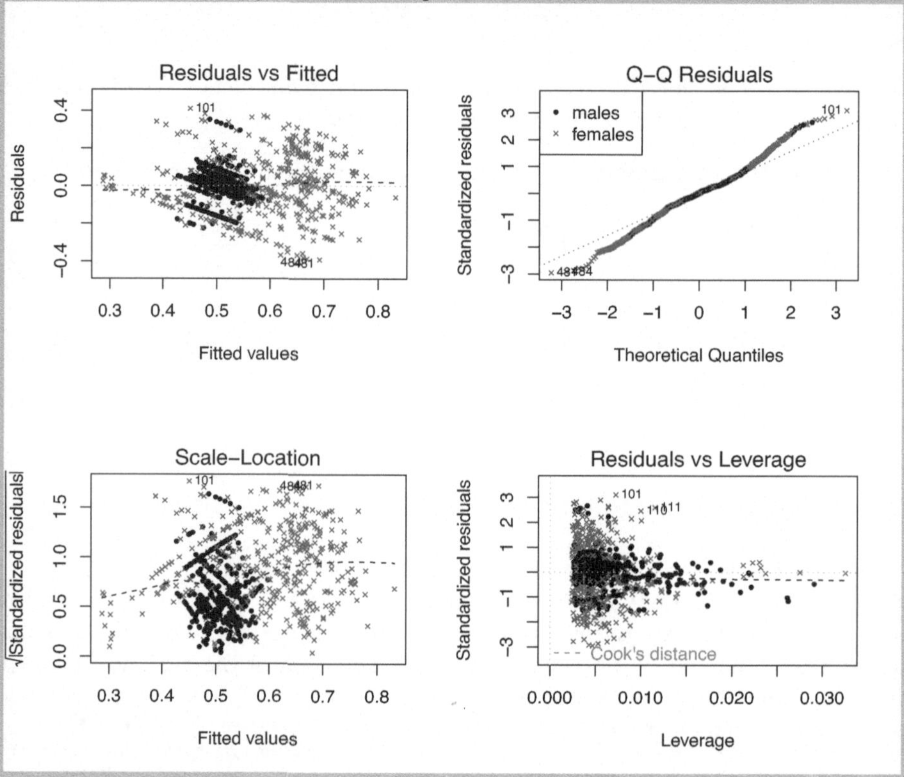

In the top left figure, we see the effect of clothing level acting as a double–bounded variable (between 0 and 1). The top right figure indicates that there is some deviation from the normal distribution (dotted diagonal line) though it is not too serious. The lower left figure indicates a slight increase in variance with higher fitted values. The lower right figure does not indicate any very influential points. There are points with high leverage (an often used limit for high leverage is $2p/n$; here $2 \cdot 5/803 = 0.012$), but these have a low standardized residual and are therefore are not marked a influential (Cooks distance<0.5 for all points). There is also an indication that residual for women have higher variance than residual for males. This can be investigated by plotting residual as a function of explanatory variables.

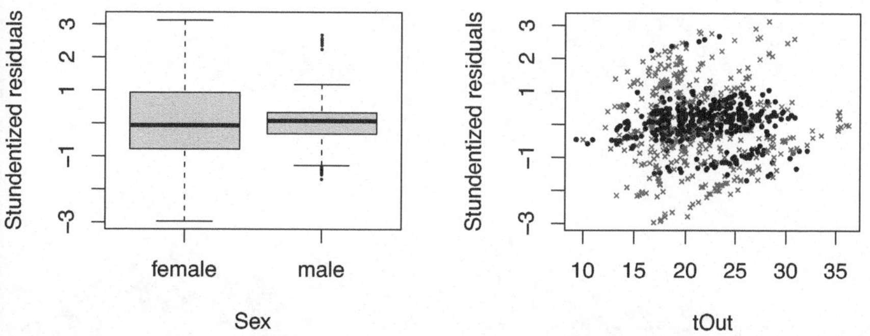

We now clearly see that there is a difference in variance. In conclusion, there are some issues that should be dealt with; we will get back to this in Section 3.12.

In the example above, we have seen how the residual analysis can pinpoint problems in the model formulation that should be solved. In general, we should use the residuals in model development. Figure 3.1 shows generalized residual plots. The action required depends on what is on the x-axis (fitted values, time or a (possible) explanatory variable). Table 3.1 gives an overview of the actions required. Some of the actions involve higher order terms in the model or transformations of the dependent variable. We will get back to these solutions in in Section 3.12.

3.4.4 Partitioning of variation

In order to construct (or derive) tests, we need distributions of test statictics; the first step in this construction (derivation) is partitioning of the variation. The deviance of the residuals can be split into two parts as given below.

Remark 3.1 (Partitioning of the variation) *The deviance between data and the expected value can be split in two parts*

$$D(\boldsymbol{y}; \boldsymbol{X}\boldsymbol{\beta}) = D(\boldsymbol{y}; \boldsymbol{X}\widehat{\boldsymbol{\beta}}) + D(\boldsymbol{X}\widehat{\boldsymbol{\beta}}; \boldsymbol{X}\boldsymbol{\beta}). \tag{3.47}$$

Table 3.1: Actions required corresponding to Figure 3.1

x-axis	Figure 1	Figure 2	Figure 3	Figure 4
$\hat{\mu}_i$	Acceptable	Variance grows with $\hat{\mu}_i$. Do a weighted analysis (see Section 3.10) or transform the Y's	Lack of constant term. Regression possibly erroneously forced trough zero	Bad model, try transforming the Y's
$x_{j,i}$	Acceptable	Variance grows with x_i. Perform weighted analysis or transform the Y's.	Error in the computations	Lack of quadratic term in x_j
Time	Acceptable	The variance grows with time. Do a weighted analysis	Lack of term of the form $\beta \cdot$ time	Lack of term of the form $\beta_1 \cdot$ time^2

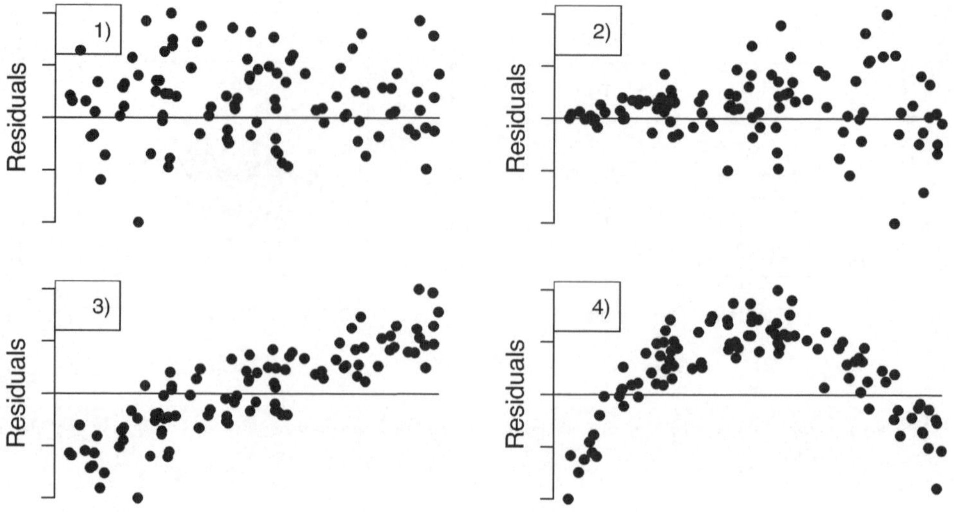

Figure 3.1: Residuals as a function of fitted values, explanatory variables or time (see Table 3.1).

In the remark above, the left–hand side is the true variation (i.e. difference between mean value structure and observation), whereas the first term on the right–hand side is the deviance between the fitted model and observed residuals (i.e. what we observe). The last term is the difference between the estimated model and the true mean value structure. Each of these terms in (3.47) can be assigned as a probability distribution.

Remark 3.2 (**χ^2-distribution of individual contributions in** (3.47)) *Under H_0, as in* (3.27)*, it follows from the normal distribution of $\boldsymbol{Y}$ that*

$$D(\boldsymbol{y}; \boldsymbol{X\beta}) = (\boldsymbol{y} - \boldsymbol{X\beta})^T \Sigma^{-1} (\boldsymbol{y} - \boldsymbol{X\beta}) \sim \sigma^2 \chi^2_n. \tag{3.48}$$

Furthermore, it follows from the normal distribution of $\boldsymbol{r}$ and of $\widehat{\boldsymbol{\beta}}$ that

$$\begin{aligned} D(\boldsymbol{y}; \boldsymbol{X}\widehat{\boldsymbol{\beta}}) &= (\boldsymbol{y} - \boldsymbol{X}\widehat{\boldsymbol{\beta}})^T \Sigma^{-1} (\boldsymbol{y} - \boldsymbol{X}\widehat{\boldsymbol{\beta}}) \sim \sigma^2 \chi^2_{n-k} \\ D(\boldsymbol{X}\widehat{\boldsymbol{\beta}}; \boldsymbol{X\beta}) &= (\widehat{\boldsymbol{\beta}} - \boldsymbol{\beta})^T \boldsymbol{X}^T \Sigma^{-1} \boldsymbol{X} (\widehat{\boldsymbol{\beta}} - \boldsymbol{\beta}) \sim \sigma^2 \chi^2_k. \end{aligned} \tag{3.49}$$

Moreover, the independence between $\boldsymbol{r}$ and $\boldsymbol{X}\widehat{\boldsymbol{\beta}}$ implies that $D(\boldsymbol{X}\widehat{\boldsymbol{\beta}}; \boldsymbol{X\beta})$ and $D(\boldsymbol{y}; \boldsymbol{X}\widehat{\boldsymbol{\beta}})$ are independent.

Thus, (3.47) *represents a partition of the $\sigma^2\chi^2_n$–distribution on the left side into two independent χ^2-distributed variables with $n - k$ and k degrees of freedom, respectively.*

The proof can be given through Cochran's theorem (see Madsen and Thyregod (2011)). The importance of the remarks given above lies in the fact that this can be generalised further into dispersion between nested models. Since we have χ^2 distributions, test statistics formulated as ratios will be F-distributed. This will be given a more precise formulation in the next section.

3.5 Likelihood ratio tests

In previous sections, we looked at marginal test where one parameter is tested. It is, however, often useful to test multiple parameters at the same time. For example, in factorial experiments/models where the effect of several covariates is given by multiple parameters, we should aim at testing if all parameters related to one effect could be zero at the same time.

The theorem below gives the general result for comparing two models that are nested.

Theorem 3.5 (A test for model reduction) *The likelihood ratio test statistic for testing*

$$H_0 : \boldsymbol{\mu} \in \Omega_0 \quad \textit{against the alternative} \quad H_1 : \boldsymbol{\mu} \in \Omega_1 \setminus \Omega_0,$$

is a monotone function

$$F(\boldsymbol{y}) = \frac{D(p_1(\boldsymbol{y}); p_0(\boldsymbol{y}))/(m_1 - m_0)}{D(\boldsymbol{y}; p_1(\boldsymbol{y}))/(n - m_1)}, \tag{3.50}$$

where $p_1(\boldsymbol{y})$ and $p_0(\boldsymbol{y})$ denote the projections of $\boldsymbol{y}$ on Ω_1 and Ω_0, respectively.

Under H_0, we have

$$F \sim F(m_1 - m_0, n - m_1), \tag{3.51}$$

i.e. large values of F reflect a conflict between the data and H_0, and hence lead to rejection of H_0. The p-value of the test is found as $p = P[F(m_1 - m_0, n - m_1) \geq F_{obs}]$, where F_{obs} is the observed value of F given the data (3.50).

The theorem is formulated as general projections, but we can think of the projections in terms of our linear models, and the null hypothesis is formed by removing a number of columns from the design matrix or equivalent by fixing a number of parameters to zero.

Example 3.5 (Clothing insulation level (cont.)) *In Example 3.3 above, we tested a number of different models using marginal tests. The initial model was*

$$y_i = \beta_0 + x_{i1}\beta_1 + x_{i2}\beta_2 + \beta_3 x_{i1} x_{i2} + \epsilon_i; \quad \epsilon_i \sim N(0, \sigma^2), \tag{3.52}$$

where x_1 and x_2 refer to the outdoor and indoor operative temperatures respectively. Based on the marginal test in the initial model, we might suggest the following null-model

$$y_i = \beta_0 + x_{i1}\beta_1 + \epsilon_i; \quad \epsilon_i \sim N(0, \sigma^2), \tag{3.53}$$

i.e. fixing two parameters to zero (or removing two columns from the design matrix). In R, the test can be performed by the function `anova` *(reduced R-output);*

```
modLMExNull <- lm(clo ~ tOut, data = dfLMEx1)
anova(modLMExNull,modLMEx11)

## Analysis of Variance Table
##
## Model 1: clo ~ tOut
## Model 2: clo ~ tOut * tInOp
##   Res.Df    RSS Df Sum of Sq      F  Pr(>F)
## 1    801 16.222
## 2    799 16.124  2  0.098172 2.4324 0.08848 .
## [...]
```

In this case, we get a different result using the F-test than the sequential marginal t-test, and using significance level $\alpha = 0.05$, we also get different models.

Traditionally, the partitioning of the variation is presented in a Deviance table (or an *ANalysis Of VAriance table*, ANOVA) as illustrated in Table 3.2, and in the R-output in Example 3.5.

3.5.1 Coefficient of determination, R^2

The so–called coefficient of determination (R^2) is given by

$$R^2 = \frac{D(p_1(\boldsymbol{y}); p_{null}(\boldsymbol{y}))}{D(\boldsymbol{y}; p_{null}(\boldsymbol{y}))} = 1 - \frac{D(\boldsymbol{y}; p_1(\boldsymbol{y}))}{D(\boldsymbol{y}; p_{null}(\boldsymbol{y}))}, \tag{3.54}$$

and measure how much the variance is reduced by the model defined by the projection $p_1()$, relative to the simplest model (only an intercept). The coefficient of determination is sometimes used as a goodness–of–fit because it is intuitively appealing. There is, however, not any generally accepted threshold for what would be a good or a bad value, as this will be entirely dependent on the stochastic component of the system, for example in the clothing example used in this chapter, there is a very dominating individual effect and hence we will not expect R^2-values close to one; for systems that are driven by physical laws, we would expect much higher R^2-values.

Another problem with the R^2-value is that it is an increasing function of model complexity (assuming nested models), i.e. when adding parameters, the coefficient of determination will increase, regardless of the significance of the parameter. To compensate for this, the coefficient of determination can be penalised by the number of parameters to get the *adjusted*

Table 3.2: Deviance table corresponding to a test for model reduction as specified by H_0. For $\Sigma = \boldsymbol{I}$, this corresponds to an analysis of variance table, and then "Deviance" is equal to the "Sum of Squared deviations" (SS).

Source	f	Deviance	Test statistic, F
Model versus hypothesis	$m_1 - m_0$	$\Vert p_1(\boldsymbol{y}) - p_0(\boldsymbol{y})\Vert^2$	$\dfrac{\Vert p_1(\boldsymbol{y})-p_0(\boldsymbol{y})\Vert^2/(m_1-m_0)}{\Vert \boldsymbol{y}-p_1(\boldsymbol{y})\Vert^2/(n-m_1)}$
Residual under model	$n - m_1$	$\Vert \boldsymbol{y} - p_1(\boldsymbol{y})\Vert^2$	
Residual under hypothesis	$n - m_0$	$\Vert \boldsymbol{y} - p_0(\boldsymbol{y})\Vert^2$	

coefficient of determination

$$R^2_{adj} = 1 - \frac{D(\boldsymbol{y}; p_1(\boldsymbol{y}))/(n-k)}{D(\boldsymbol{y}; p_{null}(\boldsymbol{y}))/(n-1)} \,. \tag{3.55}$$

where the variation of the model residuals is penalised by the number of parameters (k).

3.6 Tests for model reduction

An often recommended way to develop models is by starting with a comprehensive model that contains all terms that could be hypothesised from an the initial set up of the model. In general, the end result will depend on the initial model, and the chosen path through the model reduction. The variation of data can be partitioned in different ways, and the partitioning will partly determine the end result; three different types are often used:

Type I: The partitioning is determined before the reduction start and reductions are evaluated relative to the initial model, and removing all higher effect (used by `anova`).

Type II: The partitioning is obtained as the partial test for all factors excluding higher order interactions, or terms (relative to the one being tested, used by `drop1`).

Type III: The partitioning is obtained as the partial test for all factors.

While there is no generally accepted partitioning, we recommend to use Type II or Type III, at least when data is analysed in an explorative way as we have seen in the clothing example above.

Example 3.6 *Getting back to our clothing insulation level example, suppose we assume the following initial model (formulated in two different ways);*

```
m01 <- lm(formula = clo ~ tOut * sex + tInOp * sex +
              tOut * tInOp, data = dfLMEx1)
m02 <- lm(formula = clo ~ tOut * sex + tOut * tInOp +
              tInOp * sex, data = dfLMEx1)
```

The Type I partitioning would result in the following tables:

```
anova(m01)

## Analysis of Variance Table
##
## Response: clo
##              Df  Sum Sq Mean Sq  F value    Pr(>F)
## tOut          1  4.3294  4.3294 245.6928 < 2.2e-16 ***
## sex           1  1.2721  1.2721  72.1933 < 2.2e-16 ***
## tInOp         1  0.1457  0.1457   8.2676  0.004144 **
## tOut:sex      1  0.7020  0.7020  39.8361 4.583e-10 ***
## sex:tInOp     1  0.0612  0.0612   3.4753  0.062660 .
## tOut:tInOp    1  0.0151  0.0151   0.8544  0.355604
## Residuals   796 14.0263  0.0176
## [...]
```

and

```
anova(m02)

## [...]
## Response: clo
##              Df  Sum Sq Mean Sq  F value    Pr(>F)
## tOut          1  4.3294  4.3294 245.6928 < 2.2e-16 ***
## sex           1  1.2721  1.2721  72.1933 < 2.2e-16 ***
## tInOp         1  0.1457  0.1457   8.2676  0.004144 **
## tOut:sex      1  0.7020  0.7020  39.8361 4.583e-10 ***
## tOut:tInOp    1  0.0187  0.0187   1.0595  0.303636
## sex:tInOp     1  0.0576  0.0576   3.2701  0.070930 .
## Residuals   796 14.0263  0.0176
## [...]
```

where we see that the p-values are different depending on the order in which we enter the effects. On the other hand, using type II results in

```
drop1(m01, test = "F")

## Single term deletions
##
## Model:
## clo ~ tOut * sex + tInOp * sex + tOut * tInOp
##            Df Sum of Sq  RSS   AIC F value   Pr(>F)
## <none>                  14.0 -3236
## tOut:sex    1     0.341 14.4 -3219   19.35 0.000012 ***
## sex:tInOp   1     0.058 14.1 -3235    3.27    0.071 .
## tOut:tInOp  1     0.015 14.0 -3237    0.85    0.356
## ---
## [...]
```

We see that the p-values correspond to entering the effects last in the Type I partitioning.

When reducing models, it is not recommended to remove main effects before interaction effects since the resulting models might not be invariant to linear transformation of the regressors. We can illustrate this by a small hypothetical example; suppose that we are estimating the following simple linear regression model

$$Y = \beta_0 + \beta_1 T + \epsilon, \tag{3.56}$$

where we use the usual assumption on the residuals and T is temperature measured in Celsius, after using a Type III partitioning, we find that β_0 is not significant and hence we have the model

$$Y = \beta_1 T + \epsilon. \tag{3.57}$$

Now suppose we are asked to give the model when temperature is measure in Fahrenheit (a linear transformation), in this case we would need another model structure (namely including the intercept). Therefore we should include lower order terms (this also applies in polynomial regression).

3.7 Collinearity

When $\boldsymbol{X}^T\boldsymbol{X}$ is close to singular, there is collinearity. The simplest form of collinearity is when two regressors have a very high empirical correlation, and there are various ways to detect collinearity:

i) Examination of the correlation matrix for the estimates may reveal strong pairwise collinearities.

ii) Considering the change in standard error of the estimate of other parameters when removing a particular parameter.

If there is strong collinearity, the interpretation of the parameters might be impossible, and we might not be able to discover the real effects in the system.

Example 3.7 *Let us again consider our clothing example; only now suppose we have two sensors for the outdoor temperature (the second one being simulated for the purpose of illustration)*

```
dfLMEx1$tOut2 <- dfLMEx1$tOut + rnorm(length(dfLMEx1$tOut))
cor(dfLMEx1$tOut, dfLMEx1$tOut2)

## [1] 0.9783434
```

So, there is a large correlation between the two temperatures.

```
summary(lm(clo ~ tOut + tOut2, data = dfLMEx1))

## [...]
## Coefficients:
##              Estimate Std. Error t value Pr(>|t|)
## (Intercept)  0.891569   0.024049  37.073   <2e-16 ***
## tOut        -0.006999   0.005257  -1.331   0.1834
## tOut2       -0.008786   0.005060  -1.736   0.0829 .
## [...]
```

We see that the correlation between the regressors is very high and that none of the coefficients are significant. Removing one coefficient will lead to a highly significant parameter.

As seen from Example 3.7, collinearity can result in models that cannot be given the obvious interpretation, i.e. unstable estimates and also high standard errors. In the example, the effect of outdoor temperature is actually the sum of the two parameters related to outdoor temperature. Collinearity should of course be identified and dealt with. For the identification part, we might use graphical methods, like scatter–plots of the regressors and correlation matrix for the regressors.

For graphical methods and the use of the correlation matrix, we will only discover pairwise collinearity, and hence for detecting more general patterns, we need other methods. As the problem can be traced back to the properties of $(\boldsymbol{X}^T\boldsymbol{X})^{-1}$, another method relate to the condition number of the matrix $(\boldsymbol{X}^T\boldsymbol{X})^{-1}$ (Lattin et al. (2003) suggest that the condition number should be below 30).

Example 3.8 *In the example above, we get*

```
X <- cbind(1, dfLMEx1$tOut, dfLMEx1$tOut2) ## Design matrix
kappa(solve(t(X) %*% X)) ## Condition number

## [1] 32357
```

Hence condition number is in this case much larger than 30.

We may also simply investigate how parameters change when one parameter is removed from the regression; if this results in large changes of one or more parameters, this suggests that there is a multicollinearity problem. A quite simple method is also to use the so–called variance inflation factor (VIF). The basic idea is to investigate how much of the variation in x_i is explained by the other explanatory variables, using models describing each explanatory variable by the others, i.e.

$$VIF_i = \frac{1}{1 - R_i^2}, \tag{3.58}$$

where R_i^2 is the coefficient of determination from a general linear model describing x_i by the other explanatory variables. If VIF is low (usually 5 or 10 is used), there is not a multicollinearity problem. In our simple example with only two regressors, VIF is simply

$$VIF_i = \frac{1}{1 - r_{12}^2}, \tag{3.59}$$

where r_{12} is the empirical correlation between the 2 temperatures; in our example, the result is

```
1/(1-cor(dfLMEx1$tOut,dfLMEx1$tOut2)^2)

## [1] 23
```

and more generally we may use the `car` package (Fox and Weisberg, 2019) to find VIF of each of the explanatory variables

```
## library(car)
vif(lm(clo~tOut + tOut2, data = dfLMEx1))

##  tOut tOut2
##    23    23
```

There are a number of different approaches that can be used to solve collinearity problems; the simplest method is just to remove one regressor (in the example above it would make sense to only use one censor), also in the simple end, we could average the regressors (point-wise), again a sensible approach in the example above. In the more advanced end for counter measures is principal component regression, where the regressors are parameterised such that the design matrix is orthogonal. While very effective, the interpretation of the parameters becomes harder. Ridge regression is another way to solve the problem; here, the matrix $\boldsymbol{X}^T\boldsymbol{X}$ is simply regularised by adding some number to the diagonal; this is an effective method when the number of parameters is very large (compared to the number of observations); the price is that we introduce a bias in the parameters (this is mostly relevant when the number of parameters is large compared to the number of observation, and the reader is refered to other sources, eg. Madsen and Thyregod (2011).

Related to the problem of orthogonal parameterisation of the design matrix, we have the following theorem.

Theorem 3.6 *The Type I and Type III partitioning of the deviance are identical if the design matrix* $\boldsymbol{X}$ *is orthogonal.*

The above theorem is relevant for experimental design with controlled conditions, but in our example with clothing insulation, it is hardly relevant since we cannot control the outside temperature.

3.8 Confidence intervals and confidence regions

We have already seen that a $100(1-\alpha)\%$ confidence interval for β_j is found as

$$\widehat{\beta}_j \pm t_{1-\alpha/2}(n-m_0)\widehat{\sigma}\sqrt{(\boldsymbol{X}^T\boldsymbol{\Sigma}^{-1}\boldsymbol{X})^{-1}_{jj}}. \tag{3.60}$$

It follows from the distribution of $\widehat{\boldsymbol{\beta}}$ (see eq. (3.34)) that we can construct a confidence region for two or more model parameters by

$$(\widehat{\boldsymbol{\beta}}-\boldsymbol{\beta})^T(\boldsymbol{X}^T\boldsymbol{\Sigma}^{-1}\boldsymbol{X})(\widehat{\boldsymbol{\beta}}-\boldsymbol{\beta}) \leq m_0\widehat{\sigma}^2F_{1-\alpha}(m_0, n-m_0). \tag{3.61}$$

Such a region will be an ellipsoid in $\mathbb{R}^{m_0}$, which can be visualised in two dimensions at the time; we can of course check a specific set of hypothesised parameters in any dimension.

Example 3.9 *Referring to Example 3.3, we can test the hypothesis that* `tInOp` *and the interaction between sex and* `tInOp` *are both 0 at the same time by*

```
## DEsign matrix
X <- model.matrix(modLMEx13)[ ,c(4,6)]
## Estimated variance
sigma.sq <- summary(modLMEx13)$sigma^2
## Parameter estimates
beta.hat <- coef(modLMEx13)[c(4,6)]
## Covariance matrix for parameters
Cov <- sigma.sq * summary(modLMEx13)$cov.unscaled
## Test statistics
Q <- (beta.hat %*% solve(Cov[c(4, 6), c(4, 6)]) %*%
      beta.hat) / 2
## p-values
1 - pf(Q, df1 = 2, df2 = 797)

##        [,1]
## [1,] 0.023
```

Since p-value $= 0.023 < 0.05$*, both parameters should not be set equal to zero at the same time.*

The test above can be generalised to more general hypothesis, see Bingham and Fry (2010) of the form

$$\boldsymbol{B\beta} = \boldsymbol{c}, \tag{3.62}$$

that e.g. include hypothesis of the form $\beta_i = \beta_j$.

3.9 Prediction

When making predictions of future values of the system, the point predictions are made directly from our model and is as such straightforward, but we will also need some kind of uncertainty evaluation of the prediction. Here we distinguish between prediction and confidence intervals; while the first is related to a single new prediction, the confidence interval relates to the uncertain information we have on the true underlying regression line.

Theorem 3.7 (Prediction in the general linear model) *Assume that the unknown parameters* $\boldsymbol{\theta}$ *in the linear model are estimated by the least squares method* (3.29)*; then the* prediction *is*

$$\widehat{Y}_{n+\ell} = E[Y_{n+\ell}|\boldsymbol{X}_{n+\ell} = \boldsymbol{x}_{n+\ell}] = \boldsymbol{x}_{n+\ell}^T\widehat{\boldsymbol{\theta}}. \tag{3.63}$$

The variance of the prediction error $e_{n+\ell} = Y_{t+\ell} - \widehat{Y}_{n+\ell}$ *becomes*

$$V[e_{n+\ell}] = V[Y_{n+\ell} - \widehat{Y}_{n+\ell}] = \sigma^2[1 + \boldsymbol{x}_{n+\ell}^T(\boldsymbol{X}^T\boldsymbol{X})^{-1}\boldsymbol{x}_{n+\ell}]. \tag{3.64}$$

A $100(1-\alpha)\%$ *prediction interval for the future* $Y_{n+\ell}$ is given as

$$\widehat{Y}_{n+l} \pm t_{\alpha/2}(n-k)\sqrt{V[e_{n+l}]} = \widehat{Y}_{n+l} \pm t_{\alpha/2}(n-k)\widehat{\sigma}\sqrt{1 + \boldsymbol{x}_{n+l}^T(\boldsymbol{X}^T\boldsymbol{X})^{-1}\boldsymbol{X}_{n+l}} \tag{3.65}$$

and a $100(1-\alpha)\%$ confidence interval is given by

$$\widehat{Y}_{n+l} \pm t_{\alpha/2}(n-k)\sqrt{V[\boldsymbol{x}_{n+\ell}^T\widehat{\boldsymbol{\theta}}]} = \widehat{Y}_{n+l} \pm t_{\alpha/2}(n-k)\widehat{\sigma}\sqrt{\boldsymbol{x}_{n+l}^T(\boldsymbol{X}^T\boldsymbol{X})^{-1}\boldsymbol{X}_{n+l}}. \tag{3.66}$$

Note that the confidence interval width will converge to zero as we get larger samples, while the prediction interval width will not converge to zero, as $n \to \infty$.

Example 3.10 (Clothing insulation level (cont.)) *In Example 3.3, we found a model using a combination of forward and backward selection (see Section 2.9) and also plotted a set of confidence intervals. The figure below shows the confidence and prediction intervals together.*

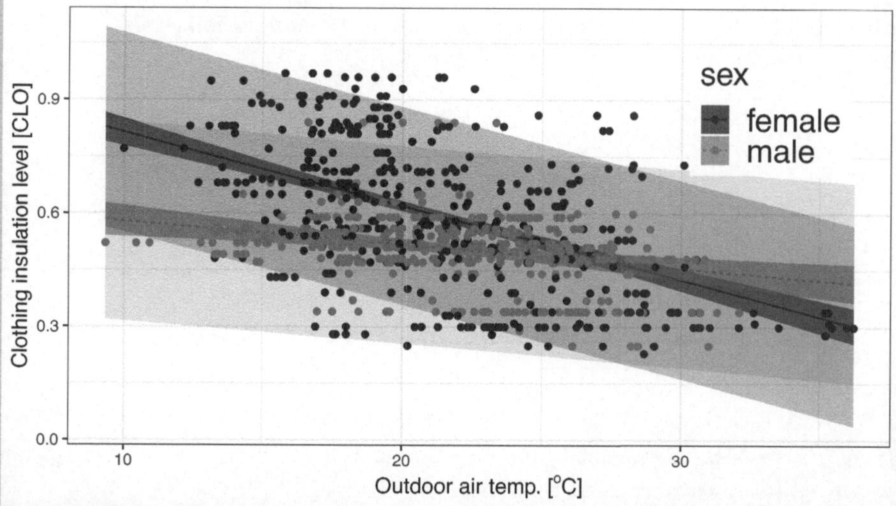

The darker areas are 95% confidence intervals while the lighter areas are the 95% prediction intervals, the wide prediction intervals reflect that there is a large underlying variance between individuals. The narrow confidence intervals reflect that we have many observations and therefore we get a good hold on the underlying regression line (assuming the assumptions are correct).

3.10 Weighted analysis

In the example, above we saw that the residual variation was much larger for women than for men; this points towards a weighted analysis. In the beginning of the chapter, we introduced the normal model as

$$\boldsymbol{Y} \sim N_n(\boldsymbol{\mu}, \sigma^2 \boldsymbol{\Sigma}) \text{ for } \boldsymbol{y} \in \mathbb{R}^n \tag{3.67}$$

but throughout the analysis, we have assumed that $\boldsymbol{\Sigma} = \boldsymbol{I}$, which indicate that the residuals are independent and identically distributed and therefore, for example the variance is independent of the regressors. If $\boldsymbol{\Sigma}$ is known, then the solution to the linear (weighted) regression problem ($\boldsymbol{\mu} = \boldsymbol{X}\boldsymbol{\beta}$) is given by (3.29)

$$\widehat{\boldsymbol{\beta}} = (\boldsymbol{X}^T \boldsymbol{\Sigma}^{-1} \boldsymbol{X})^{-1} \boldsymbol{X}^T \boldsymbol{\Sigma}^{-1} \boldsymbol{y}, \tag{3.68}$$

which is a closed–form solution. If $\boldsymbol{\Sigma}$ is not known, but we can parametrise it in some way (i.e. $\boldsymbol{\Sigma} = \boldsymbol{\Sigma}(\boldsymbol{\psi})$) then we can find the likelihood estimates of the parameters.

The step of such an optimisation would be:

1. For some value of $\boldsymbol{\psi}$, calculate $\boldsymbol{\Sigma}$.
2. For this $\boldsymbol{\Sigma}$, find $\widehat{\boldsymbol{\beta}}$,
3. Find $\hat{\sigma}^2$, using Theorm 3.3.
4. Calculate the likelihood.
5. Repeat until convergence.

The last point can be carried out with a general porpuse optimiser. We can illustrate this with our clothing insulation example:

Example 3.11 *Our residual analysis points to different variance between men and women (see the box-plot in Example 3.4), i.e.*

$$Y_i \sim N(\mu_i, \sigma_i^2), \tag{3.69}$$

where σ_i^2 depends on the sex of the subject. In this case, we could parametrise $\mathbf{\Sigma}$ such that $\mathbf{\Sigma}_{ii} = 1$ if the subject is a female and $\mathbf{\Sigma}_{ii} = 1/\gamma$ if the subject is a male. We can also get a good idea of γ directly from our model fit

```
var(residuals(modLMEx14)[dfLMEx1$sex == "female"]) /
    var(residuals(modLMEx14)[dfLMEx1$sex == "male"])

## [1] 3.6
```

Hence the residual variation of males is about $1/3.6 \cdot 100\% = 28\%$ that of females. To set up the likelihood, we need the design matrix

```
X <- model.matrix(modLMEx14)
```

The negative log-likelihood function is given by

```
nll.weight <- function(gamma,X,y, full){
    n <- length(y) ## no. of obs
    p <- dim(X)[2] ## no. of parameters
    ## Setup Sigma (step 1)
    W <- rep(1, n)
    W[X[ ,"sexmale"] == 1] <- gamma
    Sigma <- diag(1 / W)
    ISigma <- diag(W)
    ## Step 2
    beta <- solve((t(X) %*% ISigma %*%X)) %*% t(X) %*% ISigma %*% y
    ## Step 3
    r <- y - X %*% beta
    ## Moment/marginal likelihood estimate
    sig.sq <- sum(r^2 * W) / (n - p)
    ## Step 4
    ll <- sum(dnorm(y, mean = X %*% beta, sd = sqrt(sig.sq / W),
                    log = TRUE))
    ## If full TRUE, the parameters and the likelihood is
    ## returned otherwise only likelihood is returned
    if(full){
        return(list(gamma=gamma, beta = as.numeric(beta),
                    sig.sq = sig.sq, logLik = ll))
    }
    - ll ## negative log-likelihood returned
}
```

We may compare with our implementation with the log-likelihood using `lm`

```
-nll.weight(1,X=X, y=dfLMEx1$clo,full=FALSE)

## [1] 483

logLik(modLMEx14)

## 'log Lik.' 483 (df=6)
```

To find the optimal γ, we use `optimise`

```
(w1 <-optimise(nll.weight, lower = 0.01, upper = 10, X = X,
               y = dfLMEx1$clo, full = FALSE))

## $minimum
## [1] 3.6
##
## $objective
## [1] -559
```

We see that the optimal γ is $1/3.6 = 0.278$ (very close to the value calculated directly from the residuals). Also the likelihood is increased from 483.5 to 558.7 (which is a highly significant improvement); we can compare the parameters for the two different implementations by

```
cbind(coef(modLMEx14),
      nll.weight(w1$minimum, X, y=dfLMEx1$clo, full=TRUE)$beta)

##                   [,1]     [,2]
## (Intercept)    1.17812  1.11254
## tOut          -0.01949 -0.02027
## sexmale       -0.36665 -0.37161
## tInOp         -0.00628 -0.00323
## tOut:sexmale   0.01321  0.01348
```

We see that the parameters are mostly the same except for the effect of `tInOp`*, where the effect size is almost halved. We could of course go on and calculate standard errors for all the parameters and do the residual analysis, but we will leave this as an exercise for the reader.*

It should be noted that the parameter estimators for ψ are usually biased, and one would usually use the so–called Restricted Maximum likelihood estimation (see Chapter 5).

3.11 Autocorrelation

If the data is given as time–series, we should examine if there is autocorrelation present; we will treat this more formally in Chapter 8; for now, we will just say that it can by calculated in R by `acf`, and let it be part of the diagnostics. We can again turn to our clothing insulation level example.

Example 3.12 (Autocorrelation) *So far, we have treated the data as independent measurements, but it does actually consist of small time series, with a number of measurements on a subject during a day (typically 6). An example for 1 subject that was measured on 3 different days is shown below; on the right, we see the residuals from our linear model.*

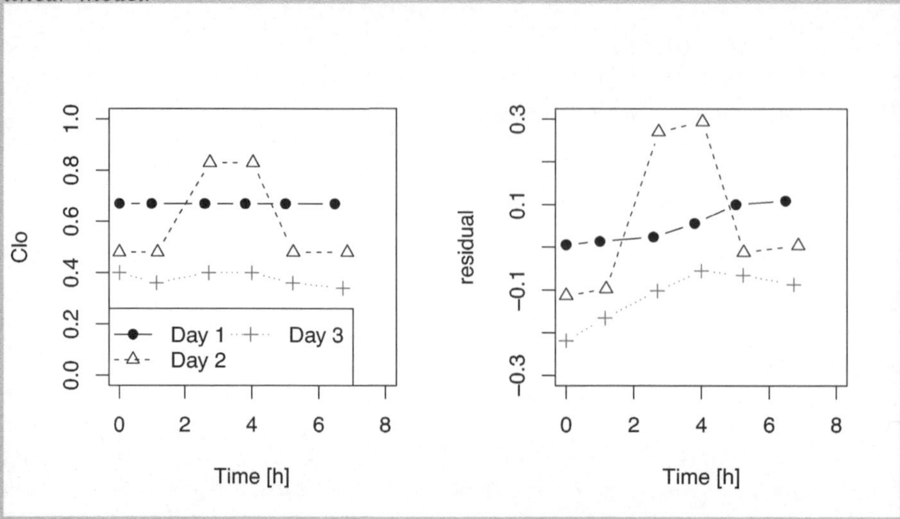

From the figures we may suspect that the residuals are not independent, however, as the timeseries are very short it does not make much sense to plot the autocorrelation function (again see Chapter 8). Instead we could turn to the AR(1) coefficient ($cor(r_i, r_{i+1})$) and check if this could be zero, in order to do so we need to calculate the AR(1) coefficient for each subject on each day

```
dat.tmp <- dfLMEx1
dat.tmp$res <- residuals(modLMEx14)
## Calculate within day residual autocorrelation
cor <- aggregate(res ~ subjId + day, data = dat.tmp,
                 function(x){acf(x,plot=FALSE)$acf[1]})[ ,3]
summary(cor)

##    Min. 1st Qu.  Median    Mean 3rd Qu.    Max.
##   -0.59    0.17    0.39    0.28    0.46    0.59

length(cor)

## [1] 136
```

We see that all quartiles are above zero, but also that the range of the AR(1) coefficients contain zero. The distribution of the correlations will not be Gaussian. However, since the number of correlations calculated is large, we can still use the usual t-test to investigate if the correlations are significantly different from zero;

```
## Test
t.test(cor)

##
##  One Sample t-test
##
## data:  cor
## t = 13, df = 135, p-value <2e-16
## alternative hypothesis: true mean is not equal to 0
## 95 percent confidence interval:
##  0.24 0.33
## sample estimates:
## mean of x
##     0.285
```

Which they definitely are. Hence the assumptions of the model are not fulfilled and we should therefore be careful with the use of the model.

When the assumptions are not fulfilled, we should be somewhat careful, and either use a more appropriate model for the data; in this case, for example mixed effect type models (see Chapter 5), or aggregate data in such a way that the correlation is no longer relevant (e.g. taking average value within subjects).

Example 3.13 *To solve the problem of autocorrelation, we saw in the last example we could simply average clothing levels within days using;*

```
agrdat <- aggregate(. ~ subjId + day, data = dfLMEx1[ ,c(2:5,6,8)],
                    mean)
```

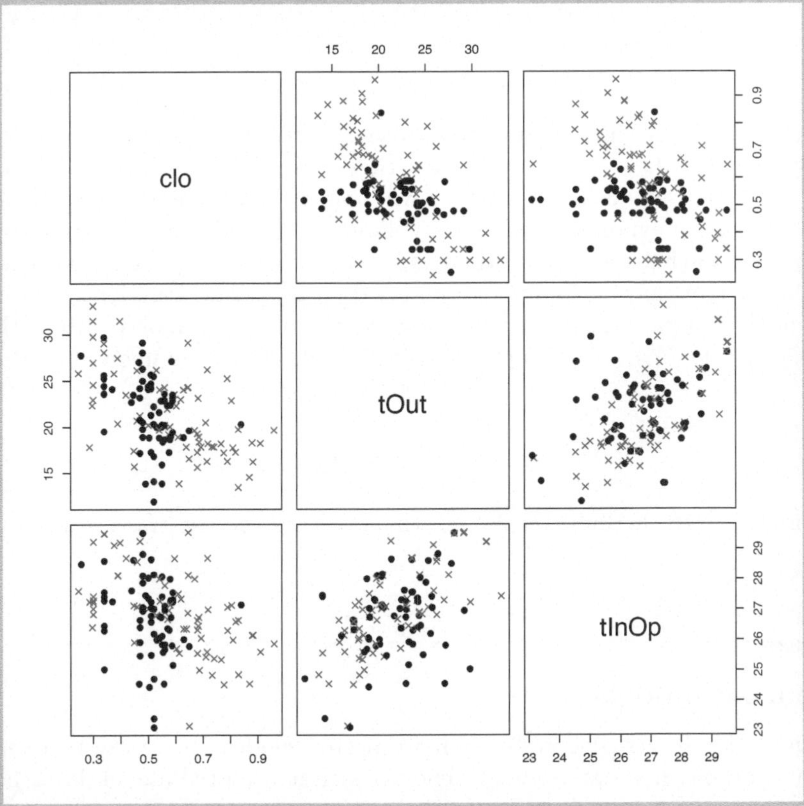

and do the modelling on the aggregated data

```
summary(lm(formula = clo ~ tOut * sex + tInOp, data = agrdat))

## [...]
## Coefficients:
##                  Estimate Std. Error t value Pr(>|t|)
## (Intercept)      1.528516   0.225348   6.783 3.66e-10 ***
## tOut            -0.020617   0.003757  -5.488 2.03e-07 ***
## sexmale         -0.377225   0.113289  -3.330  0.00113 **
## tInOp           -0.018382   0.009524  -1.930  0.05576 .
## tOut:sexmale     0.013543   0.005184   2.612  0.01005 *
## ---
## [...]
```

We see that most parameters are approximately the same (except for the intercept and the dependence on the indoor temperature). All p-value are now larger, which is a consequence of fewer datapoints (because we use the aggregated level).

3.12 Polynomial regression and transformations

The prediction intervals that we saw in Example 3.10 above are somewhat unsatisfactory. We also saw that the residuals have some quite strange shapes. These observations call for some action. If we stay in the framework of the general linear model, we either transform the response variable, the regressors, or both. Hence a quite general form would be (only one regressor)

$$\psi(Y) = \beta_0 + \beta_1 s_1(x_1) + ... + \beta_p s_p(x_1) + \epsilon; \quad \epsilon \sim N(0, \sigma^2) \tag{3.70}$$

and we could of course include similar function for other regressor. The most common choice for $s_j(\cdot)$ are polynomials $s_j(x) = x^j$ and as such these resemble non-linear functions (we can think of this as Taylor expansion of an unknown non-linear function). Even though the most common choice are polynomials, we are not bounded by this and one could use e.g. log–transformations of the regressors, or if there is a natural periodicity (like wind–direction or hour of the day) trigonometric functions (sine and cosine). The function $\psi(Y)$ change the distribution of the outcome and scale the variance, the most common choice is the log-transformation $\psi(Y) = \log(Y)$, but also power transformations are common. The Box-Cox transformation

$$Y^{(\lambda)} = \frac{Y^\lambda - 1}{\lambda}; \quad \lambda \neq 0, \tag{3.71}$$

is also a common choice. In the limit for $\lambda \to 0$, this becomes the log–transform.

3.13 Comparing models

In previous sections, we discussed how to compare nested models; if we use regression on different functions of the regressors, we might end up with non-nested model; in such cases, we can use the so-called information criteria to compare models. The most used information criteria are Akaike's Information Criterium (AIC) and the Bayesian Information Criterium (BIC); see also Section 2.8, defined by

$$\begin{aligned} AIC &= -2\log(L) + 2p \\ BIC &= -2\log(L) + p\log(n), \end{aligned} \tag{3.72}$$

where L is the likelihood value of the model, p is the number of parameters in the model and n is the number of observation used for the modelling. AIC and BIC can be used directly when the output is not transformed (i.e. $\psi(y) = Y$). Both AIC and BIC are negatively oriented (i.e. smaller values are better).

If we use a non-linear transformation of the data, we cannot compare AIC and BIC directly across transformation. We can, however, use the transformation formula for random variables (Theorem 2.3) to get a correction term that will give the likelihood on the original scale

$$\log L(\boldsymbol{\theta}; \boldsymbol{y}) = \log L(\boldsymbol{\theta}; \psi(\boldsymbol{y})) + \sum_i \log(|\psi'(y_i)|. \tag{3.73}$$

We might even parameterise the transformation with one or more parameter (i.e. $\psi(y_i, \lambda)$) and find the best transformation within some class of transformation (e.g. `boxcox` function in the R-package `MASS` (Venables and Ripley, 2002)).

Example 3.14 *Again getting back to the clothing example, we might suspect that the clothing insulation level is not a linear function of the temperature. One option is to include* log *of temperature.*

```
modLMEx15 <- lm(clo ~ tOut * sex + I(log(tOut)) * sex  +
                 tInOp, data = dfLMEx1)

##                     Df   AIC F value Pr(>F)
## <none>                 -3237
## tInOp                1 -3235    4.22   0.04 *
## tOut:sex             1 -3238    0.37   0.54
## sex:I(log(tOut))     1 -3237    2.13   0.14
## ---
## [...]
```

We see that we should exclude the interaction between outdoor temperature and sex, and actually it turns out that the model should not be reduced further. Below we compare the models by AIC and BIC (see Example 3.3 for `modLMEx14`*).*

```
modLMEx16 <- lm(clo ~ tOut  + I(log(tOut)) * sex  +
                 tInOp, data = dfLMEx1)

c(AIC(modLMEx14),AIC(modLMEx16))

## [1] -954.9054 -957.6105

c(BIC(modLMEx14),BIC(modLMEx16))

## [1] -926.7753 -924.7920
```

The model should not be reduced further (this is not shown here), and we see that we have a better AIC, but if we consider BIC we should choose the simpler one.

As indicated above we might also benefit from transformation of the data. Applying the R-function `boxcox` *indicate that, using the family of Box-Cox transformations we should choose* $\lambda = 0.5$ *(i.e. square root), the resulting model (after reduction) become;*

```
modLMEx17 <- lm(clo.sqrt ~ tOut + I(log(tOut)) * sex ,
                data = dfLMEx1)
```

The models could be compared graphically (i.e. examine diagnostic plots), or we can adjust AIC and BIC for the model

```
AIC17 <- AIC(modLMEx17) - 2 * sum(log(dfLMEx1$clo)) -
    0.5 * length(dfLMEx1$clo)
BIC17 <- BIC(modLMEx17) - 2 * sum(log(dfLMEx1$clo)) -
    0.5 * length(dfLMEx1$clo)
c(AIC(modLMEx14),AIC(modLMEx16),AIC17)

## [1] -955 -958 -960

c(BIC(modLMEx14),BIC(modLMEx16),BIC17)

## [1] -927 -925 -932
```

Both AIC and BIC improve. As the observed clothing level is a number in the interval $(0, 1)$*, it is natural to try the logit transformation, i.e*

$$\psi(y) = \log\left(\frac{y}{1-y}\right). \tag{3.74}$$

After reducing the model, we get the following model for logit transformed data

```
dfLMEx1$clo.logit <- log(dfLMEx1$clo / (1 - dfLMEx1$clo))
modLMEx18 <- lm(clo.logit ~ tOut + I(log(tOut)) * sex  + tInOp,
                data = dfLMEx1)

AIC18 <-  AIC(modLMEx18) + 2 * sum(log(dfLMEx1$clo * (1 - dfLMEx1$clo)))
BIC18 <-  BIC(modLMEx18) + 2 * sum(log(dfLMEx1$clo * (1 - dfLMEx1$clo)))

##       LMEx14 LMEx16 LMEx17 LMEx18
## AIC    -955   -958   -960   -844
## BIC    -927   -925   -932   -811
```

We see that even though the logit transformation is natural, in this case it leads to a large reduction in the likelihood and hence we should not use this transformation of the data; one could consider other transformations with the same properties (like log-log, c log-log, *probit and others). Here we will, instead, investigate a family of asymmetrical logit transformations*

$$\psi(y;\gamma) = \frac{1}{\gamma}\log\left(\frac{y^\gamma}{1-y^\gamma}\right); \quad \gamma > 0, \tag{3.75}$$

for $\gamma = 1$, we get the standard logit transformation. In the code below, γ is optimised for the initial structure of the right–hand side of the model

```
## Profile likelihood of transformation
lP.lambda <- function(lambda){
    dat <- dfLMEx1
    M <- lm(I(trans(clo,lambda)$y) ~ tOut * sex + I(log(tOut)) * sex +
              tInOp, data = dat)
    n <- dim(dat)[1]
    -log(summary(M)$sigma^2) * n / 2 +
        sum(abs(log(trans(dat$clo,lambda)$dy)))
}

## Transformation
trans<-function(x,g){
    y <- log(x^g / (1 - x^g)) / g
    dy <- 1 / (x * (1 - x^g))
     return(list(y = y, dy = dy))
}
(opt <- optimize(lP.lambda, c(0, 40), maximum = TRUE))

## $maximum
## [1] 6.6
##
## $objective
## [1] 1642
```

We see that the optimal choice for γ is 6.6, and performing the usual model reduction with that γ we get

```
modLMEx19 <- lm(I(trans(clo, 6.6)$y) ~ tOut + I(log(tOut)) * sex,
                data = dfLMEx1)

AIC19 <- AIC(modLMEx19) - 2 * sum(log(trans(dfLMEx1$clo, 6.6)$dy))
BIC19 <- BIC(modLMEx19) - 2 * sum(log(trans(dfLMEx1$clo, 6.6)$dy))

##      LMEx14 LMEx16 LMEx17 LMEx18 LMEx19
## AIC    -955   -958   -960   -844   -998
## BIC    -927   -925   -932   -811   -970
```

Hence we see improvements in both AIC and BIC using the optimised value of γ. The effect of applying transformations to temperature and clothing level is illustrated (on the original scale) in the figure below. The presented models are `modLMEx14` *(top left, Example 3.3),* `modLMEx16` *(top right, Example 3.14),* `modLMEx17` *(bottom left, Example 3.14) and* `modLMEx19` *(bottom right, Example 3.14).*

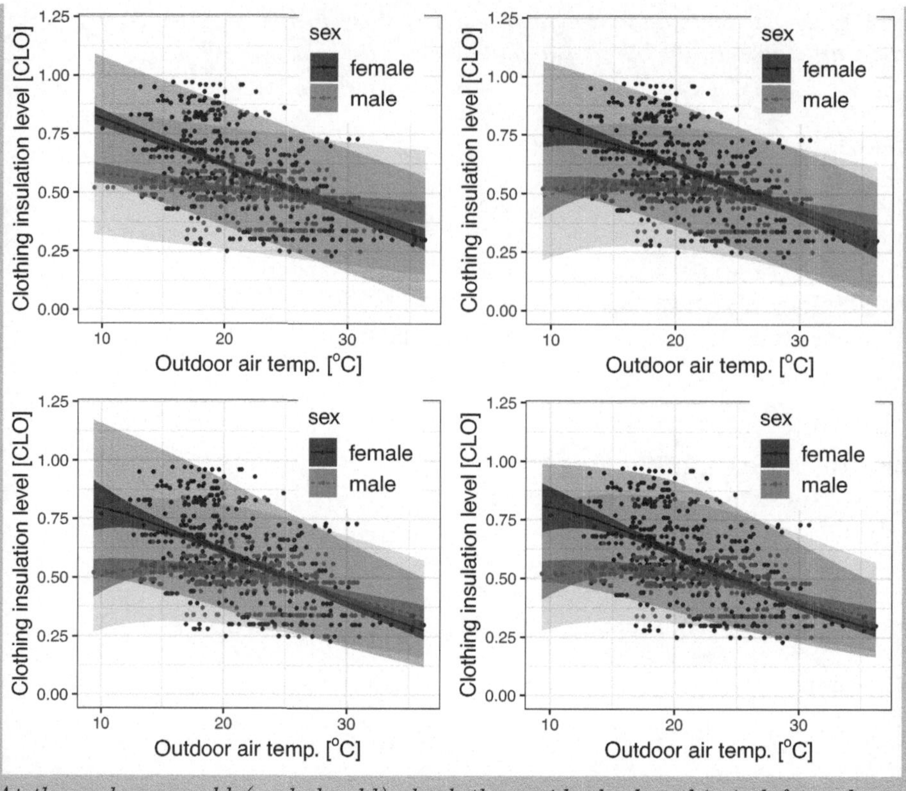

At the end, we could (and should) check the residuals, but this is left to the reader to do.

3.14 Summary and concluding remarks

In this chapter, we have covered the general linear model and a few extensions. The general linear model is unique among the models presented in this book in that closed form–solutions for the parameter estimates are available. This feature makes estimation very efficient, but the price is that the models are restricted to normal and linear models. If it is actually reasonable to assume normality, the structure also allows for explicit formulation of the likelihood ratio test, whereas in other situations, we need to rely on asymptotic results.

To some extent, the linearity assumption can be relaxed using polynomial regression, and the normality assumption can, again to some extent, be relaxed by transformation of the outcome.

As illustrated in the chapter, it is possible to get quite a long way using the general linear model, and although conclusions about mean effect may be quite robust to deviation from normality or other violations of the basic assumption, they are crucial if the models should be applied for predictions.

4

Generalized linear models

This chapter is a revision of a similar chapter in Madsen and Thyregod (2011) but tailored to occupant behaviour and the remaining chapters of the present book.

4.1 Introduction

Generalized linear models are an extension of the concept of general linear models. They were introduced by Nelder and Wedderburn (1972). In this class of models, we give up the assumption of normally distributed response variables and with it (the assumption of) a linear mean value. Instead, the class of response variables is generalized to the exponential dispersion family. The response variable $\boldsymbol{Y}$ can be considered as one of three components of a generalized linear model. The other two are a linear combination of explanatory variables $\boldsymbol{X\beta}$ (also linear predictor or systematic component) and a link function, which connects the response variable and the linear predictor in the following way: Through transformation by the link function, the mean of the response variable becomes linear and can be explained by the linear predictor. Because all information about $\boldsymbol{Y}$ is inherited in $\boldsymbol{X\beta}$, the model is considered a generalized linear model rather than a non-linear model. One example for an application in occupant behaviour modelling is the occupants' heating set-point preferences as a function of physical, psychological and social drivers (Fabi et al., 2013).

Parameters in generalized linear models are estimated by likelihood methods, and hence we could use the exact same techniques as in Chapter 2. However, generalized linear model are implemented as standard method in most statistical software packages and therefore we will treat this standard set up here.

4.1.1 Type of response variables

In models related to occupant behaviour, we consider a large variety of data types. Models of window opening behaviour, presence, thermostat preferences, lighting, shading or thermal comfort require a list of various approaches. Generalized linear models are applicable to different types of situations as they offer modelling approaches for a wide range of response variables. In the following, types of response variables that can be modelled by standard generalized linear models (i.e. using `glm`)and that we might come across in practice are given together with their potential underlying statistical distribution.

1. Count data (e.g. $y_1 = 4$, $y_2 = 25$, ...); Poisson distribution
   - Number of people in a room
   - Number of window openings in a day
2. Binary (e.g. $y_1 = 0$, $y_2 = 1$, ...) or quantal (e.g. $y_1 = 35/120$, $y_2 = 75/110$, ...) response variables; Binomial distribution.

DOI: 10.1201/9781003340812-4

- Occupant present or absence
- Window open or closed
- Proportion of open windows
- Proportion of shading covering a window

3. Count data, waiting times; Negative binomial distribution.
   - Time (measured in time steps) until occupant arrives
4. Count data, multiple ordered categories (e.g. "Very unsatisfied", "Unsatisfied", "Neutral", "Satisfied", "Very satisfied"); Multinomial distribution.
   - Thermal comfort
   - Questionnaire data
5. Count data, multiple categories.
6. Continuous responses, constant variance (e.g. $y_1 = 20.35$, $y_2 = 19.4, \ldots$); normal distribution (i.e. the general linear model).
   - Indoor air temperature
7. Continuous positive responses with constant coefficient of variation, (e.g. empirical variances, $s_1^2 = 3.82$, $s_2^2 = 2.56$, ..., $s_n^2 = 2.422$); Gamma distribution.
   - CO_2 level in ppm
8. Continuous positive highly skewed; Inverse Gaussian.

In order to get a better understanding, we take a look at the following example:

Example 4.1 (Window opening) *One facet of occupant behaviour modelling is the modelling of window opening (and closing) behaviour. This can be done, for example by observing the window status dependent on environmental variables such as outdoor air temperature, occurrence of precipitation, or relative humidity. Let us assume that a window can only be completely open or closed. Then we can consider the relative frequencies y of a window being open, given a number of observations, as a random variable (the response variable of the model). Figure 4.1 shows the relative frequencies of window openings dependent on the outdoor air temperature in an appartment between 18.00 and 21.00 (i.e. have the window been open for some time in this interval), the data is an aggregated version of one of the appartments presented in (Andersen et al., 2013).*

Figure 4.1 also includes a linear fit to the data, already from the figure we see some issues, e.g.:

- *The model will predict negative frequencies for low temperatures (below app. -1 degrees)*
- *The deviation from the model is not independent of the temperature (e.g. mostly above for high temperatures)*

In addition to the problems mentioned above, there are more issues where we have to take a closer look at data to understand it. In the table below, the actual numbers are given.

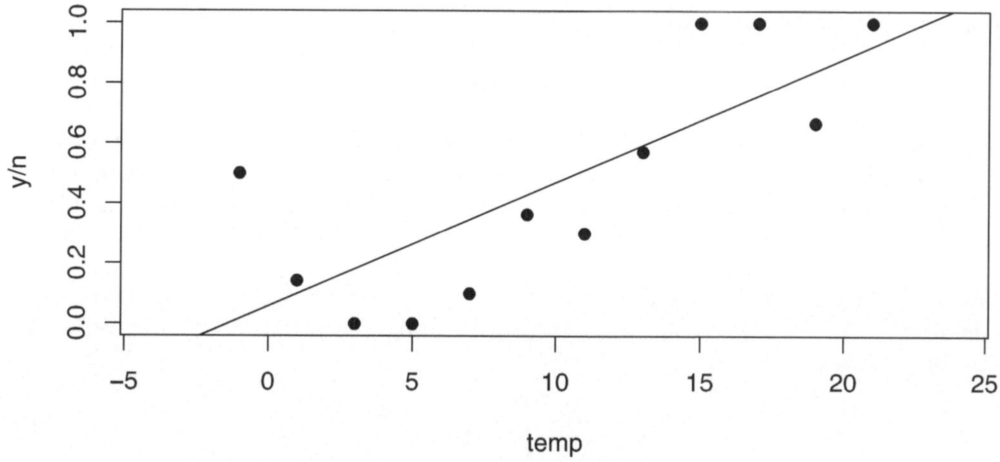

Figure 4.1: Fit of Classical GLM

```
##    -1 1 3 5  7  9 11 13 15 17 19 21
## y   1 1 0 0  1  4  3  4 10  6  2  4
## n   2 7 1 5 10 11 10  7 10  6  3  4
```

The first row gives the temperatures, the second row is the number of days where the window opened, and the last row is the number of days within the temperature range.

By inspecting the data above, we see that the number of datapoints for low temperatures is small, for example it seems from Figure 4.1 that the frequency of opening the window when the temperature is below zero is around 0.5, but this is only based on 2 observations while the observed frequency of around 0.4 at 9 degrees is based on 11 observations, that is we would be more certain about this number.

The discussion in the example above shows that we need other models that take care of the boundaries, the difference in weights, and the non-linearity in the problem. The most obvious possibility in this case is to use the binomial distribution.

The probability mass function of the binomial distribution is given by

$$f_{\text{bin}}(k, n, p) = \binom{n}{k} p^k (1-p)^{n-k}. \tag{4.1}$$

The normal distribution and the binomial distribution both belong to a generalized class of distributions, which is closely related to the concept of generalized linear models (GLMs), namely the exponential dispersion family. The idea of generalized linear models is that a transformed version of the mean value lies in a linear subspace of $\mathbb{R}^n$. This transformation is called the link function, and is provided by the distribution's representation in the exponential dispersion family.

In the case of a normally distributed random variable, the canonical link is given by the identity $\theta(\mu) = \mu$ which leads to the "untransformed" classical linear model. For the binomial distribution, the canonical link is given by

$$\theta(\mu) = \log\left(\frac{\mu}{1-\mu}\right), \tag{4.2}$$

which we refer to as the *logit* function in the following. A regression model with *logit* as link function is called logistic regression.

In a binomial model, we would write

$$Y_i \sim Binom(\mu_i, n_i), \tag{4.3}$$

meaning that the ith observation follows a binomial distribution with success probability $p_i = \mu_i$ and n_i trails with that particular success probability. The link function establishes the link between a non-linear domain (here probability) and the linear predictor, e.g. $\theta(\mu_i) = \beta_0 + \beta_i x_i$.

Example 4.2 (Window opening (continued)) *In the case of the window opening example, the logistic regression model with outdoor air temperature x as the only explanatory variable is*

$$\log\left(\frac{\mu}{1-\mu}\right) = \beta_0 + \beta_1 x. \tag{4.4}$$

With methods introduced later in this chapter, we are able to find maximum likelihood estimates $\hat{\beta}_0$ and $\hat{\beta}_1$ of the model parameters. The following figure shows the fitted model.

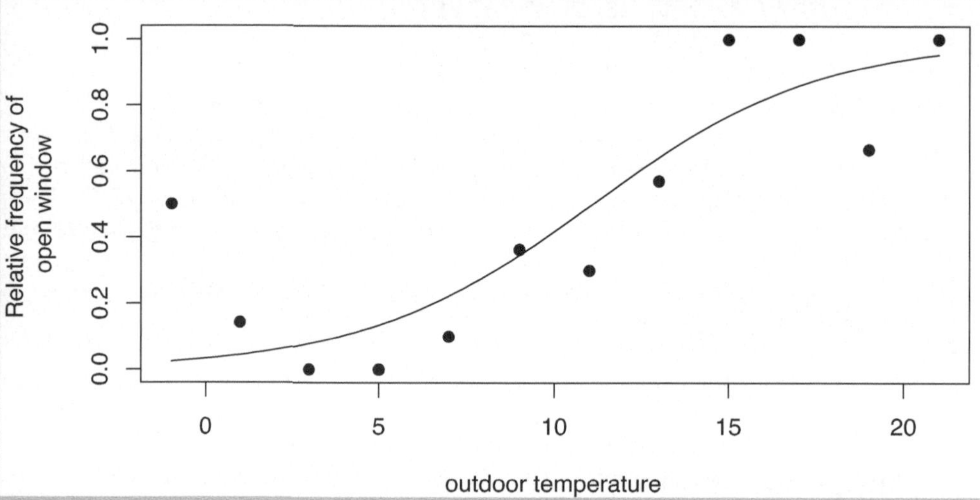

We can see that the logistic regression model fits the data much better than the linear regression model. We will get back to the analasis of the residuals later in this chapter.

Andersen et al. (2013) investigate window opening and closing behaviour in Danish dwellings divided into groups according to criteria such as building ownership and type of ventilation. Furthermore, they distinguish between seasons and room type. As an example of their findings, (4.5) *shows the fitted model of window opening behaviour for owner-occupied, naturally ventilated buildings in spring*

$$\log\left(\frac{p}{1-p}\right) = -23.04 + 1.87 \cdot \log(CO_2) + 0.163 \cdot t_i + 0.501 \log(Rad + 1), \tag{4.5}$$

where p is the probability of opening a window within the next 10 *minutes, CO_2 is the CO_2 concentration in ppm, t_i is the indoor temperature in $°C$ and Rad is the solar radiation in W/m^2.*

4.2 Exponential families of distributions

The assumption in generalized linear models is that the *pdf* of an observation can be written in terms of the *exponential family* or the *exponential dispersion family.*

Definition 4.1 (A natural exponential family) *A family of probability densities that can be written on the form*

$$f_Y(y;\theta) = c(y)\exp(\theta y - \kappa(\theta)), \quad \theta \in \Omega \tag{4.6}$$

is called a natural exponential family *of distributions. The function $\kappa(\theta)$ is called the* cumulant generator. *The representation (4.6) is called the* canonical parameterisation *of the family, and the parameter θ is called the* canonical parameter. *The parameter space, Ω, is a subset of the real line.*

In the definition above, the function $c(y)$ is a constant, i.e. not dependent on the parameter θ. A simple example of the exponential family is the Poisson distribution.

Example 4.3 (Poisson distribution) *The pdf of a Poisson distribution is given by*

$$f_Y(y;\mu) = \frac{\mu^y e^{-\mu}}{y!} = \frac{e^{\log(\mu)y-\mu}}{y!}; \tag{4.7}$$

hence, we see that the Poisson distribution is a member of the natural exponential family. Further we see that $\theta(\mu) = \log(\mu)$, $\kappa(\theta) = e^\theta = \mu$, and $c(y) = 1/y!$.

As we can see the Poisson distribution fits directly into the exponential family, for other distributions it is more natural to include a so-called dispersion parameter, and in this case we have the exponential dispersion family.

Definition 4.2 (An exponential dispersion family) *A family of probability densities which can be written as*

$$f_Y(y;\theta,\lambda) = c(y,\lambda)\exp(\lambda\{\theta y - \kappa(\theta)\}) \tag{4.8}$$

is called an exponential dispersion family *of distributions. The parameter $\lambda > 0$ is called the* precision parameter.

We see that the exponential dispersion family has an extra parameter (λ), that controls the dispersion (related to the variance).

Example 4.4 (Normal distribution) *The normal distribution, is a member of the exponential dispersion family. To see this, consider the pdf of the normal distribution*

$$\begin{aligned} f_Y(y;\mu,\sigma^2) &= \frac{1}{\sqrt{2\pi\sigma^2}} e^{-\frac{(y-\mu)^2}{2\sigma^2}} \\ &= \frac{1}{\sqrt{2\pi\sigma^2}} e^{\frac{1}{\sigma^2}(-y^2/2-\mu^2/2+\mu y)}, \end{aligned} \tag{4.9}$$

Hence, the normal distribution is a member of the exponential dispersion family with $\theta = \mu$, $\kappa(\theta) = \frac{1}{2}\theta^2$, $\lambda = 1/\sigma^2$, and $c(y,\lambda) = \frac{e^{-\lambda y^2}}{\sqrt{2\pi/\lambda}}$.

As we saw in the example above, the normal distribution is a member of the exponential dispersion family, and hence the generalized linear models generalise the class of general linear models. In the normal model, the dispersion parameter is a free parameter that should be estimated, while in other models, the precision parameter is given by the experimental condition.

Example 4.5 *The binomial distribution is a member of the exponential dispersion family (with $q = y/n$)*

$$\begin{aligned} f_Y(q; p, n) &= \binom{n}{nq} p^{nq}(1-p)^{n(1-q)} \\ &= \binom{n}{nq} \exp\{nq\log(p) + n(1-q)\log(1-p)\} \\ &= \binom{n}{nq} \exp\left\{n\left(q\log\left(\frac{p}{1-p}\right) + \log(1-p)\right)\right\}, \end{aligned} \tag{4.10}$$

also set

$$\begin{aligned} p &= \frac{e^\theta}{1+e^\theta} \\ log(1-p) &= \log\left(1 - \frac{e^\theta}{1+e^\theta}\right) = -\log\left(1+e^\theta\right) \end{aligned} \tag{4.11}$$

hence we have $\theta = \log(p/(1-p))$, $\kappa(\theta) = \log\left(1+e^\theta\right)$ and $\lambda = n$. Here the dispersion parameter is given by the exponential conditions (the number of trials n).

In order to set up the statistical model for the exponential dispersion family, we need to have a formula for the mean value, and in order to specify uncertainties, we will of course also need the variance. The expected value can be found by differentiating the integral of the pdf (noting that $\int f dy = 1$ and hence $\frac{\partial}{\partial\theta}\int f dy = 0$)

$$\begin{aligned} 0 &= \frac{\partial}{\partial\theta}\int c(y,\lambda)\exp(\lambda\{\theta y - \kappa(\theta)\})dy \\ &= \int c(y,\lambda)\frac{\partial}{\partial\theta}\exp(\lambda\{\theta y - \kappa(\theta)\})dy \\ &= \int c(y,\lambda)\lambda\left(y\exp(\lambda\{\theta y - \kappa(\theta)\}) - \kappa'(\theta)\exp(\lambda\{\theta y - \kappa(\theta)\})\right)dy \\ &= \lambda\int\left[y f_Y(y;\theta,\lambda) - \kappa'(\theta) f_Y(y;\theta,\lambda)\right]dy \\ &= \lambda\left(E[Y] - \kappa'(\theta)\right). \end{aligned} \tag{4.12}$$

Hence we can define the mean value function as

$$E_\theta[Y] = \kappa'(\theta) = \tau(\theta) = \mu. \tag{4.13}$$

In a similar way, one can show that the variance function is

$$V_\theta[Y] = \frac{\kappa''(\theta)}{\lambda}. \tag{4.14}$$

The mean value function (4.13) implies a "natural" link between the mean value space and the natural or canonical parameters. This is called the canonical link function.

Definition 4.3 (The canonical link function) *The relation between the two parameterizations (canonical and mean value) is determined as* $\mu = \tau(\theta)$ *with* $\tau(\cdot)$ *given by* (4.13). *The inverse mapping*

$$\theta = \tau^{-1}(\mu), \tag{4.15}$$

is called the canonical link function.

Example 4.6 (Normal distribution) *In the case of the normal distribution, we can find the mean and variance as*

$$\begin{aligned} E_\theta[Y] &= \kappa'(\theta) = \theta \\ V_\theta[Y] &= \frac{\kappa''(\theta)}{\lambda} = \frac{1}{1/\sigma^2} = \sigma^2, \end{aligned} \tag{4.16}$$

Hence we also have $\tau(\theta) = \theta$*, and the canonical link is the identity.*

Example 4.7 (Binomial distribution) *In the case of the binomial distribution, we can find the mean and the variance as* ($Q = \frac{Y}{n}$)

$$\begin{aligned} E_\theta[Q] &= \kappa'(\theta) = \frac{e^\theta}{1+e^\theta} \\ V_\theta[Q] &= \frac{\kappa''(\theta)}{\lambda} = \frac{1}{n}\left(\frac{e^\theta}{1+e^\theta} - \frac{e^{2\theta}}{(1+e^\theta)^2}\right) \\ &= \frac{1}{n}\frac{e^\theta}{(1+e^\theta)^2}. \end{aligned} \tag{4.17}$$

Hence, we also have $\tau(\theta) = e^\theta/(1+e^\theta)$*, and the canonical link is* $\theta = \log(\mu/(1-\mu))$. *Also note that the variance in terms of the natural parameter is* $V_\mu[Q] = \mu(1-\mu)/n$.

Even though the distribution directly suggests a link function, we are not bounded by this and the choice of link function will, as we will discuss later, be an integrated part of the statistical modelling.

4.2.1 Exponential family densities as a statistical model

In order to estimate parameters, we need the likelihood of our model, and hence also the joint density of a set of observations. The joint density, using the canonical parameter, is

$$f(\boldsymbol{y};\boldsymbol{\theta}) = \exp\left[\sum_{i=1}^n w_i(\theta_i y_i - \kappa(\theta_i))\right]\prod_{i=1}^n c(y_i, w_i). \tag{4.18}$$

Here w_i can either be part of the modeling (as in the normal case), or be defined by the model/data (as in the case of the binomial). We are now ready to define the generalized linear model.

Definition 4.4 (The generalized linear model) *Assume that* $Y_1, Y_2, \ldots, Y_n$ *are mutually independent, and the density can be described by an exponential dispersion model (see equation* (4.8)*) with the same variance function* $V(\mu)$.

A generalized linear model *for $Y_1, Y_2, \ldots, Y_n$ describes an affine hypothesis for $\eta_1, \eta_2, \ldots, \eta_n$, where*

$$\eta_i = g(\mu_i), \tag{4.19}$$

is a transformation of the mean values $\mu_1, \mu_2, \ldots, \mu_n$.

The hypothesis is of the form

$$H_0 \ : \ \boldsymbol{\eta} - \boldsymbol{\eta}_0 \in L \, , \tag{4.20}$$

where L is a linear subspace $\mathbb{R}^n$ of dimension k, and where $\boldsymbol{\eta}_0$ denotes a vector of known off-set values.

The definition above gives a quite abstract formulation of the generalized linear model in terms of linear subspaces. We can also give this a more concrete definition in line with what we saw for the general linear model.

Definition 4.5 (Design matrix for the generalized linear model) *Consider the linear subspace $L = \mathtt{span}\{x_1, \ldots, x_k\}$, i.e. the subspace is spanned by k vectors ($k < n$) such that the hypothesis can be written*

$$\boldsymbol{\eta} - \boldsymbol{\eta}_0 = \boldsymbol{X\beta} \ \ \mathit{with} \ \boldsymbol{\beta} \in \mathbb{R}^k \, , \tag{4.21}$$

where $\boldsymbol{X}$ has full rank. The $n \times k$ matrix $\boldsymbol{X}$ is called the design matrix.

The i'th row of the design matrix is given by the model vector

$$\boldsymbol{x}_i^T = \begin{pmatrix} x_{i1} & x_{i2} & \ldots & x_{ik} \end{pmatrix} \, , \tag{4.22}$$

for the i'th observation.

Hence we have the linear model given in a matrix-vector notation and on top of that we have a transformation from the linear space into the mean value space ($\mu(\boldsymbol{X\beta}) = g^{-1}(\boldsymbol{X\beta})$).

Example 4.8 (Window opening) *In the model from Andersen et al. (2013) (see Example 4.2) we have*

$$\boldsymbol{\eta} = \boldsymbol{X\beta}, \tag{4.23}$$

where one row of the design matrix is given by $\boldsymbol{x}_i = [1, \log(CO_{2,i}), \log(Rad_i + 1)]$, and the prediction of the probability of opening a window (within the next 10 minutes) is

$$p_i = \frac{e^{\eta_i}}{1 + e^{\eta_i}}. \tag{4.24}$$

In the example above, the data suggests a natural density (here the binomial), but in other cases, the data in itself does not directly suggest a distribution. One example of such data is the clothing example in Chapter 3. There the data as such does not imply a specific distribution, and we investigated different transformation of the data to get a better fit (but staying within the normal model). In the example below, we will use the gamma model to re-investigate these data.

Example 4.9 (Clothing insulation level) *In the clothing example, the clothing level is a positive variable and therefore the standard approach would be to model the log-transformed data using the normal distribution or (equivalent) use the log-normal distribution for the original data. Another option is to use the gamma distribution (which have positive outcomes). In R, the model (using the same structure as found in Chapter 3, Example 3.14) can be estimated by*

```
modGLMEx11.gam <- glm(clo ~ tOut + I(log(tOut)) * sex + tInOp,
                      data = dfLMEx1, family = "Gamma")
summary(modGLMEx11.gam)

## [...]
## Deviance Residuals:
##     Min       1Q   Median       3Q      Max
## -0.7903  -0.1426   0.0095   0.1270   0.7487
##
## Coefficients:
##                      Estimate Std. Error t value Pr(>|t|)
## (Intercept)            4.2652     1.0990    3.88  0.00011 ***
## tOut                   0.1665     0.0260    6.40  2.6e-10 ***
## I(log(tOut))          -2.1794     0.5340   -4.08  4.9e-05 ***
## sexmale                2.3834     0.4629    5.15  3.3e-07 ***
## tInOp                  0.0202     0.0102    1.99  0.04676 *
## I(log(tOut)):sexmale  -0.7057     0.1536   -4.59  5.0e-06 ***
## ---
## (Dispersion parameter for Gamma family taken to be 0.057)
##
##     Null deviance: 69.471  on 802  degrees of freedom
## Residual deviance: 47.837  on 797  degrees of freedom
## AIC: -989.3
## [...]
```

In the code above, we see that the family is specified, but we have not specified the link function, implying that R will use the canonical link (in this case the inverse link). We see that the output is very similar to the one we got in the general linear model. The interpretation of the parameter standard errors and p-values is the same as in Chapter 3. The dispersion parameter ($1/\lambda$) *is estimated at* 0.057. *Further, we see that the AIC is better than most of the models from Chapter 3 (except for the complicated estimated transformation). The model along with its confidence intervals is shown in the figure below. The indoor operative temperature has been fixed to the observed average (i.e. 26.8 °C).*

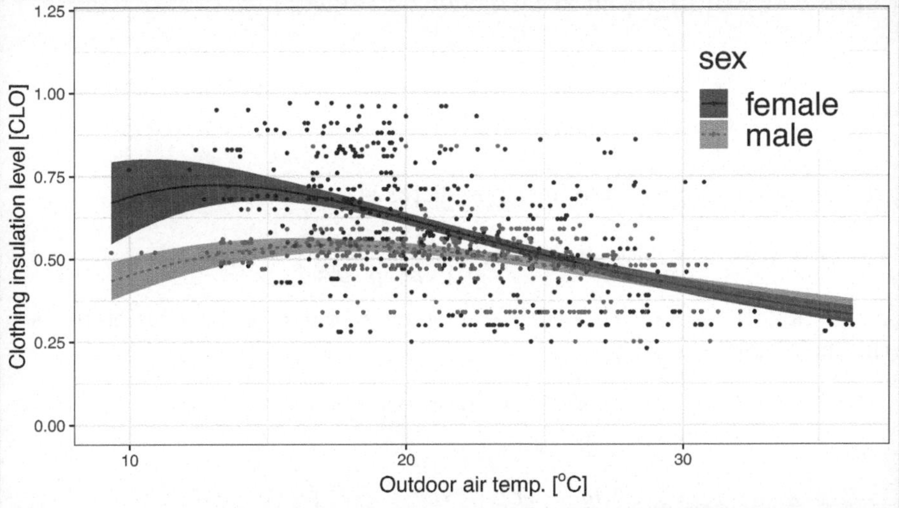

In the sections below we will discuss the parameter estimates and how standard errors are estimated.

4.3 Estimation of parameters

The parameter estimation relies on likelihood estimation, and the methods from Chapter 2 apply. As opposed to the general linear models, there are in general no closed–form solutions for the parameter estimates and we have to rely on non-linear optimisation.

In order to find the parameter estimates in the generalized linear model, we find the *score function* wrt. the canonical parameter (θ)

$$\frac{\partial}{\partial \boldsymbol{\theta}} \ell_\theta(\boldsymbol{\theta}; \boldsymbol{y}) = \text{diag}(\boldsymbol{w})(\boldsymbol{y} - \boldsymbol{\tau}(\boldsymbol{\theta})), \tag{4.25}$$

where $\boldsymbol{w}_i = \lambda_i$. Here, we aim at writing the score function wrt. the parameters, and a step towards this is to write the variance as function of the mean value.

Definition 4.6 ((Unit) variance function) *By introducing the mean value parameter in* (4.14) *we may express the variance as a function, the* (unit) variance function, *of the mean value in the distribution as*

$$V(\mu) = \kappa''(\tau^{-1}(\mu)). \tag{4.26}$$

The definition is also useful for calculating the variance of predictions; further we can write the score function as a function of the mean value

$$\begin{aligned}
\frac{\partial}{\partial \boldsymbol{\mu}} \ell_\mu(\boldsymbol{\mu}; \boldsymbol{y}) =& \frac{\partial}{\partial \boldsymbol{\theta}} \ell_\theta(\boldsymbol{\theta}; \boldsymbol{y}) \frac{\partial \boldsymbol{\theta}(\boldsymbol{\mu})}{\partial \boldsymbol{\mu}} \\
=& \text{diag}(\boldsymbol{w})(\boldsymbol{y} - \boldsymbol{\tau}(\tau^{-1}\boldsymbol{\mu})) \frac{\partial \tau^{-1}(\boldsymbol{\mu})}{\partial \boldsymbol{\mu}} \\
=& \text{diag}(\boldsymbol{w})(\boldsymbol{y} - \boldsymbol{\mu}) \frac{1}{\tau'(\tau^{-1}\boldsymbol{\mu})} \\
=& \text{diag}\left(\frac{w_i}{V(\mu_i)}\right)(\boldsymbol{y} - \boldsymbol{\mu}),
\end{aligned} \tag{4.27}$$

where $\text{diag}\left(\frac{w_i}{V(\mu_i)}\right)$ is a diagonal matrix with diagonal elements $\frac{w_i}{V(\mu_i)}$. The next step is to find the derivative wrt. to the linear prediction, i.e. we need

$$\begin{aligned}
\frac{\partial \mu_i}{\partial \eta_i} =& \frac{\partial g^{-1}(\eta_i)}{\partial \eta_i} \\
=& \frac{1}{g'(g^{-1}(\eta_i))} = \frac{1}{g'(\mu_i)}.
\end{aligned} \tag{4.28}$$

Hence, the score function wrt. the linear predictor is

$$\frac{\partial}{\partial \boldsymbol{\eta}} \ell_\eta(\boldsymbol{\eta}; \boldsymbol{y}) = \text{diag}\left(\frac{w_i}{V(\mu_i)} \frac{1}{g'(\mu_i)}\right)(\boldsymbol{y} - \boldsymbol{\mu}). \tag{4.29}$$

Based on Table 4.1, we see that the parameters are found as the solution to the equation (using the chain rule $(\frac{\partial}{\partial \theta} l = \frac{\partial \mu}{\partial \theta} \frac{\partial l}{\partial \mu})$)

$$\boldsymbol{X}(\boldsymbol{\beta})^T \boldsymbol{i}_\mu(\boldsymbol{\mu})(\boldsymbol{y} - \boldsymbol{\mu}) = \boldsymbol{0}\,, \tag{4.30}$$

where

$$\boldsymbol{X}(\boldsymbol{\beta}) = \frac{\partial \mu}{\partial \beta} = \left[\frac{d\mu}{d\eta}\right]^T \frac{\partial \eta}{\partial \beta} = \text{diag}\left\{\frac{1}{g'(\mu_i)}\right\} \boldsymbol{X}, \tag{4.31}$$

is called the local design matrix, and we see that the parameter mapping from the natural space to the linear space is included in the local design matrix.

Table 4.1: Likelihood, score function and Fisher information for generalized linear models, in terms of the canonical parameter, the mean value and the linear predictor.

	$l_i()$	$S_i()$	$i()$
θ	$\lambda_i(\theta_i y_i - \kappa(\theta_i))$	$\lambda_i(y_i - \kappa'(\theta_i))$	$\lambda_i \kappa''(\theta_i)$
μ	$\lambda_i(\tau^{-1}(\mu_i) y_i - \kappa(\tau^{-1}(\mu_i)))$	$\frac{\lambda_i}{V(\mu_i)}(y_i - \mu_i)$	$\frac{\lambda_i}{V(\mu_i)}$
η	$\lambda_i(\tau^{-1}(\mu_i) y_i - \kappa(\tau^{-1}(\mu_i)))$	$\frac{\lambda_i}{V(\mu_i) g'(\mu_i)}(y_i - \mu_i)$	$\frac{\lambda_i}{V(\mu_i)[g'(\mu_i)^2]}$

Remark 4.1 (The score function for the canonical link) *For the canonical link, the local design matrix $\boldsymbol{X}(\boldsymbol{\beta})$ is diag$\{V(\mu_i)\}\boldsymbol{X}$, and hence the score function* (4.30) *becomes*

$$\boldsymbol{X}^T \mathit{diag}\{V(\mu_i)\} \mathit{diag}\left\{\frac{w_i}{V(\mu_i)}\right\}(\boldsymbol{y} - \boldsymbol{\mu}(\boldsymbol{\beta})) = \boldsymbol{0}\,, \tag{4.32}$$

or

$$\boldsymbol{X}^T \mathit{diag}\{w_i\}\, \boldsymbol{y} = \boldsymbol{X}^T \mathit{diag}\{w_i\}\, \boldsymbol{\mu}. \tag{4.33}$$

Equation (4.33) *is called the* mean value equation, *or* the normal equation.

For an unweighted model, the mean value equation simply becomes

$$\boldsymbol{X}^T \boldsymbol{y} = \boldsymbol{X}^T \boldsymbol{\mu}(\boldsymbol{\beta}). \tag{4.34}$$

We can also find the variance-covariance matrix for the parameters as

$$\begin{aligned} \boldsymbol{i}_\beta =& \frac{\boldsymbol{\eta}}{\boldsymbol{\beta}}^T \boldsymbol{i}_\eta \frac{\boldsymbol{\eta}}{\boldsymbol{\beta}} \\ =& \boldsymbol{X}^T \boldsymbol{i}_\eta \boldsymbol{X} \end{aligned} \tag{4.35}$$

Hence, we can write the dispersion of $\boldsymbol{\beta}$ as

$$\boldsymbol{\Sigma} = D(\widehat{\boldsymbol{\beta}}) = \left(\boldsymbol{X}^T \boldsymbol{i}_\eta \boldsymbol{X}\right)^{-1}. \tag{4.36}$$

The theorem below states the Wald statistics for the ML estimator, Wald confidence intervals and Wald test for individual parameters.

Theorem 4.1 (Asymptotic distribution of the ML estimator) *Under the hypothesis $\boldsymbol{\eta} = \boldsymbol{X}\boldsymbol{\beta}$, we have asymptotically*

$$\widehat{\boldsymbol{\beta}} - \boldsymbol{\beta} \in N_k(\boldsymbol{0}, \boldsymbol{\Sigma})\,, \tag{4.37}$$

where $\boldsymbol{\Sigma}$ is the dispersion matrix for $\widehat{\boldsymbol{\beta}}$ (4.36). *Hence we can construct the usual Wald confidence intervals by*

$$\hat{\beta}_i \pm z_{1-\alpha/2}\sqrt{\boldsymbol{\Sigma}_{ii}}. \tag{4.38}$$

If we have estimated a dispersion parameter (as in the case of the gamma distribution example above), we will use the t-distribution, that is

$$\hat{\beta}_i \pm t_{1-\alpha/2}(n-p)\sqrt{\boldsymbol{\Sigma}_{ii}}. \tag{4.39}$$

In the case of the canonical link, the equations for the variance simplify since

$$g'(\mu) = \frac{1}{\tau'(\tau^{-1}(\mu))} = \frac{1}{V(\mu)}. \tag{4.40}$$

In generalized linear models, an option is also to use profile likelihood for estimating confidence intervals, this is implemented in R in the `confint` function for `glm` objects.

Example 4.10 (Window opening (binomial)) *In the window opening example (see Example 4.1 for the definition of data), the observed values consist of the number of times the window has been opened during the specified hours* (`y`)*, and at the specified temperature* (`temp`)*, and the number of times where window openings could have been observed* (`n`)*. Data is organised as*

```
##     temp  y  n
## 1     -1  1  2
## 2      1  1  7
## 3      3  0  1
## 4      5  0  5
## [...]
```

Here, `temp` *is the temperature,* `y` *is the number of window openings and* `n` *is the number of days with the specified temperature in that period. In this case $\lambda_i = n_i$, and the* log*-likelihood is*

$$l(q_i = y_i/n_i, \theta_i) = n_i(q_i\theta_i + \log(1 + e^{\theta_i})) + c(n_i, q_i), \tag{4.41}$$

with $\theta_i = \log(p_i/(1 - p_i))$ and $\kappa(\theta_i) = \log(1 + e^{\theta_i})$. We also have

$$\kappa'(\theta_i) = \frac{e^{\theta_i}}{1 + e^{\theta_i}}, \tag{4.42}$$

and hence the canonical link is $g(\mu_i) = \log(\mu_i/(1 - \mu_i))$; further

$$g'(\theta) = \frac{1 - \mu_i}{\mu_i}\left(\frac{1}{1 - \mu_i} + \frac{\mu_i}{(1 - \mu_i)^2}\right) \tag{4.43}$$

$$= \frac{1 - \mu_i}{\mu_i}\left(\frac{1}{(1 - \mu_i)^2}\right) \tag{4.44}$$

$$= \frac{1}{\mu_i(1 - \mu_i)}, \tag{4.45}$$

and $V(\mu_i) = \mu_i(1 - \mu_i)/n_i$. Using the canonical link we can implement the model in R by;

```
mod.binom <- glm(cbind(y, n - y) ~ temp, data = WindowDat1,
                 family = binomial())

## [...]
## Coefficients:
##             Estimate Std. Error z value Pr(>|z|)
## (Intercept)  -3.3869     0.8362   -4.05 0.000051 ***
## temp          0.3057     0.0721    4.24 0.000022 ***
## ---
## [...]
```

Notice that the observation is the number of times the window opened and the numbers of times the window did not open. The prediction of the probability at a given temperature is

$$p_i = \frac{e^{-3.4+0.31x_i}}{1+e^{-3.4+0.31x_i}}. \tag{4.46}$$

The confidence intervals for the parameters can be calculated by;

```
## Profile likelihood CI
confint(mod.binom)

## Waiting for profiling to be done...

##                2.5 % 97.5 %
## (Intercept) -5.23   -1.92
## temp         0.18    0.47

## Wald CI
coef(mod.binom)+1.96 * cbind(-sum.binom$coefficients[ ,2],
          sum.binom$coefficients[ ,2])

##               [,1]   [,2]
## (Intercept) -5.03 -1.75
## temp         0.16  0.45
```

Notice that there is a small difference in the CIs due to the two different methods that were used in this case (profile likelihood and Wald confidence intervals).

4.4 Fitted values and predictions

As we have seen so far, we have predictions in different spaces, namely the linear space, i.e.

$$\widehat{\eta} = \boldsymbol{X}\widehat{\beta}, \tag{4.47}$$

and the response space

$$\widehat{\mu} = \mu(\widehat{\eta}), \tag{4.48}$$

both of which are asymptotically normally distributed with variance-covariance given by

$$\begin{aligned} D[\widehat{\eta}] &\approx \widehat{\sigma}^2 \boldsymbol{X\Sigma X}^T \\ D[\widehat{\mu}] &\approx \widehat{\sigma}^2 \left[\frac{\partial \mu}{\partial \eta}\right]^2 \boldsymbol{X\Sigma X}^T. \end{aligned} \tag{4.49}$$

From the dispersion matrices above, we can directly calculate confidence intervals for the predictions.

Theorem 4.2 (Confidence interval for linear predictor) *An approximate* $100(1-\alpha)\%$ *confidence interval for the linear prediction is obtained as*

$$\widehat{\eta}_i \pm z_{1-\alpha/2}\sqrt{\widehat{\sigma^2\widehat{\sigma}_{ii}}}, \tag{4.50}$$

with $\widehat{\eta}_i$ *given by* (4.21), *and* $\widehat{\sigma}_{ii}$ *denoting the i'th diagonal element in* $\boldsymbol{X\Sigma X}^T$ *where* $\boldsymbol{\Sigma}$ *given by* (4.36). $z_{1-\alpha/2}$ *is the* $1-\alpha/2$ *quantile of a standard normal distribution.*

For reporting, it is often more useful to find confidence intervals in the natural or original domain; the definitions are given below and both are exemplified below in Examples 4.11 and 4.12 (the first one including the specific calculations).

Theorem 4.3 (Confidence interval for response) *An approximate* $100(1-\alpha)\%$ *confidence interval for the response is obtained as*

$$\widehat{\mu}_i \pm z_{1-\alpha/2}\sqrt{D[\widehat{\boldsymbol{\mu}}]_{ii}}, \tag{4.51}$$

with $\widehat{\mu}_i$, *and* $z_{1-\alpha/2}$ *as in Theorem 4.2.*

If we want to make prediction intervals, the stochastic uncertainty needs to be included, the easiest way to do that is by bootstrap. That is, simulate μ_i from a normal distribution, with the estimated variance and then simulate from the assumed distribution using the realisation of the mean value.

Theorem 4.4 (Prediction intervals) *An approximate* $100(1-\alpha)\%$ *prediction interval for a new observation can be obtained by simulation in 3 steps:*

1: Simulate k-values ($k >> 1$, *e.g. 1000*) ($\mu_{i,j}$) *of* μ_i *from* $N(\widehat{\mu}_i, \widehat{\sigma}^2_{\mu,i})$.

2: For each $\mu_{i,j}$, *simulate an outcome* ($y^*_{i,j}$) *from the appropriate distribution.*

3: Find appropriate quantiles ($\alpha/2, 1-\alpha/2$) *of* $\boldsymbol{y}^*_{i,\cdot}$ *(which is the now prediction intervals).*

In the first step, there are actually more options, for example:

- *Simulate k-values* ($k >> 1$, *e.g. 1000*) ($\eta_{i,j}$) *of* η_i *from* $N(\widehat{\eta}_i, \widehat{\sigma}^2_{\eta,i})$ *and use the inverse link to get realisations of* $\mu_{i,j}$.
- *Simulate data-sets using* $\widehat{\boldsymbol{\theta}}$, *estimate parameters from those data-sets, and simulate* ($y^*_{i,j}$) *based on each parameter set.*

Example 4.11 *Continuing the clothing example, we might be interested in a prediction interval for the level of clothing for a female at an outdoor temperature of* 20 *degrees; this can be obtained by;*

```
## Step 1:
tout <- 20
X.f <- data.frame(tOut=tout, tInOp = mean(dfLMEx1$tInOp),
                  sex="female")
conf11.f <- predict(modGLMEx11.gam,newdata=X.f,type="response",
                    se.fit=TRUE)
## Confidence interval
(CI <- conf11.f$fit + 1.96 * c(-1,1) * conf11.f$se.fit)

## [1] 0.60 0.64

## Step 2: Prediction interval
k <-  100000
a <- summary(modGLMEx11.gam)$dispersion
r <- rgamma(k, scale = rnorm(k, mean = conf11.f$fit,
                                sd = conf11.f$se.fit) * a,
            shape=1/a)
## Step 3: Expected value, median, and 95% Prediction interval
c(mean(r), quantile(r,probs = c(0.5,0.025,0.975)))

##          50%  2.5% 97.5%
##  0.62   0.61  0.36  0.94
```

Hence we see that the confidence interval is $[0.60, 0.64]$ *and the prediction interval is* $[0.36; 0.94]$. *The figure below shows the prediction and confidence interval for different values of the outdoor temperature (using indoor operating temperature equal to the average of observed indoor operating temperatures).*

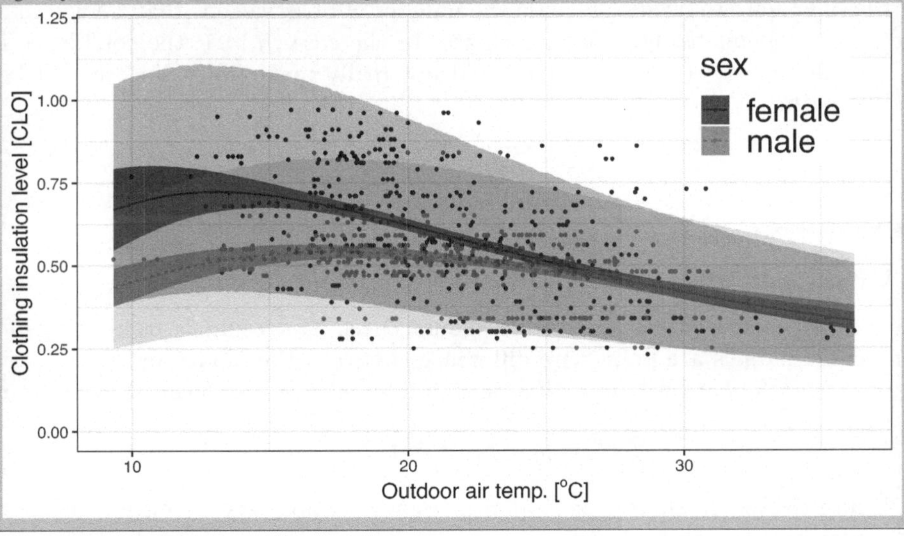

Example 4.12 *In the case of the binomial model, it is also straightforward to make the confidence intervals. The interpretation of the confidence interval is that the true probability lies in the given interval. For a prediction interval, we would need to specify the total number of observation at a specific temperature, e.g.* $n_i = 20$ *as in the figure below (light grey shows the prediction interval, and dark grey shows the confidence interval).*

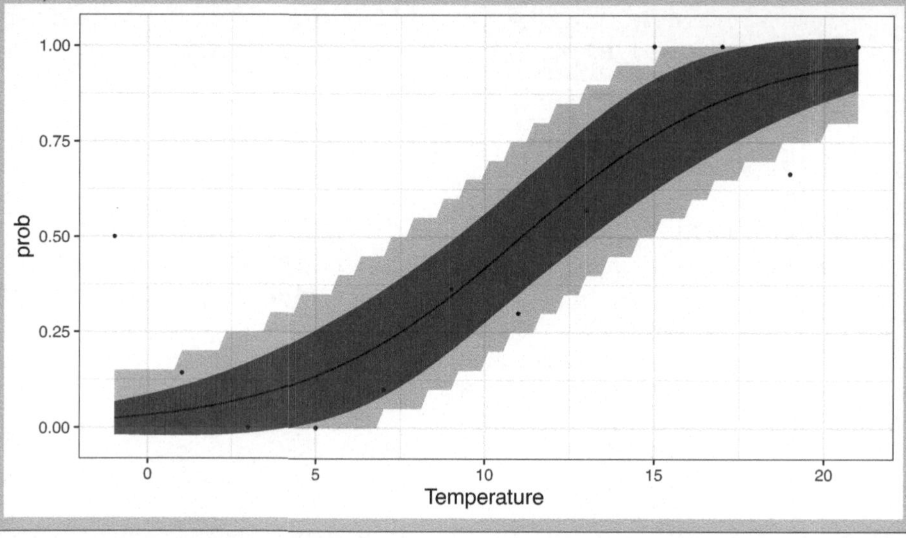

The examples above illustrate how the interpretation of prediction and confidence intervals depends on the model: In the case of the gamma distribution, prediction intervals can be interpreted in the same way as prediction intervals for Gaussian models, while in the case of the binomial, we need to include the number of observations that we will make, and the width will depend strongly on this choice. In the case $n_i = 1$, the confidence interval (which is really a discrete set of points) will not really make sense since it will be either $\{0\}$, $\{1\}$ or both.

4.5 Residuals and model diagnostics

Just as for the general linear model, we should check the residuals from the generalized linear model (residuals are point-wise differences between the model and the observation). The most straightforward way to look at such a difference is the so-called *response residual*

$$r_i^R = y_i - \widehat{\mu}_i \ , \ i = 1, 2, \ldots, n, \tag{4.52}$$

although intuitively appealing, the response residual is not very useful for diagnostic, because the variance is a function of the mean value. Therefore, we cannot check if, for example the variation is constant. The *Pearson residuals* are obtained by dividing the residuals with the standard deviation given by the model, i.e.

$$r_i^P = \frac{y_i - \widehat{\mu}_i}{\sqrt{V(\hat{\mu}_i)/w_i}} \ , \ i = 1, 2, \ldots, n. \tag{4.53}$$

This is similar to what we did for the general linear model (where we divided with the variance of the individual observations to the standardized residuals); however, in the case of generalized models, this is much more important since the residuals can have very different variances originating from the stochastic uncertainty (i.e. from the model, not just the sampling).

Finally, we may use the so-called *deviance residuals*; the idea is to consider the likelihood contribution from each observation. The log-likelihood of the exponential dispersion model can be written as (ignoring constant terms)

$$l(\boldsymbol{y}; \boldsymbol{\theta}) = \sum_{i=1}^{n} w_i(\theta_i y_i - \kappa(\theta_i)). \tag{4.54}$$

Now, we define the residual deviance as

$$D(\boldsymbol{\mu}; \boldsymbol{y}) = -2l(\boldsymbol{y}; \boldsymbol{\mu}) = \sum_{i=1}^{n} w_i d(y_i, \hat{\mu}_i). \tag{4.55}$$

Hence, the contribution from observation i to the log-likelihood is $w_i d(y_i, \hat{\mu}_i)$, and from this we can define the deviance residuals as

$$r_i^D = \sqrt{w_i d(y_i, \widehat{\mu}_i)}\texttt{sign}(y_i, \widehat{\mu}_i)). \tag{4.56}$$

Note that for the normal distribution the deviance and the Pearson residuals are the same.

In addition to the residuals defined above, one will also sometimes encounter working residuals, which are however primarily used for estimation and will be omitted here.

Both the deviance and the Pearson residuals come in standardized (and studentized) forms

$$r_i^{P,S} = \frac{r_i^P}{\sqrt{1 - h_{ii}}}; \quad r_i^{D,S} = \frac{r_i^D}{\sqrt{1 - h_{ii}}}, \tag{4.57}$$

where $h_{ii} = \mathbf{H}_{ii}$, and the matrix $\mathbf{H}$ is given by

$$\boldsymbol{H} = \boldsymbol{W}^{1/2}\boldsymbol{X}(\boldsymbol{X}^T\boldsymbol{W}\boldsymbol{X})^{-1}\boldsymbol{X}^T\boldsymbol{W}^{1/2}. \tag{4.58}$$

Since $\boldsymbol{W} = \text{diag}\{w_i\}$ is a diagonal matrix, the square root is well defined and easy to calculate.

4.5.1 Model diagnostics

Model diagnostics is done in the same way as for the general linear model, but we should use either Pearson residuals or deviance residuals rather than response residuals.

Example 4.13 (Clothing) *The figure below shows the standard (from R) residual plot[a] for the clothing insulation level model, discussed above.*

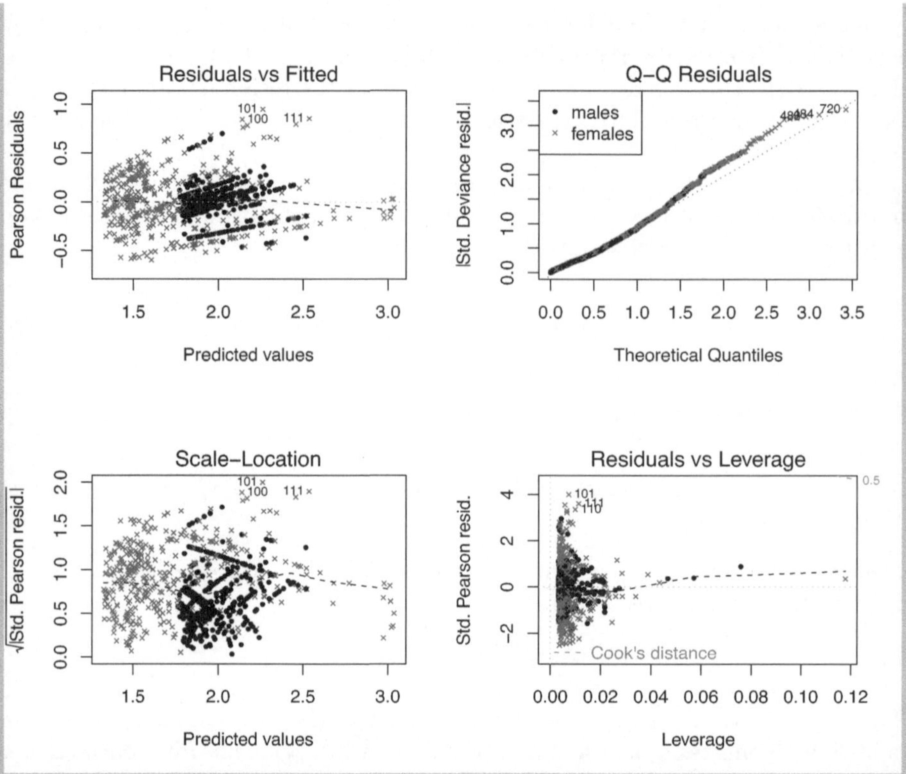

There are some issues with the model diagnostics, e.g. we still have some patterns in the residuals vs. fitted. The absolute value of the standardized deviance residuals is compared to the half normal distribution, and some issues are seen. In particular, we see that the residuals for females are much larger than those for mmales, which is the same pattern we saw for the general linear model (Example 3.4, on page 56).

[a]It should be noted that these plots may change over time with updates of R. The basic patterns that should be observed however remains.

Example 4.14 (Window) *Diagnostic plot for the window opening example is shown below.*

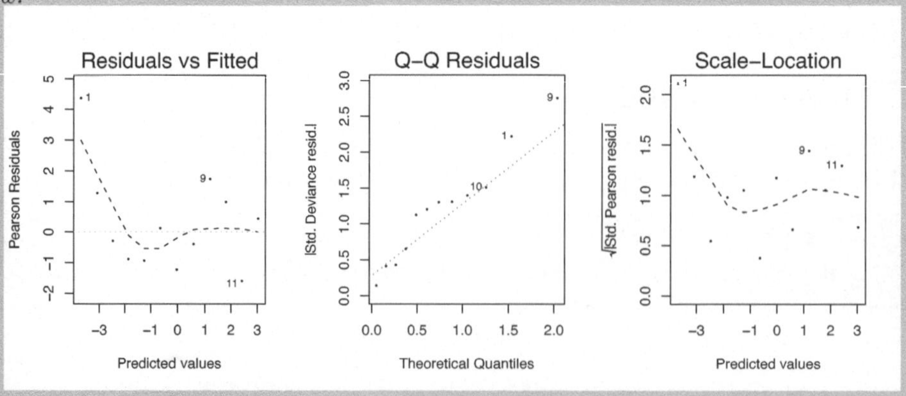

Here, we see some pattern in residuals vs. fitted and also in the scaled location plot: these figures suggest that the functional relationship could be improved.

The diagnostic plots above rely on the normal approximation of the residuals, and, if such an approximation is reasonable, depend on the data and the model at hand. As a particular example, the normal approximation of binomial data relies on the number of trails. Hence, if we had not aggregated the window opening data, we would (almost) get the same model but the diagnostic plot would suggest large problems with the model (see Example 4.15 below).

Example 4.15 (Window) *The figure below shows the binary data along with binomial model predictions, fitted to aggregated data and to binary data; it is quite clear that the models are basically identical.*

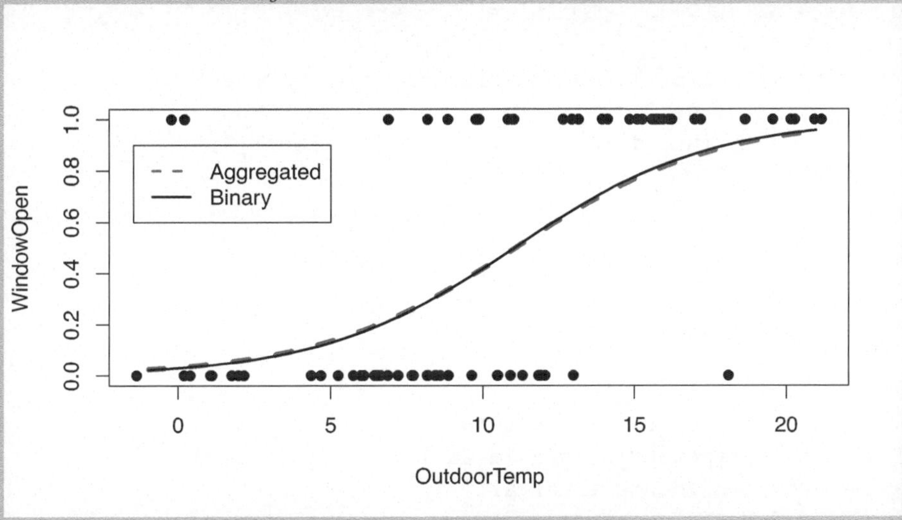

However, if we look at the diagnostic plot, we see very clear patterns that stem from the way data is sampled (i.e. 0, 1).

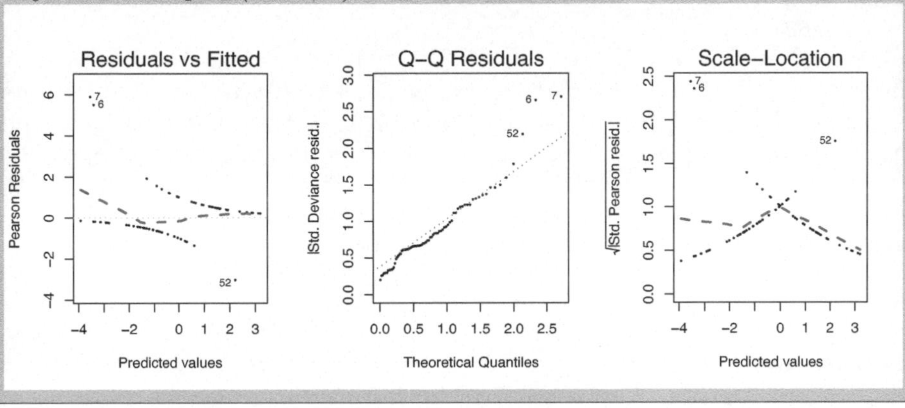

4.5.2 Independence

In the model diagnostics above we have assumed independence, and this can often be assumed from data; however if data is obtained as time–series or multiple time–series, we should check for such independence. This is most easily done by considering the autocorrelation (we will get back to the autocorrelation function in Chapter 8) of the residuals. Here, we will restrict our selves to consider the lag 1 autocorrelation ($\rho(1)$), that is

$$\rho(1) = cor(r_i, r_{i-1}), \tag{4.59}$$

with r_i being either Pearson or deviance residuals. If the residuals are independent, the autocorrelation should be 0 (in all lags).

Example 4.16 (Clothing) *The clothing data is actually taking over a number of days where each subject reports their clothing insulation level at specific time points (usually around 6 times per day), and one could suspect that the measurements taken on the same day are actually correlated. This can be checked by considering the autocorrelation of the residuals within each day: we should use the Pearson or the deviance residuals for such an analysis. The R-code for the analysis is illustrated below.*

```
dat.tmp <- dfLMEx1
dat.tmp$res <- residuals(modGLMEx11.gam, type = "pearson")
cor <- aggregate(res ~ subjId + day, data = dat.tmp,
                 function(x){acf(x,plot=FALSE)$acf[2]})[ ,3]

## Test
t.test(cor)

##
##  One Sample t-test
##
## data:  cor
## t = 11, df = 135, p-value <2e-16
## alternative hypothesis: true mean is not equal to 0
## 95 percent confidence interval:
##  0.220 0.316
## sample estimates:
## mean of x
##     0.268
```

In this case, the test shows that there is actually a positive correlation (of around 0.27), as we also found in Example 3.12 (page 70), in the residuals, which should be taken into account in the modelling. We will return to this subject in Chapter 6.

The example above illustrates that we need to consider our data carefully. A wrong independence assumption might invalidate our results.

4.6 Link functions

As mentioned above, the canonical link is only one possible link function, and basically the choice is free and part of the modeller's job is to find the most suited link function for a particular problem. Some link functions are, however, more natural for a given family than others. A natural starting point is the canonical link; further, it is natural to choose a link function that takes the real axis into the domain of the model, e.g. for binomial models, it is natural to choose logit, probit, cloglog, etc. It is natural to choose log or power link for models that have positive outcomes, for example Poisson or gamma models. Some commonly used link functions are listed in Table 4.2.

When comparing models with different link functions, we can use AIC or BIC to compare the models.

Table 4.2: Some commonly used link functions.

Name	Link function $\eta = g(\mu)$	$\mu = g^{-1}(\eta)$
identity	μ	η
log	$\log(\mu)$	$\exp(\eta)$
logit	$\log(\mu/(1-\mu))$	$\exp(\eta)/[1+\exp(\eta)]$
inverse	$1/\mu$	$1/\eta$
power	μ^k	$\eta^{1/k}$
sqrt	$\sqrt{\mu}$	η^2
probit	$\Phi^{-1}(\mu)$	$\Phi(\eta)$
log-log	$\log(-\log(\mu))$	$\exp(-\exp(\eta))$
cloglog	$\log(-\log(1-\mu))$	$1-\exp(-\exp(\eta))$

Example 4.17 (Clothing) *For the gamma model, the link functions that are implemented in the* `glm` *method are the Inverse link,* log-*link and identity link. These are compared by AIC below.*

```
modGLMEx11.gam.log <- glm(formula = clo ~ tOut +
                              I(log(tOut)) * sex + tInOp,
                          family = Gamma(link=log),
                          data = dfLMEx1)
modGLMEx11.gam.ident <- glm(formula = clo ~ tOut +
                              I(log(tOut)) * sex + tInOp,
                            family = Gamma(link=identity),
                            data = dfLMEx1)
AIC(modGLMEx11.gam,modGLMEx11.gam.log,modGLMEx11.gam.ident)

##                      df  AIC
## modGLMEx11.gam        7 -989
## modGLMEx11.gam.log    7 -990
## modGLMEx11.gam.ident  7 -987
```

The difference between the inverse link and the log-link is quite small, but with preference to the log-link while the identity is worse. Hence we should choose the log-link.

In the example above, we see that the log-link gives a better result than the canonical link, and it illustrates that the link function should be an integrated part of our modelling.

Example 4.18 (Window) *For the binomial model, we test the link functions that take the real axis into the interval (0,1) and are implemented in R; these are given below.*

```
##                    df  AIC
## mod.binom           2 37.8
## mod.binom.probit    2 39.2
## mod.binom.cauchit   2 33.3
## mod.binom.cloglog   2 36.7
## mod.binom.loglog    2 43.5
```

Here, it is quite clear that we should choose the Cauchit link; some diagnostic plots are shown below.

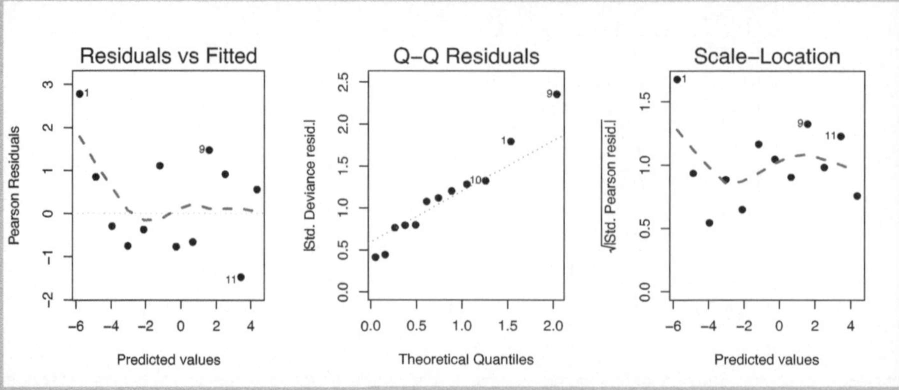

We see that in particular the scale-location plot (last column) has improved. For completeness, we include predictions with two of the models.

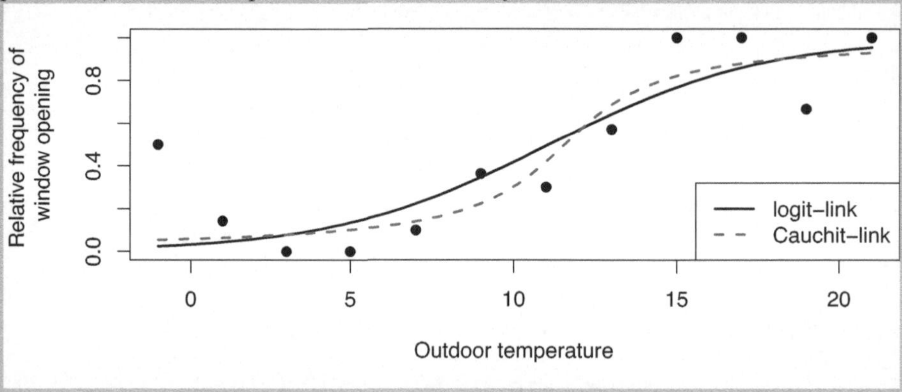

We see that the reponse is quite different in the two cases. It should be noted that only the clog-log link is implemented directly in R, but we can implement the log-log link by shifting the output, i.e. change the output from number of window opening to number of not opening windows.

4.7 Including an offset

In some cases, it is natural to include an offset in the model; the most obvious case is for Poisson model where the parameter is given as a rate, i.e. intensity of events over unit time or distance. Let us, for example say we have observations of number of events over one hour from one population, and number of events over 2 hours from another population: in order to compare the number of events from the two populations, we would need to take the different observation periods into account. In the case of Poisson distributed data, this would imply

$$\frac{\mu_i}{t_i} = \lambda_i, \tag{4.60}$$

where t_i is the time used to get observation i. Taking the log on both side gives

$$\log(\lambda_i) = \boldsymbol{x}_i^T \boldsymbol{\beta} + \log(t_i), \tag{4.61}$$

we use our clothing data to illustrate this.

Example 4.19 (Change of clothes) *The clothing data used in Example 4.16 (and other examples of this book) is actually taken as small time–series, where each subject is in the office on a number of days and records the clothing insulation level at a number number of time-points (usually 6) during the day. Hence we could count number of times that subjects change their clothing insulation level during a day. The output below shows a small subset of the data;*

```
##    subjId day clo time nobs    sex delta.t
## 1      11   1   0  6.7    5 female     1.3
## 2      17   1   0  6.7    5 female     1.3
## 3      19   1   1  6.8    5 female     1.4
## 4      29   1   0  6.6    5 female     1.3
## 5      35   1   0  6.8    5   male     1.4
## [...]
```

We see that the data contains an identifier for the subject, day (1-4, but varying between subjects), the number of times the subject change clothing insulation level during the day (`clo`*), the total time of observations (*`time`*), the number of records on the day (*`nobs`*), average time between measurements* `delta.t` *and sex of the subject. Some key numbers are;*

```
##         clo              time             nobs           delta.t
##  Min.    :0.0   Min.    :1.5   Min.    :1.0   Min.    :1.20
##  1st Qu.:0.0   1st Qu.:6.7   1st Qu.:5.0   1st Qu.:1.33
##  Median :0.0   Median :6.7   Median :5.0   Median :1.34
##  Mean    :0.6   Mean    :6.6   Mean    :4.9   Mean    :1.35
##  3rd Qu.:1.0   3rd Qu.:6.8   3rd Qu.:5.0   3rd Qu.:1.35
##  Max.    :4.0   Max.    :7.2   Max.    :5.0   Max.    :1.70
```

We see that most subjects do not change clothing insulation level during the day and that none changes clothing insulation level more than 4 times per day. Most subjects use approximately 6.6 hours a day, but a few use significantly less. Further most have 5 record of changes in insulation level (i.e. 6 observations of insulation level). Also, note that the number of observations and the total time of observation is closely connected, and the average time between observations is approx. 1.34 hours, with quite small variation (IQR=[1.33,1.35]). Given the total number of observation, it seems natural to fit a binomial model (we will take into account sex here);

```
fit.binom <- glm(cbind(clo, nobs - clo) ~ sex, data = dfcount,
                 family = binomial)
exp(cumsum(coef(fit.binom))) / (1 + exp(cumsum(coef(fit.binom))))

## (Intercept)     sexmale
##        0.17        0.06
```

Hence, we see that the probability of changing insulation level between two measurements (app. 1.34 hours) is about 17% for women and 6% for males. Even though the data is obtained by recording insulation level at specific time points, we could model this as the number of changes in clothing insulation level over a day using the Poisson model, ignoring that a "day" represents different time intervals, which would result in the following model;

```
fit.pois <- glm(clo ~ sex, data = dfcount, family = poisson)
exp(cumsum(coef(fit.pois)))

## (Intercept)     sexmale
##        0.86        0.29
```

That is, a rate of changing insulation level in a day. However, here we do not take into account that the length of day differs considerably; hence, we should include an offset;

```
fit.poisOff <- glm(clo ~ sex + offset(log(time)), data = dfcount,
                   family = poisson)
exp(cumsum(coef(fit.poisOff)))

## (Intercept)     sexmale
##       0.128       0.044
```

Here, we have the rate of change in clothing insulation level for men and women during 1 time unit (here 1 hour). We can compare the models by e.g. AIC

```
AIC(fit.binom,fit.pois,fit.poisOff)

##             df AIC
## fit.binom    2 274
## fit.pois     2 271
## fit.poisOff  2 270
```

Hence, we see that the Poisson model gives a better fit than the binomial, and also that including the offset results in a slightly better fit (likelihood) than when ignoring it. The figure below compares the probability of changing clothing insulation level 0–5 times during a standard day (6.7 hours and 5 records) using the Poisson and the binomial model.

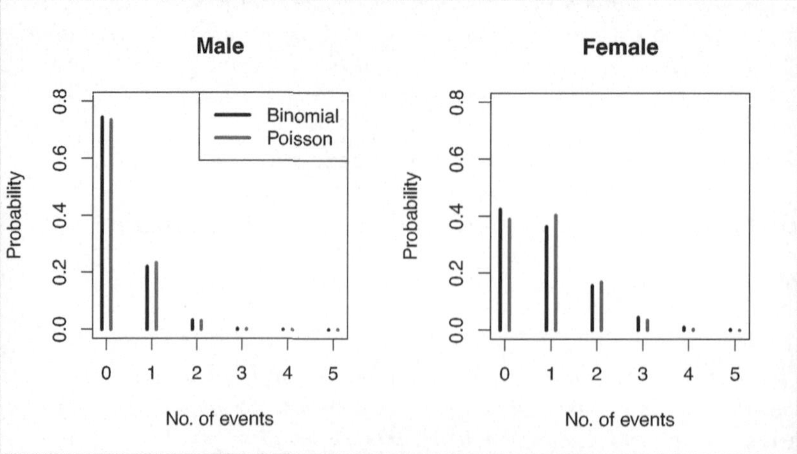

We see that the models (binomial and Poisson) are very similar for both males and females. The figures above do not allow for a direct comparison with the data as n and time are not constant over the data.

4.8 Test for model reduction – nested models

For nested models, we can compare the models by means of likelihood ratio test, and as long as we do not estimate a dispersion parameter, the methods are exactly the same as described in Chapter 2. In the clothing insulation example, we use the data to estimate the dispersion parameters, and in such cases we should also take the parameter uncertainty into account in the test procedure. This implies using the F-distribution (like we saw in Chapter 3) rather than the χ^2-distribution (see Brockhoff et al. (2015) for an introduction to sampling distributions).

Example 4.20 *In the clothing example, we might test if the indoor temperature has a significant effect.*

```
modGLMEx11.gam.log <- glm(clo ~ tOut + I(log(tOut)) * sex +
                              tInOp, family = Gamma(link=log),
                          data = dfLMEx1)

modGLMEx12.gam.log <- glm(clo ~ tOut + I(log(tOut)) * sex,
                          family = Gamma(link=log),
                          data = dfLMEx1)

anova(modGLMEx12.gam.log,modGLMEx11.gam.log,test="F")

## Analysis of Deviance Table
##
## Model 1: clo ~ tOut + I(log(tOut)) * sex
## Model 2: clo ~ tOut + I(log(tOut)) * sex + tInOp
##   Resid. Df Resid. Dev Df Deviance    F Pr(>F)
## 1       798       48.0
## 2       797       47.8  1    0.146 2.59   0.11
```

Hence, with a p-value of 0.1, we should disregard the indoor temperature in the model.

When the dispersion parameter is not estimated (like in the binomial and Poison models), we use the χ^2-distribution for testing nested model.

4.8.1 Initial goodness of fit

In cases where we have not estimated a dispersion parameter, we might access the goodness of the fit from an analysis of the deviance table. This will give a test to evaluate if the model can at all be maintained. In case it cannot we would have to choose a different structure; this may include another distribution assumption, including random effects (see Chapters 5 and 6), more/other explanatory variables.

Remark 4.2 (Test for model "sufficiency") *Consider the generalized linear model from Definition 4.4 on page 83 and assume that the dispersion is* $\sigma^2 = 1$. *Let* H_{full} *denote the* full, *or* saturated *model allowing each observation to have its own, freely varying mean value, i.e.* $H_{\text{full}}: \boldsymbol{\mu} \in \mathbb{R}^n$ *(restricted only by restrictions on mean values), and consider the hypothesis*

$$H_0: \boldsymbol{\eta} \in L \subset \mathbb{R}^k, \tag{4.62}$$

with L *parameterised as* $\boldsymbol{\eta} = \boldsymbol{X}_0\boldsymbol{\beta}$.

Then, as the residual deviance under H_{full} *is 0, the test statistic is the residual deviance* $D(\boldsymbol{\mu}(\widehat{\boldsymbol{\beta}}))$. *If* H_0 *is true, the test statistic is distributed as* $\chi^2(n-k)$. *The test rejects for large values of* $D(\boldsymbol{\mu}(\widehat{\boldsymbol{\beta}}))$.

Example 4.21 *In the window opening example (Example 4.18), we test if the Cauchit-link model is appropriate for the data.*

```
anova(mod.binom.cauchit, test = "Chisq")

## [...]
##       Df Deviance Resid. Df Resid. Dev Pr(>Chi)
## NULL                     11       50.1
## temp  1     35.4         10       14.7  2.6e-09 ***
## [...]
```

In this case, the residual deviance is 14.7, which should be compared to a χ^2*-distribution with 10 degrees of freedom and hence the p-value is*

```
1 - pchisq(14.7, df = 10)

## [1] 0.14
```

Hence, we conclude that the model can be maintained (on a 5% level), and of course also that temperature is highly significant (p-value $\approx 2.6 \cdot 10^{-9}$*)*

Example 4.22 *In the case of the changes in clothing insulation level, the goodness of fit is given by*

```
1 - pchisq(summary(fit.poisOff)$deviance, df = 134)

## [1] 0.18
```

Hence, the test does not suggest over-dispersion, but we could make the hypothesis that there is a variation due to differences between the subjects,

```
fit.poisOffSat <- glm(clo ~ factor(subjId) + offset(log(time)),
    data = dfcount, family = poisson)
```

Comparing the two models gives

```
anova(fit.pois,fit.poisOffSat,test="Chisq")

## Analysis of Deviance Table
##
## Model 1: clo ~ sex
## Model 2: clo ~ factor(subjId) + offset(log(time))
##   Resid. Df Resid. Dev Df Deviance Pr(>Chi)
## 1       134      150.0
## 2        89       63.5 45     86.6  0.00019 ***
## [...]
```

Hence, there is a clear effect of subjId, which should be included in some way, but the way chosen here only allows us to conclude about the specific persons included in this study. A better way is to include the subjects as random effects: we will discuss this in Chapters 5 and 6.

4.9 Over-dispersion

Suppose that we have estimated a comprehensive model and that it fails the goodness–of–fit test. In order to fix this, we should try different link functions before modelling over-dispersion. Again, suppose we have done so, but still it fails the goodness–of–fit; in such cases we might introduce a so-called over-dispersion parameter, implying that there is more variance/dispersion in the data than would be anticipated from the model. Such over-dispersion may be due to grouping structures in data that is not included in the model (see Chapters 5 and 6), but for now we will simply estimate it.

As we have seen, the exponential dispersion family can be written as

$$f_Y(y;\theta) = c(y,\lambda)\exp(\lambda\{\theta y - \kappa(\theta)\}). \tag{4.63}$$

The dispersion parameter might be a natural part of the modelling (e.g. gamma and Normal models); in other cases, it is given directly by the experimental conditions (e.g. binomial and Poisson models). The formulation implies that the estimation of the dispersion parameter is separated from the estimation of the mean value (defined by the cumulant generator, $\kappa(\theta)$).

In the case of the binomial, the precision parameter is

$$\lambda = n, \tag{4.64}$$

but we could replace this by

$$\lambda = n/\sigma^2, \tag{4.65}$$

or more generally

$$\lambda = w/\sigma^2, \tag{4.66}$$

where σ^2 is a dispersion parameter to be estimated. The problem is now that the dispersion parameter is a part of the constant $c(y,\lambda)$ and hence a direct calculation of the likelihood is in general not possible (we can do it for, e.g. the gamma and the normal model). An alternative is to use the observed dispersion to estimate the over–dispersion parameter.

Remark 4.3 (Approximate moment estimate for the dispersion parameter) *The over-dispersion parameter can be estimated by the deviance using the estimator*

$$\widehat{\sigma}^2_{dev} = \frac{D(\boldsymbol{y}; \boldsymbol{\mu}(\widehat{\boldsymbol{\beta}}))}{n-k}, \tag{4.67}$$

which is asymptotically unbiased for σ^2.

Alternatively, one could use the corresponding Pearson goodness–of–fit test statistic

$$X^2 = \sum_{i=1}^{n} w_i \frac{(y_i - \widehat{\mu}_i)^2}{V(\widehat{\mu}_i)}, \tag{4.68}$$

and use the estimator

$$\widehat{\sigma}^2_{Pears} = \frac{X^2}{n-k}. \tag{4.69}$$

Example 4.23 *As we have seen earlier, there is not really any evidence of over-dispersion in the window opening example, but for illustrative purposes, we will estimate the dispersion parameter here. In* `glm`*, this is done using the family* `quasibinomial`*;*

```
mod.quasibinom <- glm(formula = cbind(y, n - y) ~ temp,
                      family = quasibinomial(link=cauchit),
                      data = WindowDat1)
summary(mod.quasibinom)

## [...]
## Coefficients:
##              Estimate Std. Error t value Pr(>|t|)
## (Intercept)    -5.335      2.540   -2.10    0.062 .
## temp            0.462      0.220    2.09    0.063 .
## (Dispersion parameter for quasibinomial family taken to be 1.7)
##
##     Null deviance: 50.114  on 11  degrees of freedom
## Residual deviance: 14.671  on 10  degrees of freedom
## [...]
```

In this case, the dispersion parameter (σ^2) is estimated at 1.7.

Since the residual deviance has been used for estimating the dispersion, there is no goodness–of–fit test in over-dispersed models. We can see in the example above that there is no AIC; this is because the likelihood cannot be evaluated directly (due to the normalisation constant), and therefore we cannot use, for example the profile likelihood to evaluate the confidence interval for the dispersion parameter. As an alternative, we can calculate the bootstrap p-value.

Example 4.24 *An example of such a bootstrap p-value is given below. The procedure is that observations are simulated using the estimated mean value model (with no over-dispersion) and same "experimental" conditions (i.e. n). For each realisation, the over–dispersion parameter is estimated and the estimated dispersion parameter is compared to the simulated distribution (i.e. under the assumption that the true dispersion parameter is equal to 1);*

```
n <- WindowDat1$n
temp <- WindowDat1$temp
y <- WindowDat1$y

pred <- predict(mod.binom.cauchit, type= "response")
k <- 1000
disp <- numeric(k)
for(i in 1:k){
    y <- rbinom(length(n), prob = pred, size = n)
    disp[i] <- summary(glm(cbind(y, n - y) ~ temp,
                           family = quasibinomial(link = cauchit))
    )$dispersion
}

sum(disp > summary(mod.quasibinom)$dispersion) / k

## [1] 0.12
```

Hence the one-sided p-value is 0.12 and we cannot reject that the true dispersion parameter is 1.

4.10 Modelling dispersion

The focus of modelling is usually (and should be) on the mean value, while the dispersion is a nuisance parameter we estimate in order to give meaning to the statistical model. However, in order predict new data points and construct reliable confidence and prediction intervals, we need accurate models, also for the dispersion. For example, the dispersion might depend on levels or regressors, as in the clothing insulation level example (Example 3.11) where the residual variation for males and females is quite different. We could therefore consider the dispersion as a parameter that could be modelled in the same way as the mean values.

Example 4.25 *Again, consider the clothing insulation example (Example 4.17). The dispersion parameter is, in this case, an integrated part of the model estimation, and there is therefore no goodness-of-fit test. However, the data suggested that there is a difference in dispersion between men and women. In order to compare this, we should use either the Pearson or the deviance residuals; here, we use the Pearson residuals and consider the ratio of the sum of squared residuals for males and females;*

```
rP <- residuals(modGLMEx12.gam.log,type="pearson")
(ratio <- mean(rP[dfLMEx1$sex=="female"]^2)/
     mean(rP[dfLMEx1$sex=="male"]^2))

## [1] 2.8
```

As we can see the ratio is around 2.8, hence the variation of the residuals is around 2.8 times higher for females than for males.

In order to model this, we should set up the likelihood including a model for the dispersion, i.e. the model

$$Y_i \sim G(\lambda_i, \mu_i), \tag{4.70}$$

where μ_i *is modelled by*

$$\log(\boldsymbol{\mu}) = \boldsymbol{\eta} = \boldsymbol{X}\boldsymbol{\beta}, \tag{4.71}$$

and λ_i *is modelled by*

$$\log(\boldsymbol{\lambda}) = \boldsymbol{\eta}_\lambda = \boldsymbol{X}_\lambda\boldsymbol{\beta}_\lambda, \tag{4.72}$$

where $\boldsymbol{X}_\lambda$ *is a design matrix corresponding to a male/female model. In our particular model, this means that we can write the dispersion for males as* λ *and the dispersion for women as* $\lambda_w = \lambda\gamma$ *and a reasonable null hypothesis is that* $\gamma = 1$*. The model is implemented below, using the usual R-implementation of the gamma distribution, that is*

$$\begin{aligned} \texttt{shape}_i &= 1/a_i \\ \texttt{scale}_i &= \mu_i a_i. \end{aligned} \tag{4.73}$$

The likelihood function is implemented by;

```
## Negative log.likelihood function
nll <- function(pars,X.mu,X.disp,y){
    ## mean value parameters
    beta <- pars[1:dim(X.mu)[2]]
    ## Dispersion
    a <- exp(X.disp %*% pars[-c(1:dim(X.mu)[2])])
    ## Linear predictor
    eta <- X %*% beta
    ## mean value
    mu <- exp(eta)
    ## scale
    s <- mu * a
    - sum(dgamma(y, shape = 1 / a, scale = s, log = TRUE))
}
```

the design matrices are defined by;

```
## Design matrix for mean value model
X <- model.matrix(modGLMEx12.gam.log)
## Design matrix for dispersion model
X.disp <- matrix(1,nrow=dim(X)[1],ncol=2)
X.disp[ ,2] <- dfLMEx1$sex=="female"
```

and finally the MLEs are estimated using `nlminb`*;*

```
## Initial paramters
pars <- rep(0,dim(X)[2]+dim(X.disp)[2])
## Find optimal paramters
opt.gam <- nlminb(pars, nll, X.mu = X, X.disp = X.disp, y = dfLMEx1$clo)
## Log-likelihood
-opt.gam$objective

## [1] 551

## Campare with gamma model
logLik(modGLMEx12.gam.log)

## 'log Lik.' 501 (df=6)
```

We see that we have an increase in log-likelihood of about 50 on 1 degree of freedom, which is significant on any reasonable level. We can get the standard error and hence test statistics from the hessian of the likelihood;

```
## library(numDeriv)
H <- hessian(nll ,opt.gam$par, X.mu = X, X.disp = X.disp,
             y = dfLMEx1$clo)
se <- sqrt(diag(solve(H)))
par.tab <- cbind(opt.gam$par, se,opt.gam$par / se)
colnames(par.tab) <- c("Est", "se", "z.value")
rownames(par.tab) <- c(names(coef(modGLMEx12.gam.log)),
                       "log(lambda)", "log(gamma)")
round(par.tab, digits=3)

##                           Est    se z.value
## (Intercept)            -1.169 0.576    -2.0
## tOut                   -0.071 0.012    -5.8
## I(log(tOut))            0.708 0.273     2.6
## sexmale                -1.527 0.238    -6.4
## I(log(tOut)):sexmale    0.457 0.078     5.8
## log(lambda)            -3.486 0.072   -48.5
## log(gamma)              1.021 0.099    10.3
```

Hence a 95% confidence interval for γ is $e^{1.021\pm0.099 z_{0.975}} = [2.29, 3.37]$, i.e. the dispersion for females is between 2.29 and 3.37 that of males.

4.11 Ordinal and nominal regression

Ordinal and nominal regression are used when the outcome does not have an numerical interpretation. When there is a natural ordering of the outcome (e.g. unsatisfied, satisfied, very satisfied) the regression is called ordinal regression, while if this outcome does not have a natural ordering (e.g. red, blue, green), it is called nominal regression. The multinomial model can be given as a generalized linear model, but we would need a more general definition of the generalized linear model than the one used in this book; we will omit this here and focus on the model formulation. We will use the example presented below to illustrate the methods. The modelling of such data is based on the multinomial distribution with pmf

$$f(\boldsymbol{y}|n) = \frac{n!}{y_1!y_2!\cdots y_k!}p_1^{y_1}p_2^{y_2}\cdots p_k^{y_k}, \tag{4.74}$$

where $\boldsymbol{y} = [y_1, ..., y_k]^T$ and $\sum_{i=1}^k y = n$, $\sum_{i=1}^k p_i = 1$, and $p_i > 0$. Hence the log-likelihood is

$$l((\boldsymbol{y}|n) = \sum_{i=1}^{k} y_i \log(p_i). \tag{4.75}$$

We will start with a motivating example.

Example 4.26 (Ceiling fan speed) *This example deals with the usage of a ceiling fan. Ceiling fans are in regions with warm outdoor conditions a common adaptive opportunity leading to a higher air velocity and consequently a reduced thermal strain (Schweiker et al., 2014). At the same time, we can make the hypothesis that the chosen speed of a ceiling fan depends on the thermal strain, i.e. with a higher thermal strain, the likelihood of switching the ceiling fan to a higher speed increases. In addition, a moderating effect can be expected by the characteristics of the air velocity field created by different types of ceiling fans.*

In order to investigate the hypotheses that a) the chosen speed of a ceiling fan depends on the thermal strain, and b) there is a moderating effect of the type of ceiling fan, a subset of the data from (Schweiker et al., 2014) can be used. The authors conducted an experimental study, where 21 subjects were working for two days in a realistic, but thermally controlled office environment (the LOBSTER). On these days, the type of ceiling fan differed: one was creating a dominant downstream, the other a dominant upstream of air. Note that in reality, it was the same ceiling fan, which was set on one day to its normal mode and on the other day on reverse mode.

Let us assume that we were not able to get our hands on the physical indoor conditions such as the air temperature and air velocity, but we were able to assess the data from the thermal sensation votes (TSV) on a seven-point scale from cold to hot, collected from each subject up to six times per working day, and the fan speed at the time of vote binned into three levels from 1, i.e. off, to 3.

In order to test the above hypothesis, we will build a model that consists of a multinomial distributed outcome variable, the fan speed, one 3-level fator explanatory variable, the thermal sensation vote, and one 2 level factor explanatory variable representing the type of ceiling fan. The type of outcome variable calls for a generalized linear model, and initially we will ignore the subjId.

The data is summarised in the table below;

```
## fanSpeed TSV          fanType
## 0:145    0:137   downstream:120
## 1: 52    1: 74   upstream  :117
## 2: 40    2: 26
```

We see that the fan speed and thermal sensation vote each have 3 levels, while fan type have two levels. If fan type is ignored, we can test if there is a significant impact of TSV on the preferred fan speed. In this case, we can use a simple contingency table (and the resulting χ^2-test) as illustrated below.

```
tab <- aggregate(fanSpeed ~ TSV, data = dfCeilingFan,table)
colnames(tab) <- c("TSV","fanSpeed")
tab

##   TSV fanSpeed.0 fanSpeed.1 fanSpeed.2
## 1   0         97         20         20
## 2   1         40         24         10
## 3   2          8          8         10

(chi.tab <- chisq.test(tab[ ,-1]))

##
##  Pearson's Chi-squared test
##
## data:  tab[, -1]
## X-squared = 23, df = 4, p-value = 0.0001

## Warning message:
## In chisq.test(tab[, -1]) : Chi-squared approximation may be incorrect
```

The conclusion is that ($0.0001 < 0.05$) *there is a difference in the distribution of fan speed with difference in TSV. Formally, the hypothesis that we test is*

$$H_0: \quad P(Fanspeed = i|TSV = j) = P(Fanspeed = i|TSV = j'). \tag{4.76}$$

That is, the distribution of the fanspeed is independent of the TSV. The warning in the above appears because the expected number (under the null hypothesis) in one of the cells is small (< 5).

We could also use a likelihood ratio test to test independence and the probabilities and the likelihood under the alternative hypothesis is;

```
## Observed proportions
p1 <- tab[1,-1] / sum(tab[1,-1])
p2 <- tab[2,-1] / sum(tab[2,-1])
p3 <- tab[3,-1] / sum(tab[3,-1])
(l1 <- sum(tab[1,-1] * log(p1)) + sum(tab[2,-1] * log(p2)) +
     sum(tab[3,-1] * log(p3)))

## [1] -211
```

and the likelihood under the null hypothesis is

```
p <- colSums(tab[ ,-1]) / sum(colSums(tab[ ,-1]))
(l0 <- sum(tab[1,-1] * log(p)) + sum(tab[2,-1] * log(p))+
     sum(tab[3,-1] * log(p)))

## [1] -221
```

and we can calculate the p-value for the test by

```
(Q <- 2 * (11 - 10))

## [1] 22

1 - pchisq(Q, df = 6-2)

## [1] 0.00025
```

we see that the p-values are somewhat different because we have used different methods. The method we present here is based on the last result (i.e. the likelihood), while the `chisq.test` *is based on normal approximation of the Pearson residuals.*

In the example above, we saw that we can make a simple test to check if there is a difference between probabilities depending on the TSV. This test relies on the fact that TSV takes discrete value, if TSV had been on a continuous scale, we would have to collect into interval in order to construct the table. We should also be able to fit more general structures, for example hypothesis of linear dependence. Such models are described below. The models can formulated in different ways; here we follow the formulation used in the R-package `ordinal` (Christensen, 2022).

4.11.1 Nominal regression

In order to avoid complicated restrictions, the cumulative distribution is modeled

$$\pi_j(\boldsymbol{x}) = P(Y \leq j|\boldsymbol{x}); \quad j = 1, ..., k-1. \tag{4.77}$$

In this case, a natural link function is

$$\eta_j = \log\left(\frac{\pi_j}{1-\pi_j}\right); \quad j = 1, ..., k-1, \tag{4.78}$$

which implies the inverse function

$$\pi_j = \frac{e^{\eta_j}}{1+e^{\eta_j}} \quad j = 1, ..., k-1 \tag{4.79}$$

$$\pi_k = 1 - \sum_{j=1}^{k-1} \pi_j, \tag{4.80}$$

the regression model is build around the same principle as we have seen in previous chapters, i.e.

$$\eta_j(\boldsymbol{x}) = \theta_j + \boldsymbol{x}^T\boldsymbol{\beta}_j; j = 1, ..., k-1. \tag{4.81}$$

Hence the cumulative probabilities are modelled independently.

Example 4.27 (Ceiling fan speed) *In this case, there is actually a natural ordering, but for illustrative purposes, we will start with a nominal analysis. Below the R-code for usual backward selection is given.*

```
fit1.nom <- clm(fanSpeed ~ 1, nominal = ~ TSV * fanType,
                data = dfCeilingFan)
fit2.nom <- clm(fanSpeed ~ 1, nominal = ~ TSV + fanType,
                data = dfCeilingFan)
fit3.nom <- clm(fanSpeed ~ 1, nominal = ~ TSV,
                data = dfCeilingFan)
fit4.nom <- clm(fanSpeed ~ 1, nominal = ~ 1,
                data = dfCeilingFan)
```

The analysis of deviance table shown below shows that we should choose the model including TSV, *but not* fanSpeed;

```
anova(fit1.nom, fit2.nom, fit3.nom, fit4.nom)

## Likelihood ratio tests of cumulative link models:
##
##          no.par AIC logLik LR.stat df Pr(>Chisq)
## fit4.nom      2 447   -221
## fit3.nom      6 433   -211   21.53  4    0.00025 ***
## fit2.nom      8 431   -208    5.70  2    0.05786 .
## fit1.nom     12 436   -206    3.24  4    0.51882
## [...]
```

We see that the result in terms of p-value and test statistics is identical to Example 4.26 (when using the likelihood ratio test). The difference is that the contingency table uses Pearson residuals while the regression model uses likelihood ratio test (i.e. deviance residuals).

*The parameters and their standard errors are given in the summary from R (*fit.nom3*)*

```
## [...]
## Threshold coefficients:
##                    Estimate Std. Error z value
## 0|1.(Intercept)      0.8858     0.1879    4.71
## 1|2.(Intercept)      1.7664     0.2420    7.30
## 0|1.TSV1            -0.7233     0.2995   -2.41
## 1|2.TSV1             0.0899     0.4173    0.22
## 0|1.TSV2            -1.6968     0.4646   -3.65
## 1|2.TSV2            -1.2964     0.4702   -2.76
## [...]
```

Hence the probabilities estimate of fan speeds when the thermal sensation vote is 0 is

$$\begin{aligned} \pi_1(0) &= \frac{1}{1+e^{-0.88583}} = 0.708 \\ \pi_2(0) &= \frac{1}{1+e^{-1.7664}} = 0.85, \end{aligned} \tag{4.82}$$

Which is equal to the observed frequency from the previous example (97/137, and (97+20)/137).

We see the result in terms of p-value and test statistics is almost identical to Example 4.26.

4.11.2 Ordinal regression

In the case of ordinal regression the linear prediction is formulated as

$$\eta_j(\boldsymbol{x}) = \theta_j + \boldsymbol{x}^T\boldsymbol{\beta}. \tag{4.83}$$

Hence the same linear regression parameters are used for all probabilities (but different intercepts). In this case, we can write the odds as

$$ODDS_j(\boldsymbol{x}) = \frac{\pi_j(\boldsymbol{x})}{1-\pi_j(\boldsymbol{x})} \tag{4.84}$$

$$= e^{\theta_j - \boldsymbol{x}\boldsymbol{\beta}}, \tag{4.85}$$

and the odds ratio as

$$OR_j(\boldsymbol{x}_1, \boldsymbol{x}_2) = e^{(\boldsymbol{x}_2-\boldsymbol{x}_1)^T\boldsymbol{\beta}}; \tag{4.86}$$

hence the odds ratio only depends on the difference in the regressors (not on the probability).

Example 4.28 *We continue with the example above and to simplify, we will only consider a model with TSV;*

```
fit3.ord <- clm(fanSpeed ~ TSV, data = dfCeilingFan)

## [...]
## Coefficients:
##      Estimate Std. Error z value Pr(>|z|)
## TSV1    0.564      0.288    1.96     0.05 .
## TSV2    1.606      0.410    3.92 0.000089 ***
##
## Threshold coefficients:
##     Estimate Std. Error z value
## 0|1    0.834      0.188    4.44
## 1|2    2.045      0.230    8.89
```

Further, we could assume that the TSV could be measured on a continuous scale.

```
dfCeilingFan$numTSV <- as.numeric(dfCeilingFan$TSV) - 1
fit32.ord <- clm(fanSpeed ~ numTSV, data = dfCeilingFan)

## Likelihood ratio tests of cumulative link models:
##
##             no.par AIC logLik LR.stat df Pr(>Chisq)
## fit32.ord        3 433   -214
## fit3.ord         4 434   -213    0.65  1       0.42
```

Hence, there is no evidence that we should not use the TSV on a continuous scale, and the parameters of the model are;

```
summary(fit32.ord)

## [...]
## Coefficients:
##        Estimate Std. Error z value Pr(>|z|)
## numTSV    0.739      0.189    3.92 0.00009 ***
##
## Threshold coefficients:
##     Estimate Std. Error z value
## 0|1    0.883      0.180    4.91
## 1|2    2.088      0.225    9.29
```

That is, in this case we have ($\pi_j(i) = P(\texttt{fanSpeed} \leq i|\texttt{TSV} = j)$)

$$\pi_0(TSV = 0) = \frac{1}{1 + e^{-0.883}} \tag{4.87}$$

$$\pi_0(TSV = 1) = \frac{1}{1 + e^{-2.088}} \tag{4.88}$$

$$\pi_1(TSV = 0) = \frac{1}{1 + e^{-0.883+0.739}} \tag{4.89}$$

$$\vdots \tag{4.90}$$

$$\pi_1(TSV = 2) = \frac{1}{1 + e^{-2.088+0.739\cdot 2}} \tag{4.91}$$

or more generally

$$\pi_i(TSV) = \frac{1}{1 + e^{-\theta_i+0.739\cdot TSV}} \tag{4.92}$$

the odds can be calculated by

$$ODDS_i(TSV) = \frac{\pi_i(TSV_1)}{1 - \pi_i(TSV_1)} \tag{4.93}$$

$$= \frac{\frac{1}{1+e^{-\theta_i+0.739 \cdot TSV}}}{1 - \frac{1}{1+e^{-\theta_i+0.739 \cdot TSV}}} \tag{4.94}$$

$$= \frac{1}{e^{-\theta_i+0.739 \cdot TSV}} \tag{4.95}$$

$$= e^{\theta_i-0.739 \cdot TSV} \tag{4.96}$$

and we can calculate the odds ratio by

$$OR(TSV_1, TSV_2) = \frac{\pi_i(TSV_1)}{1 - \pi_i(TSV_1)} \frac{1 - \pi_i(TSV_2)}{\pi_i(TSV_2)} \tag{4.97}$$

$$= e^{\theta_i-0.739 \cdot TSV_1} e^{-\theta_i+0.739 \cdot TSV_2} \tag{4.98}$$

$$= e^{-0.739 \cdot (TSV_1 - TSV_2)} \tag{4.99}$$

hence only depending on the difference in TSV.

The figure below illustrates the three different models.

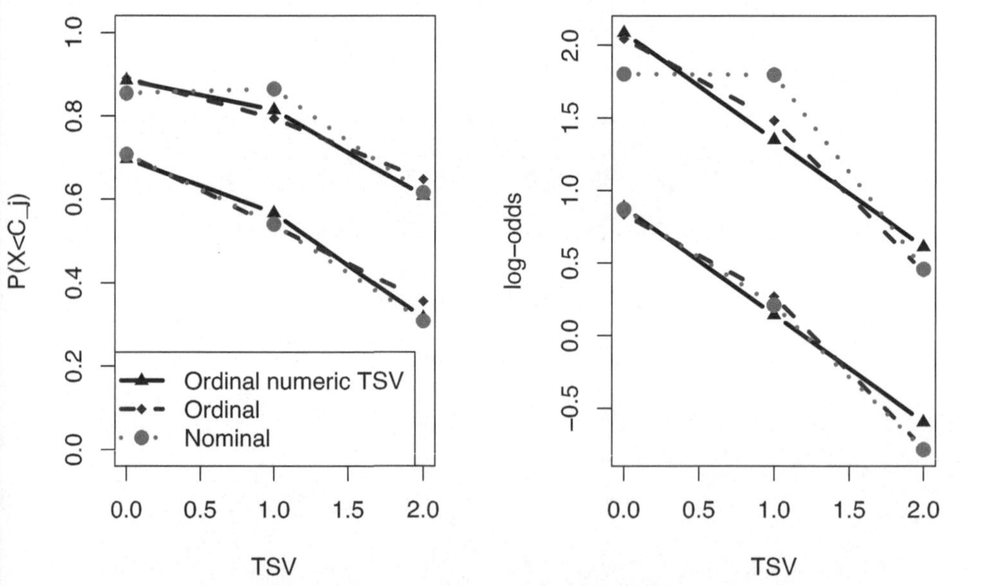

We see that in the simple case (numerical TSV), the lines for log-odds have the same slope implying that the log odds ratio is constant.

As for the generalized linear models we have discussed earlier, the results for ordinal regression can be applied using other link functions and generalised in other ways, see Christensen (2022) for details.

The ordinal package has been applied for thermal satisfaction modelling in a recent field study by Lassen et al. (2021) studying the thermal satisfaction in office building, and by Favero et al. (2022) in a controlled laboratory setting, in the latter the difference between ordinal and a continuous voting scale (using the Beta-distribution) was also investigated. In both settings the models were generalized to mixed effect settings (see Section 6.2.2).

4.12 Summary and conclusion

In this chapter, we have presented generalized linear models; the primary benefit here is that a wider range of distribution assumptions are now available. These range from Poisson and multinomial distributions to, for example the gamma distribution. From a modelling point of view, the benefits are clear. We can choose the distribution that fits the data at hand, the price is that the results regarding tests and residuals analysis rely on asymptotic results, and prediction intervals become a bit more tricky to handle.

When estimating the parameters, the difference between the general and generalized linear models is usually not a concern as both are implemented in standard software, and the difference will only be visible in very large data-sets/models.

The generalized and general linear models also form the basis for the further development in the next two chapters, where additional structures are added in order to include random effect models, which are able to handle population parameters that relate to e.g. between person variability.

5

Linear mixed effects models

This chapter is a revision of a similar chapter in Madsen and Thyregod (2011) but tailored to occupant behaviour and the remaining chapters of the present book. Occupant models often deal with observations from different observational units (persons, dwellings etc.), if we are not interested in the individual observational units, but rather the variation betweeen observational units, then mixed effect models is the right approach. In Chapter 3, we looked at general linear models; in this chapter, we will discuss models where one or more parameters are considered as random variables, the simplest version of such a model is

$$Y_{ij} = \mu + \alpha_i + \epsilon_{ij}, \tag{5.1}$$

where $\alpha_i \sim N(0, \sigma_i^2)$ and $\epsilon_{ij} \sim N(0, \sigma^2)$, both iid (see Section 2.1). The only difference between this model and the general linear model as described in Chapter 3 is that α_i is now a random variable rather than a fixed (but estimated) number. The interpretation of the model does change since the model now describes a population parameter rather than actual individuals.

Example 5.1 *To illustrate principles, we turn again to our clothing insulation level example. In Chapter 3, we ignored the fact that there are actually different individuals. We could have started with a model where we only included the variation between individuals i.e.* (5.1) *(but with α_i fixed). There are 47 subjects in the data-set, and the model is estimated by*

```
fit.aov <- lm(clo ~ subjId, data = dfMEMEx1)
```

The model has 47 parameters, which can be interpreted as the mean level of clothing for each of the individuals, but cannot be generalised to other individuals from the population. If we are interested in the population, we should estimate the parameters of a random effect model such as (`lmer` is a method from the R-package `lme4` (Bates et al., 2015));

```
fit0.lme <- lmer(clo  ~ 1 + (1| subjId), data = dfMEMEx1)
```

The output from this last model is

DOI: 10.1201/9781003340812-5

```
summary(fit0.lme)

## Linear mixed model fit by REML ['lmerMod']
## Formula: clo ~ 1 + (1 | subjId)
##    Data: dfMEMEx1
##
## REML criterion at convergence: -1275
##
## Scaled residuals:
##    Min     1Q Median     3Q    Max
## -3.006 -0.477 -0.050  0.480  3.277
##
## Random effects:
##  Groups   Name        Variance Std.Dev.
##  subjId   (Intercept) 0.01567  0.1252
##  Residual             0.00978  0.0989
## Number of obs: 803, groups:  subjId, 47
##
## Fixed effects:
##             Estimate Std. Error t value
## (Intercept)   0.5511     0.0186    29.6
```

Hence the model is

$$Y_{ij} = 0.55 + \alpha_i + \epsilon_{ij} \tag{5.2}$$

with $\epsilon_{ij} \sim N(0, 0.099^2)$ *and* $\alpha_i \sim N(0, 0.125^2)$.

This chapter will cover models like the one presented in the example above and generalisations of it.

5.1 Data used in this chapter

In this chapter, the main data–set will be the clothing insulation level data that we looked at before (Chapters 3 and 4). We will also look at sub-samples and aggregated versions of the data set, as explained in the example below.

Example 5.2 (Data for the Chapter) *The variables for the clothing insulation level data are summarised in Table 5.1. Further, some key numbers for selected variables are given below;*

```
##        clo             tOut          tInOp
##  Min.   :0.23   Min.   : 9   Min.   :22
##  1st Qu.:0.47   1st Qu.:18   1st Qu.:25
##  Median :0.54   Median :21   Median :27
##  Mean   :0.55   Mean   :22   Mean   :27
##  3rd Qu.:0.64   3rd Qu.:25   3rd Qu.:28
##  Max.   :0.97   Max.   :36   Max.   :32
```

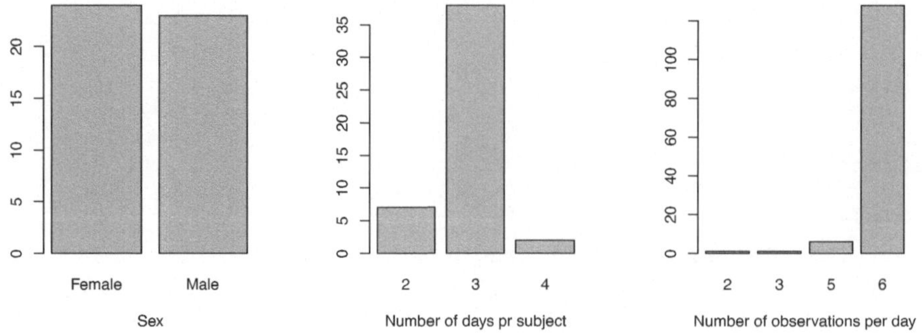

Figure 5.1: Summary of variables related to structure of random effects in the clothing insulation data

Table 5.1: Variables in the clothing insulation data.

Variable	Explanation	Type
`clo`	Clothing insulation level	Real
`tOut`	Outdoor temperature	Real
`tInOp`	Indoor operating temperature	Real
`sex`	Sex of subject (male/female)	Factor
`SubjId`	Subject identifier (47 levels)	Factor
`day`	Day identifier (within subject)	Factor
`tOutCen`	Centred outdoor temperature	Real
`tInOpCen`	Centred indoor operating temperature	Real
`obs.no`	The observation number within the day	Factor (ordered)

That is we have the clothing insulation level (column `clo`*) that we aim at modelling, outdoor temperature (*`tOut`*), indoor operative temperature (tInOp), and sex of the subject. In addition, we will model the variations between subjects, the variations between days and finally the variation between observations taken on the same day (within the same subject). Most subjects visit the facility on 3 separate days and report 6 measurements of the clothing insulation level per day (see Figure 5.1).*

In the first part of the chapter, we will look at an aggregated version of the data-set only including females visited the facility on 3 days and reported 6 observations on each day. All variables are reported as the average within the day.

There are in total 21 females who visited the facility 3 times; this will be used as simple illustration of the method in the first part of the chapter.

Finally we will use a data-set with within day aggregation (average) of all variables (`dat.aggr`*).*

To summarise, we will use the following data-sets:

1 The full data-set containing structures related to subjects, days and within day variation (`dfLMEx1`).

2 An aggregated version containing daily averages, hence only structures related to subjects can be identified (`dat.aggr`).

3 A subset of the aggregated data-set with only females who visited the lab 3 times; hence, this is a balanced design (`dat.aggrFemale`).

We will analyse all 3 data-sets, starting with the simplest one, and in the end of the chapter, discuss and implement models for the full data-set.

5.2 One-way random effects models

The model that we introduced in the example above is called an one-way random effect model; more formally, it is defined as

Definition 5.1 (One-way model with random effects) *Consider the random variables $Y_{ij}, i = 1, 2, \ldots, k$; $j = 1, 2, \ldots, n_i$.*

A one-way random effects model *for Y_{ij} is a model such that*

$$Y_{ij} = \mu + U_i + \epsilon_{ij} \ , \tag{5.3}$$

wwhere $\epsilon_{ij} \sim N(0, \sigma^2)$ and $U_i \sim N(0, \sigma_u^2)$, and ϵ_{ij} are mutually independent, and also the U_i's are mutually independent, and finally the U_i's are independent of ϵ_{ij}.

We shall put

$$N = \sum_{i=1}^{k} n_i. \tag{5.4}$$

When all groups are of the same size, $n_i = n$, we shall say that the model is balanced.

Example 5.3 *In the aggregated data for clothing insulation, the simplest mixed effect model we can consider is a model with a mean level and a random effect of subject, each day a subject visits the lab is a new observation, and the R-output below shows the number of subjects visiting the lab at least 1, 2, 3 and 4 days;*

```
table(dat.aggr$day)

##
##  1  2  3  4
## 47 47 40  2

table(dat.aggrFemale$day)

##
##  1  2  3
## 21 21 21
```

So in the larger of the two data-sets we have 47 subjects, all visit the lab on two days, 40 visit the lab on 3 different days and only two visit the lab on 4 different days. For the reduced data–set, we have 21 subjects all visiting the lab on 3 different days. Here, days are repetitions; hence, the first data–set is not balanced while the second is.

In order to exemplify the simple setup with a balanced design, we start by the balanced dataset. The result for a standard ANOVA analysis for the simple model,

$$Y_{ij} = \mu + \alpha_i + \epsilon_{ij}; \quad \epsilon_{ij} \sim N(0, \sigma^2) \tag{5.5}$$

is;

```
anova(lm(clo ~ subjId, data = dat.aggrFemale))

## Analysis of Variance Table
##
## Response: clo
##           Df Sum Sq Mean Sq F value  Pr(>F)
## subjId    20  1.670  0.0835    7.18 4.5e-08 ***
## Residuals 42  0.488  0.0116
## ---
## [...]
```

Hence, we can conclude that there is a significant difference between the subjects (p–value $<< 0.05$*).*

The model presented in the example above is a fixed effect model as discussed in Chapter 3. In this chapter, we treat some effects as random variables (e.g. the subject effect). In order to formulate the likelihood, we need the marginal distribution of the observations. Since all variables are additive constants or additive Gaussian random variables, the marginal distribution will also be Gaussian. The covariance between two observations, from the model (5.3), can be calculated by (u_i and ϵ_{kl} are independent)

$$\begin{aligned} Cov[Y_{ij}, Y_{kl}] =& Cov[\mu + U_i + \epsilon_{ij}, \mu + U_k + \epsilon_{kl}] \\ =& Cov[U_i, U_k] + Cov[\epsilon_{ij}, \epsilon_{kl}]. \end{aligned} \tag{5.6}$$

The first term is equal to zero if $i \neq k$ and equals σ_u^2 if $i = k$. The second term is zero if $(i, j) \neq (k, l)$, and it is equal to σ^2 if $(i, j) = (k, l)$. Hence, we have the following theorem:

Theorem 5.1 (Marginal distributions) *The marginal distribution of* $\boldsymbol{Y}$ *is a multivariate normal distribution with*

$$E[Y_{ij}] = \mu \tag{5.7a}$$

$$Cov[Y_{ij}, Y_{hl}] = \begin{cases} \sigma_u^2 + \sigma^2 & \text{for } (i,j) = (h,l) \\ \sigma_u^2 & \text{for } i = h,\ j \neq l \\ 0 & \text{for } i \neq h. \end{cases} \tag{5.7b}$$

From Theorem 5.1, we can find the marginal variance-covariance matrix for one observation, that is

$$\boldsymbol{Y}_i \sim N(\boldsymbol{\mu}, \boldsymbol{V}_i), \tag{5.8}$$

where $\boldsymbol{\mu} = \boldsymbol{1}\mu$, $(\boldsymbol{V}_i)_{kl} = Cov[Y_{ik}, Y_{il}]$, and $\boldsymbol{Y}_i = [Y_{i1}, ..., Y_{in_i}]$. The collection of all observations $\boldsymbol{Y} = [\boldsymbol{Y}_1, ..., \boldsymbol{Y}_k]$, and hence the likelihood can be written as

$$\begin{aligned} L(\boldsymbol{\theta}; \boldsymbol{y}) =& p(\boldsymbol{y}; \boldsymbol{\theta}) \\ =& \frac{1}{(2\pi)^{N/2}\sqrt{|\boldsymbol{V}(\boldsymbol{\theta})|}} \exp\left(-\frac{(\boldsymbol{y}-\boldsymbol{\mu})^T \boldsymbol{V}^{-1}(\boldsymbol{y}-\boldsymbol{\mu})}{2}\right), \end{aligned} \tag{5.9}$$

where $\boldsymbol{V}$ is a block diagonal matrix with diagonal block i being $\boldsymbol{V}_i$ (see Figure 5.2, on page 121). With the remark below, this can be simplified to a series of problems of smaller dimensions.

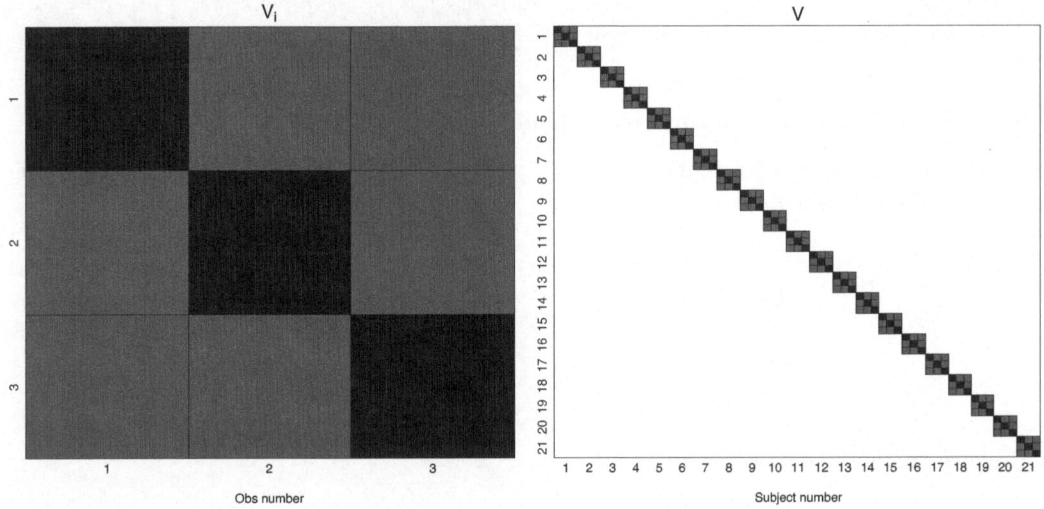

Figure 5.2: Plot of correlation matrix for the simple random effect model presented in Example 5.4.

Remark 5.1 (Observation vector for a group) *When we consider the set of observations corresponding to the i'th group as an n_i-dimensional column vector,*

$$\boldsymbol{Y}_i = \begin{bmatrix} Y_{i1} \\ Y_{i2} \\ \vdots \\ Y_{in_i} \end{bmatrix}, \tag{5.10}$$

the correlation matrix in the marginal distribution of $\boldsymbol{Y}_i$ is an equicorrelation matrix of the form

$$\boldsymbol{E}_{n_i} = (1-\rho)\boldsymbol{I}_{n_i} + \rho \boldsymbol{J}_{n_i} = \begin{bmatrix} 1 & \rho & \dots & \rho \\ \rho & 1 & \dots & \rho \\ \vdots & \vdots & \ddots & \vdots \\ \rho & \rho & \dots & 1 \end{bmatrix}, \tag{5.11}$$

where $\boldsymbol{J}_{n_i}$ is an $n_i \times n_i$-dimensional matrix consisting solely of 1's.

Thus, the set of observations $\boldsymbol{Y}_i$, $i = 1, 2, \dots, k$ may be described as k independent observations *of an n_i dimensional variable $\boldsymbol{Y}_i \sim N_{n_i}(\boldsymbol{\mu}, \sigma^2 \boldsymbol{I}_{n_i} + \sigma_u^2 \boldsymbol{J}_{n_i})$, with $\boldsymbol{Y}_i$ and $\boldsymbol{Y}_j$ independent for $i \neq j$ i.e. the dispersion matrix for $\boldsymbol{Y}_i$ is*

$$\begin{aligned} \boldsymbol{V}_i &= D[\boldsymbol{Y}_i] \\ &= E[(\boldsymbol{Y}_i - \boldsymbol{\mu})(\boldsymbol{Y}_i - \boldsymbol{\mu})^T] \\ &= \begin{bmatrix} \sigma_u^2 + \sigma^2 & \sigma_u^2 & \dots & \sigma_u^2 \\ \sigma_u^2 & \sigma_u^2 + \sigma^2 & \dots & \sigma_u^2 \\ \vdots & \vdots & \ddots & \vdots \\ \sigma_u^2 & \sigma_u^2 & \dots & \sigma_u^2 + \sigma^2 \end{bmatrix}. \end{aligned} \tag{5.12}$$

Such a matrix is denoted a compound symmetrical matrix.

Example 5.4 *In the clothing insulation example (simple setup), the marginal variance has a block diagonal structure as explained above and exemplified in Figure 5.2.*

5.2.1 Likelihood function for simple one-way setup

With the theorem and the remark above, we can write the likelihood function of the simple one-way random effect model as

$$\begin{aligned} L(\boldsymbol{\theta};\boldsymbol{y}) =& \prod_{i=1}^{k} p(\boldsymbol{y}_i;\boldsymbol{\theta}) \\ =& \prod_{i=1}^{k} \frac{1}{(2\pi)^{n_i/2}\sqrt{|\boldsymbol{V}_i(\boldsymbol{\theta})|}} \exp\left(-\frac{(\boldsymbol{y}_i-\boldsymbol{\mu})^T\boldsymbol{V}_i^{-1}(\boldsymbol{y}_i-\boldsymbol{\mu})}{2}\right), \end{aligned} \tag{5.13}$$

and hence the log-likelihood is given by

$$l(\boldsymbol{\theta};\boldsymbol{y}) = -\frac{1}{2}\sum_{i=1}^{k} \log|\boldsymbol{V}_i(\boldsymbol{\theta})| - \frac{(\boldsymbol{y}_i-\boldsymbol{\mu})^T\boldsymbol{V}_i^{-1}(\boldsymbol{y}_i-\boldsymbol{\mu})}{2}, \tag{5.14}$$

with $\boldsymbol{\mu} = \mu\boldsymbol{1}$.

Even though the main example we use in this chapter is not balanced, we will, for illustrative purposes, go through the balanced case ($n_i = n$) in details here, and find the closed–form solution for the parameters. Readers more interested in implementation and interpretation rather than mathematical derivations may skip directly to Theorem 5.2 on page 125.

The within subject variance-covariance of the simple one-way random effect model can be written as

$$\boldsymbol{V}_i(\boldsymbol{\theta}) = \sigma^2\left(\boldsymbol{I}_n + \frac{\sigma_u^2}{\sigma^2}\boldsymbol{J}_n\right). \tag{5.15}$$

Further, the inverse and the determinant of $\boldsymbol{V}_i(\boldsymbol{\theta})$ can be written as (Demidenko, 2004)

$$\begin{aligned} \boldsymbol{V}_i^{-1}(\boldsymbol{\theta}) =& \frac{1}{\sigma^2}\left(\boldsymbol{I} - \frac{1}{\frac{\sigma^2}{\sigma_u^2}+n}\boldsymbol{J}\right) \\ |\boldsymbol{V}_i(\boldsymbol{\theta})| =& \sigma^{2n}\left(1+n\frac{\sigma_u^2}{\sigma^2}\right). \end{aligned} \tag{5.16}$$

This simple form can be used to derive the estimators for the parameters.

5.2.2 Estimator for the mean

Hence the log-likelihood can, using (5.14)–(5.16), be written as

$$\begin{aligned} l(\boldsymbol{\theta};\boldsymbol{y}) =& -\frac{k}{2}\left(n\log(\sigma^2) + \log\left(1+n\frac{\sigma_u^2}{\sigma^2}\right)\right) - \\ & \frac{1}{2\sigma^2}\sum_{i=1}^{k}(\boldsymbol{y}_i-\mu\boldsymbol{1})^T\left(\boldsymbol{I} - \frac{1}{\frac{\sigma^2}{\sigma_u^2}+n}\boldsymbol{J}\right)(\boldsymbol{y}_i-\mu\boldsymbol{1}) \end{aligned} \tag{5.17}$$

The derivative of the log-likelihood wrt. μ is in this case

$$\begin{aligned}
\frac{\partial l(\boldsymbol{\theta};\boldsymbol{y})}{\partial \mu} =& -\frac{1}{2\sigma^2}\sum_{i=1}^{k} 2\mu\mathbf{1}^T\left(\boldsymbol{I}-\frac{1}{\frac{\sigma^2}{\sigma_u^2}+n}\boldsymbol{J}\right)\mathbf{1}-\boldsymbol{y}_i^T\left(\boldsymbol{I}-\frac{1}{\frac{\sigma^2}{\sigma_u^2}+n}\boldsymbol{J}\right)\mathbf{1}- \\
&\mathbf{1}^T\left(\boldsymbol{I}-\frac{1}{\frac{\sigma^2}{\sigma_u^2}+n}\boldsymbol{J}\right)\boldsymbol{y}_i \\
=& -\frac{1}{2\sigma^2}\sum_{i=1}^{k} 2n\mu - 2\frac{\mu}{\frac{\sigma^2}{\sigma_u^2}+n}\mathbf{1}^T\boldsymbol{J}\mathbf{1} - 2n\bar{y}_i + \frac{1}{\frac{\sigma^2}{\sigma_u^2}+n}\mathbf{1}^T\boldsymbol{J}\boldsymbol{y}_i + \\
&\frac{1}{\frac{\sigma^2}{\sigma_u^2}+n}\boldsymbol{y}_i^T\boldsymbol{J}\mathbf{1}.
\end{aligned} \tag{5.18}$$

The matrix-vector products may be written as $\boldsymbol{J}\mathbf{1}=n\mathbf{1}$, $\mathbf{1}^T\boldsymbol{J}\mathbf{1}=n^2$ and $\mathbf{1}\boldsymbol{y}_i=\sum_j y_{ij}=n_i\bar{y}_i$, and we get (using $\bar{\bar{y}}=\frac{1}{N}\sum_i\sum_j y_{ij}$)

$$\begin{aligned}
\frac{\partial l(\boldsymbol{\theta};\boldsymbol{y})}{\partial \mu} =& -\frac{1}{2\sigma^2}\left(\sum_{i=1}^{k} 2n\mu - 2\frac{\mu n^2}{\frac{\sigma^2}{\sigma_u^2}+n} - 2n\bar{y}_i + \frac{n\bar{y}_i}{\frac{\sigma^2}{\sigma_u^2}+n} + \frac{n\bar{y}_i}{\frac{\sigma^2}{\sigma_u^2}+n}\right) \\
=& -\frac{1}{2\sigma^2}\left(2N\mu - 2\frac{\mu k n^2}{\frac{\sigma^2}{\sigma_u^2}+n} - 2n\sum_{i=1}^{k}\bar{y}_i + \frac{2n\sum_{i=1}^{k}\bar{y}_i}{\frac{\sigma^2}{\sigma_u^2}+n}\right) \\
=& -\frac{1}{2\sigma^2}\left(2N\mu - 2\frac{\mu N n}{\frac{\sigma^2}{\sigma_u^2}+n} - 2nk\bar{\bar{y}} + \frac{2nk\bar{\bar{y}}}{\frac{\sigma^2}{\sigma_u^2}+n}\right) \\
=& -\frac{1}{\sigma^2}\left(N\mu - \frac{\mu N n}{\frac{\sigma^2}{\sigma_u^2}+n} - N\bar{\bar{y}} + \frac{N\bar{\bar{y}}}{\frac{\sigma^2}{\sigma_u^2}+n}\right) \\
=& -\frac{1}{\sigma^2}\left(N\mu - N\bar{\bar{y}}\right).
\end{aligned} \tag{5.19}$$

Hence the MLE of μ does not depend on σ^2 and σ_u^2, and is

$$\hat{\mu}=\bar{\bar{y}}, \tag{5.20}$$

which is independent of σ^2 and σ_u^2.

5.2.3 Estimator for the variance components

The profile log-likelihood (i.e. inserting $\hat{\mu}$) of $[\sigma^2,\sigma_u^2]^T$ can be written as

$$\begin{aligned}
l_p([\sigma^2,\sigma_u^2]^T;\boldsymbol{y}) =& -\frac{k}{2}\left(n\log(\sigma^2)+\log\left(1+n\frac{\sigma_u^2}{\sigma^2}\right)\right) - \\
&\frac{1}{2\sigma^2}\sum_{i=1}^{k}(\boldsymbol{y}_i-\bar{\bar{y}}\mathbf{1})^T\left(\boldsymbol{I}-\frac{1}{\frac{\sigma^2}{\sigma_u^2}+n}\boldsymbol{J}\right)(\boldsymbol{y}_i-\bar{\bar{y}}\mathbf{1}).
\end{aligned} \tag{5.21}$$

Now set

$$\begin{aligned} g([\sigma^2,\sigma_u^2]) &= -\frac{k}{2}\left(n\log(\sigma^2)+\log\left(1+n\frac{\sigma_u^2}{\sigma^2}\right)\right) \\ &= -\frac{k}{2}\left((n-1)\log(\sigma^2)+\log\left(\sigma^2+n\sigma_u^2\right)\right), \end{aligned} \tag{5.22}$$

to get

$$\begin{aligned} l_p([\sigma^2,\sigma_u^2];\boldsymbol{y}) =& g([\sigma^2,\sigma_u^2]) - \frac{1}{2\sigma^2}\sum_{i=1}^{k}(\boldsymbol{y}_i-\bar{\bar{y}}\mathbf{1})^T\boldsymbol{I}(\boldsymbol{y}_i-\bar{\bar{y}}\mathbf{1}) - \\ & \sum_{i=1}^{k}(\boldsymbol{y}_i-\bar{\bar{y}}\mathbf{1})^T\frac{1}{\frac{\sigma^2}{\sigma_u^2}+n}\boldsymbol{J}(\boldsymbol{y}_i-\bar{\bar{y}}\mathbf{1}) \\ =& g([\sigma^2,\sigma_u^2]) - \frac{1}{2\sigma^2}\left(SSE+SSB\left(1-\frac{n}{\frac{\sigma^2}{\sigma_u^2}+n}\right)\right) \\ =& g([\sigma^2,\sigma_u^2]) - \frac{1}{2}\left(\frac{SSE}{\sigma^2}+SSB\left(\frac{1}{\sigma^2+n\sigma_u^2}\right)\right), \end{aligned} \tag{5.23}$$

with

$$\begin{aligned} SSE &= \sum_{i=1}^{k}\sum_{j=1}^{n_i}(y_{ij}-\bar{y}_i)^2 \\ SSB &= \sum_{i=1}^{k} n_i(\bar{y}_i-\bar{\bar{y}})^2, \end{aligned} \tag{5.24}$$

SSE measures the within subject variation and SSB measures the between subject variation.

The derivative of the profile log-likelihood wrt. σ^2 can be written as

$$\frac{\partial l_p([\sigma^2,\sigma_u^2];\boldsymbol{y})}{\partial\sigma^2} = -\frac{k}{2}\left(\frac{n-1}{\sigma^2}+\frac{1}{\sigma^2+n\sigma_u^2}\right)+\frac{1}{2}\left(\frac{SSE}{\sigma^4}+\frac{SSB}{(\sigma^2+n\sigma_u^2)^2}\right) \tag{5.25}$$

and the derivative of the profile likelihood wrt. σ_u^2 is

$$\frac{\partial l_p([\sigma^2,\sigma_u^2];\boldsymbol{y})}{\partial\sigma_u^2} = -\frac{k}{2}\frac{n}{\sigma^2+n\sigma_u^2}+\frac{n}{2}\frac{SSB}{(\sigma^2+n\sigma_u^2)^2}. \tag{5.26}$$

Hence, we get

$$\hat{\sigma}^2+n\hat{\sigma}_u^2 = \frac{SSB}{k}. \tag{5.27}$$

Inserting in the result from the derivative wrt. σ^2, we get

$$\begin{aligned} 0 &= -\frac{k}{2}\left(\frac{n-1}{\sigma^2}+\frac{k}{SSB}\right)+\frac{1}{2}\left(\frac{SSE}{\sigma^4}+\frac{k^2SSB}{SSB^2}\right) \\ &= -\frac{k}{2}\frac{n-1}{\sigma^2}+\frac{1}{2}\frac{SSE}{\sigma^4}, \end{aligned} \tag{5.28}$$

or

$$\hat{\sigma}^2 = \frac{SSE}{k(n-1)}, \tag{5.29}$$

and we have

$$\hat{\sigma}_u^2 = \frac{SSB}{nk} - \frac{SSE}{nk(n-1)} = \frac{SSB}{N} - \frac{\hat{\sigma}^2}{n}. \tag{5.30}$$

We summarise the finding in the theorem below.

Theorem 5.2 *For the simple one-way random effect model*

$$Y_{ij} = \mu + U_i + \epsilon_{ij}, \tag{5.31}$$

with $U_i \sim N(0, \sigma_u^2)$, $\epsilon_{ij} \sim N(0, \sigma^2)$ *and the number of observation in each group equal* $n_i = n$ *the MLE's are*

$$\begin{aligned} \hat{\mu} &= \bar{\bar{y}} \\ \hat{\sigma}^2 &= \frac{SSE}{k(n-1)} \\ \hat{\sigma}_u^2 &= \frac{SSB}{N} - \frac{\hat{\sigma}^2}{n}. \end{aligned} \tag{5.32}$$

Concrete calculations in R are illustrated in the example below

Example 5.5 *For the balanced data we can now estimate the parameters by calculating manually in R:*

```
## no. of groups, obs. per group, and observations
n <- 3;
k <- length(levels(dat.aggrFemale$subjId))
N <- n * k

## estimates
mu.hat <- mean(dat.aggrFemale$clo)
SSE <- sum((n - 1) * as.vector(by(dat.aggrFemale$clo,
                                  dat.aggrFemale$subjId, var)))
SSB <- n * sum( (as.vector(by(dat.aggrFemale$clo,
                              dat.aggrFemale$subjId, mean)) -
                 mu.hat)^2)
sigma2.hat <- SSE / (k * (n-1))
sigmau2.hat <- SSB / N - SSE / (N * (n-1))

## report estimates
c(mu.hat, sqrt(sigma2.hat), sqrt(sigmau2.hat))

## [1] 0.60 0.11 0.15
```

We get the same values for the parameters, when using;

```
lmer(clo ~ 1 + (1 | subjId), data = dat.aggrFemale,
     REML = FALSE)

## Linear mixed model fit by maximum likelihood  ['lmerMod']
## Formula: clo ~ 1 + (1 | subjId)
##    Data: dat.aggrFemale
##      AIC      BIC   logLik deviance df.resid
##      -55      -49       31      -61       60
## Random effects:
##  Groups   Name        Std.Dev.
##  subjId   (Intercept) 0.150
##  Residual             0.108
## Number of obs: 63, groups:  subjId, 21
## Fixed Effects:
## (Intercept)
##       0.595
```

Note that we have to turn off the REML correction here; we will discuss REML estimation below.

5.2.4 Unbiased estimators

We are usually interested in unbiased estimators. The expected value of SSE is

$$E[SSE] = \sigma^2 k(n-1). \tag{5.33}$$

Hence, $\hat{\sigma}^2$ is an unbiased estimator in this case while the expected value of SSB is

$$E[SSB] = \sigma^2(k-1) + \sigma_u^2 n(k-1), \tag{5.34}$$

and therefore

$$E[\hat{\sigma}_u^2] = \sigma_u^2 \frac{k-1}{k} - \frac{\sigma^2}{nk}, \tag{5.35}$$

and the bias of $\hat{\sigma}_u^2$ is

$$E[\hat{\sigma}_u^2 - \sigma_u^2] = -\frac{\sigma_u^2}{k} - \frac{\sigma^2}{nk}. \tag{5.36}$$

Hence, the likelihood estimator for σ_u^2 is biased downwards so we will on average underestimate σ_u^2. The derivations above (equation (5.35) and (5.36)) also provide the unbiased estimator in the balanced case, that is

$$\tilde{\sigma}_u^2 = \frac{SSB/(k-1) - \hat{\sigma}^2}{n}. \tag{5.37}$$

Of course the argument above implies that the unbiased estimator is usually preferred.

Example 5.6 *The unbiased estimator for σ_u is calculated by*

```
sigma2.u.m <- (SSB / (k - 1) - sigma2.hat) / n
round(sqrt(sigma2.u.m), digits = 4)

## [1] 0.15
```

In R, we can get this by

```
fit <- lmer(clo ~ 1 + (1 | subjId), data = dat.aggrFemale,
            REML = TRUE)

## [1] Random effects:
##  Groups   Name        Std.Dev.
##  subjId   (Intercept) 0.155
##  Residual             0.108
## [1] Fixed effects:
##             Estimate Std. Error t value
## (Intercept)      0.6      0.036      16
```

Hence we see that `lmer` *calculate the moment estimates if the REML methods is used. Also note that we will be given usual t-test values for the fixed effect parameters. We will discuss the REML method in the next section.*

5.2.5 REML–estimation

As we saw above, estimators in mixed effect models might be biased. In the simple balanced case as we can actually find the unbiased estimator, but in more general situations, this might be difficult or impossible. A general way to deal with bias is the so-called REML-estimator (restricted or residual maximum likelihood). The idea is to introduce a penalty term in the log-likelihood function. The penalty term is given by $\frac{1}{2}\log|\boldsymbol{X}^T\boldsymbol{V}^{-1}\boldsymbol{X}|$, and the restricted log-likelihood function is given by

$$l_{RE}(\boldsymbol{\theta};\boldsymbol{y}) = l(\boldsymbol{\theta};\boldsymbol{y}) - \frac{1}{2}\log|\boldsymbol{X}^T\boldsymbol{V}^{-1}\boldsymbol{X}|. \tag{5.38}$$

In the balanced case, the REML estimates are equal to the moment estimates, and the REML estimates are generally recommended over the likelihood estimates.

Example 5.7 *Consider the simple statistical model $Y_i \sim N(\mu,\sigma^2)$ and Y_i iid. The model can be written as*

$$\boldsymbol{Y} = \boldsymbol{X}\mu + \boldsymbol{\epsilon}; \quad \boldsymbol{\epsilon} \sim N(\boldsymbol{0},\sigma^2\boldsymbol{I}), \tag{5.39}$$

where $\boldsymbol{X} = \boldsymbol{1}$, and the log-likelihood function is

$$l([\mu,\sigma^2];\boldsymbol{y}) = -\frac{n}{2}\log\sigma^2 - \frac{1}{2\sigma^2}\sum_{i=1}^{n}(y_i-\mu)^2, \tag{5.40}$$

the MLE is $\hat{\mu} = \bar{y}$ and $\hat{\sigma}^2_{ML} = \frac{1}{n}\sum_{i=1}^{n}(y_i-\bar{y})^2$. $\hat{\mu}$ is an unbiased estimator while the unbiased estimator for σ^2 is $\hat{\sigma}^2 = \frac{1}{n-1}\sum_{i=1}^{n}(y_i-\bar{y})^2$. In this case, we have $\boldsymbol{X}^T\boldsymbol{V}^{-1}\boldsymbol{X} = \frac{n}{\sigma^2}$ and therefore

$$l_{RE}(\boldsymbol{\theta};\boldsymbol{y}) = -\frac{n}{2}\log\sigma^2 - \frac{1}{2\sigma^2}\sum_{i=1}^{n}(y_i-\mu)^2 - \frac{1}{2}\log\left(\frac{n}{\sigma^2}\right) \tag{5.41}$$

$$= -\frac{n-1}{2}\log\sigma^2 - \frac{1}{2\sigma^2}\sum_{i=1}^{n}(y_i-\mu)^2 - \frac{1}{2}\log(n). \tag{5.42}$$

Since the penalty term does not include the mean value parameter μ, the estimate of μ is unaffected by the REML term (i.e. $\hat{\mu}_{REML} = \hat{\mu}_{ML} = \bar{y}$), inserting $\hat{\mu}$ and taking the derivative wrt. σ^2 gives

$$\frac{\partial l_{RE}(\sigma^2; \hat{\mu}, \boldsymbol{y})}{\partial \sigma^2} = -\frac{n-1}{2\sigma^2} + \frac{1}{2\sigma^4}\sum_{i=1}^{n}(y_i - \bar{y})^2, \tag{5.43}$$

setting equal 0 and solving for σ^2 give

$$\hat{\sigma}^2_{REML} = \frac{1}{n-1}\sum_{i=1}^{n}(y_i - \bar{y})^2, \tag{5.44}$$

which equal the unbiased estimator.

In the simple balanced one-way random effect model, the REML correction term becomes (again readers more interested in implementation and interpretation may skip directly to Section 5.2.6)

$$\begin{aligned}\log|\boldsymbol{X}^T\boldsymbol{V}^{-1}\boldsymbol{X}| &= \log|\sum_{i=1}^{k}\mathbf{1}_n^T\boldsymbol{V}_i^{-1}\mathbf{1}^T| \\ &= \log|k\mathbf{1}_n^T\boldsymbol{V}_i^{-1}\mathbf{1}^T| \\ &= \log(k) + \log|\mathbf{1}_n^T\boldsymbol{V}_i^{-1}\mathbf{1}^T|,\end{aligned} \tag{5.45}$$

and the matrix multiplication inside the log can be written as

$$\begin{aligned}\mathbf{1}_n^T\boldsymbol{V}_i^{-1}\mathbf{1}_n &= \frac{1}{\sigma^2}\mathbf{1}_n^T\left(\boldsymbol{I} - \frac{1}{\frac{\sigma^2}{\sigma_u^2}+n}\boldsymbol{J}\right)\mathbf{1}_n = \frac{1}{\sigma^2}\left(n - \frac{n^2}{\frac{\sigma^2}{\sigma_u^2}+n}\right) \\ &= \frac{n}{\sigma^2 + n\sigma_u^2},\end{aligned} \tag{5.46}$$

and hence

$$\log|\mathbf{1}_n^T\boldsymbol{V}_i^{-1}\mathbf{1}_n| = \log(n) - \log(\sigma^2 + n\sigma_u^2), \tag{5.47}$$

and the derivatives of the REML term are

$$\begin{aligned}\frac{1}{2}\frac{\partial\log|\mathbf{1}_n^T\boldsymbol{V}_i\mathbf{1}_n|}{\partial\sigma_u^2} &= \frac{1}{2}\frac{n}{\sigma^2 + n\sigma_u^2} \\ \frac{1}{2}\frac{\partial\log|\mathbf{1}_n^T\boldsymbol{V}_i\mathbf{1}_n|}{\partial\sigma^2} &= \frac{1}{2}\frac{1}{\sigma^2 + n\sigma_u^2}.\end{aligned} \tag{5.48}$$

The derivative of the profile log-likelihood is

$$\begin{aligned}\frac{\partial l_{p,REML}([\sigma^2, \sigma_u^2]; \boldsymbol{y})}{\partial\sigma_u^2} &= -\frac{k}{2}\frac{n}{\sigma^2 + n\sigma_u^2} + \frac{n}{2}\frac{SSB}{(\sigma^2 + n\sigma_u^2)^2} + \frac{1}{2}\frac{n}{\sigma^2 + n\sigma_u^2} \\ &= -\frac{k-1}{2}\frac{n}{\sigma^2 + n\sigma_u^2} + \frac{n}{2}\frac{SSB}{(\sigma^2 + n\sigma_u^2)^2}.\end{aligned} \tag{5.49}$$

This results in

$$\hat{\sigma}^2_{REML} + n\hat{\sigma}^2_{u,REML} = \frac{SSB}{k-1}, \tag{5.50}$$

and

$$\frac{\partial l_{p,REML}([\sigma^2,\sigma_u^2];\boldsymbol{y})}{\partial\sigma^2} = -\frac{k}{2}\left(\frac{n-1}{\sigma^2}+\frac{1}{\sigma^2+n\sigma_u^2}\right)+\frac{1}{2}\left(\frac{SSE}{\sigma^4}+\frac{SSB}{(\sigma^2+n\sigma_u^2)^2}\right)+ \frac{1}{2}\frac{1}{\sigma^2+n\sigma_u^2}. \tag{5.51}$$

Hence the estimating equation for σ^2 becomes

$$\begin{aligned} 0 =& -\frac{k}{2}\left(\frac{n-1}{\sigma^2}+\frac{k-1}{SSB}\right)+\frac{1}{2}\left(\frac{SSE}{\sigma^4}+\frac{(k-1)^2}{SSB}\right)+\frac{k-1}{2SSB} \\ =&\frac{1}{2\sigma^2}\left(-k(n-1)+\frac{SSE}{\sigma^2}\right)+\frac{k-1}{2SSB}\left(-k+k-1+1\right), \end{aligned} \tag{5.52}$$

and the REML estimates are

$$\begin{aligned} \hat{\sigma}^2_{REML} =& \frac{SSE}{k(n-1)} \\ \hat{\sigma}^2_{u,REML} =& \frac{SSB}{n(k-1)} - \frac{\hat{\sigma}^2_{REML}}{n} \\ =& \frac{SSB}{n(k-1)} - \frac{SSE}{kn(n-1)}, \end{aligned} \tag{5.53}$$

which are also the unbiased (moment) estimators.

The reasoning above implies that we will usually prefer the REML estimates over the maximum likelihood estimates, also in more complicated structures where we cannot write down the explicit result as above.

5.2.6 Unbalanced design

In case of an unbalanced design, the estimates can still be found analytically. However, the estimates are more complicated functions of the data (Madsen and Thyregod, 2011), and we will simply use the REML estimator that can be applied in the same way as we saw above.

Example 5.8 *The fixed effect in the simple linear model is compared with the overall mean for the simple unbalanced data as presented on page 118;*

```
fit.aggr <- lmer(clo  ~ 1 + (1|subjId), data = dat.aggr)
fixef(fit.aggr) - mean(dat.aggr$clo)

## (Intercept)
##       0.00025
```

Hence we see that in the case of unbalanced design, the fixed effect parameter is not exactly equal to the average of all clothing levels; rather, it is a weighted sum. Even though we can find closed–form solutions in this case, we will not delve into the exact formulation; just note that it is estimated by the REML method as described above.

5.2.7 Including regressors

Further regressors can be included both in the fixed effect terms and in the random effect part of the model. Formally, the model is defined by

Definition 5.2 (Linear mixed effects model) *The model*

$$Y = X\beta + ZU + \epsilon, \tag{5.54}$$

with X *and* Z *denoting known matrices, and where* $\epsilon \sim N(\mathbf{0}, \Sigma)$ *and* $U \sim N(\mathbf{0}, \Psi)$ *are independent is called a mixed general linear model. In the general case, the covariance matrices* Σ *and* Ψ *may depend on some unknown parameters,* ψ*, that also need to be estimated.*

The parameters β *are called* fixed effects *or* systematic effects *while the quantities* U *are called* random effects.

Based on the definition above, we can find the expected value and the variance-covariance of the response Y as

$$\begin{aligned} E[Y] &= X\beta \\ V[Y] &= V[ZU] + V[\epsilon] \\ &= ZV[U]Z^T + \Sigma \\ &= Z\Psi Z^T + \Sigma. \end{aligned} \tag{5.55}$$

Hence the marginal distribution for the response is $Y \sim N(X\beta, Z\Psi Z^T + \Sigma)$ and we will denote the marginal variance by $V = Z\Psi Z^T + \Sigma$.

Example 5.9 *In the full clothing insulation data (see page 118), we implement a simple mean value structure, that is* $X = \mathbf{1}$ *and the random effect is also simple. As we have a total of 47 subjects and 803 observations, the dimension of* Z *is* 803×47 *and a structure given by*

$$Z = \begin{bmatrix} 1 & 0 & \dots & 0 \\ 0 & 1 & \dots & 0 \\ \vdots & \vdots & \ddots & \vdots \\ 0 & 0 & \dots & 1 \end{bmatrix}. \tag{5.56}$$

In this case, $U_i \sim N(0, \sigma_u^2)$ *or* $\Psi = \sigma_u^2 I_{47}$*. If we include random slopes in the model, then the random design matrix becomes*

$$Z = \begin{bmatrix} Z_1 & 0 & \dots & 0 \\ 0 & Z_2 & \dots & 0 \\ \vdots & \vdots & \ddots & \vdots \\ 0 & 0 & \dots & Z_{47} \end{bmatrix}, \tag{5.57}$$

with

$$Z_i = \begin{bmatrix} 1 & z_{i1} \\ \vdots & \vdots \\ 1 & z_{in_i} \end{bmatrix}, \tag{5.58}$$

where z_{ij} *is some explanatory variable, for example indoor operative temperature. In this case, we would have* $U_i \sim N(\mathbf{0}, \Psi_i)$*, with*

$$\Psi_i = \begin{bmatrix} \sigma_{u_1}^2 & \sigma_{u_1,u_2} \\ \sigma_{u_1,u_2} & \sigma_{u_2}^2 \end{bmatrix}, \tag{5.59}$$

and therefore $\boldsymbol{\Psi}$ *becomes a block diagonal matrix*

$$\boldsymbol{\Psi} = \begin{bmatrix} \boldsymbol{\Psi}_1 & 0 & \dots & 0 \\ 0 & \boldsymbol{\Psi}_1 & \dots & 0 \\ \vdots & \vdots & \ddots & \vdots \\ 0 & 0 & \dots & \boldsymbol{\Psi}_1 \end{bmatrix} = \boldsymbol{I} \otimes \boldsymbol{\Psi}_1. \tag{5.60}$$

In actual implementations we will usually estimate the correlation between the random effect ($\rho_{u_1,u_2} = \frac{\sigma_{u_1,u_2}}{\sigma_{u_1}\sigma_{u_2}} \in (-1,1)$) *rather than the covariance.*

5.2.8 Estimation of fixed effects and variance parameters

As we saw above, the marginal distribution for $\boldsymbol{Y}$ can be written as

$$\boldsymbol{Y} \sim N(\boldsymbol{X\beta}, \boldsymbol{V}(\boldsymbol{\psi})). \tag{5.61}$$

Using equations (3.67) and (3.68) on page 67 the estimates for the fixed effect parameters for any given value of $\boldsymbol{\psi}$ are given by

$$\widehat{\boldsymbol{\beta}}(\boldsymbol{\psi}) = (\boldsymbol{X}^T\boldsymbol{V}^{-1}(\boldsymbol{\psi})\boldsymbol{X})^{-1}\boldsymbol{X}^T\boldsymbol{V}(\boldsymbol{\psi})^{-1}\boldsymbol{y}. \tag{5.62}$$

Hence, the profile log-likelihood for the variance parameters, $\boldsymbol{\psi}$, is

$$\ell_p(\boldsymbol{\psi}) = -\tfrac{1}{2}\log|\boldsymbol{V}| - \tfrac{1}{2}(\boldsymbol{Y} - \boldsymbol{X}\widehat{\boldsymbol{\beta}})^T\boldsymbol{V}^{-1}(\boldsymbol{Y} - \boldsymbol{X}\widehat{\boldsymbol{\beta}}). \tag{5.63}$$

As discussed above, usually the REML term is added to get the *modified profile log-likelihood*

$$\ell_m(\boldsymbol{\psi}) = -\tfrac{1}{2}\log|\boldsymbol{V}| - \tfrac{1}{2}\log|\boldsymbol{X}^T\boldsymbol{V}^{-1}\boldsymbol{X}| - \tfrac{1}{2}(\boldsymbol{Y} - \boldsymbol{X}\widehat{\boldsymbol{\beta}})^T\boldsymbol{V}^{-1}(\boldsymbol{Y} - \boldsymbol{X}\widehat{\boldsymbol{\beta}}). \tag{5.64}$$

Example 5.10 *We continue with the aggregated data-set with all subjects. The most general model is a model with slope depending on sex and also slopes as random effects. Below, it is tested if the random slope gives a significant contribution to the model;*

```
fit0 <- lmer(clo  ~ sex * tInOpCen + (tInOpCen| subjId),
             data = dat.aggr)

## boundary (singular) fit: see ?isSingular

fit0R <- lmer(clo  ~ sex * tInOpCen + (1| subjId),
              data = dat.aggr)
aov <- anova(fit0, fit0R)

## refitting model(s) with ML (instead of REML)

##       npar  BIC logLik Chisq Df Pr(>Chisq)
## fit0R    6 -166   97.9    NA NA         NA
## fit0     8 -158   98.5  1.33  2       0.51
```

Notice that we get a warning for the first fit, this is because the (absolute) correlation between the random intercept and the random slope becomes very close to 1 (i.e. the parameters cannot be identified). Also note that anova *will refit the models without the REML correction term. In this case, the random slopes are not significant, and the final model includes random intercepts. With the random effects in place, we test the fixed effects.*

```
drop1(fitOR,test="Chisq")

## Single term deletions
##
## Model:
## clo ~ sex * tInOpCen + (1 | subjId)
##                npar  AIC  LRT Pr(Chi)
## <none>              -184
## sex:tInOpCen      1 -176 9.28  0.0023 **
## ---
## [...]
```

We see that all fixed effect terms are significant and hence we can report the final model as (parts of the R-summary is left out);

```
## [...]
## Random effects:
##  Groups   Name        Variance Std.Dev.
##  subjId   (Intercept) 0.00629  0.0793
##  Residual             0.01002  0.1001
## Number of obs: 136, groups:  subjId, 47
##
## Fixed effects:
##                  Estimate Std. Error t value
## (Intercept)        0.5988     0.0202   29.65
## sexmale           -0.0921     0.0290   -3.18
## tInOpCen          -0.0643     0.0128   -5.01
## sexmale:tInOpCen   0.0564     0.0184    3.06
##
## [...]
```

Hence, the final model is

$$Y_{ij} = \beta_0(\texttt{sex}_i) + \texttt{tInOp}_{ij}\beta_1(\texttt{sex}_i) + U_i + \epsilon_{ij} \tag{5.65}$$

with $U_i \sim N(0, 0.079^2)$, $\epsilon_{ij} \sim N(0, 0.1^2)$, *the fixed effect are* $\beta_0(\texttt{female}) = 0.60$, $\beta_0(\texttt{male}) = 0.51$, $\beta_1(\texttt{female}) = -0.064$, *and* $\beta_1(\texttt{male}) = -0.008$; *hence, the males seem to be much less affected by the indoor operating temperature.*

5.3 Gaussian mixed effects model

The general mixed effect model can be formulated as a so-called hierarchical model, as (see (Madsen and Thyregod, 2011))

Definition 5.3 (Gaussian mixed model) *A general formulation of the* Gaussian mixed model *is*

$$\boldsymbol{Y}|\boldsymbol{U} = \boldsymbol{u} \sim N(\boldsymbol{\mu}(\boldsymbol{\beta}, \boldsymbol{u}), \boldsymbol{\Sigma}(\boldsymbol{\beta})) \tag{5.66a}$$

$$\boldsymbol{U} \sim N(\boldsymbol{0}, \boldsymbol{\Psi}(\boldsymbol{\psi})), \tag{5.66b}$$

where the dimension of $\boldsymbol{U}$ *and therefore* $\boldsymbol{\Psi}$ *might be large, but the dimension of* $\boldsymbol{\psi}$ *is generally small.*

This model formulation is also a so-called hierarchical model *where (5.66a) and (5.66b) are the first and second stage models, respectively.*

We will use the definition above to define more general models (i.e. non-Gaussian and non-linear, see Chapter 6), but for now, we will use it as a starting point for estimation of the random effect models. The formulation above gives the conditional distribution of the observation given the random effects. The random effects are not observed; hence, we will aim at finding $\boldsymbol{U}|\boldsymbol{Y} = \boldsymbol{y}$. To this end we need the joint distribution of $\boldsymbol{Y}$ and $\boldsymbol{U}$. The marginal distibution of both have been discussed above, and since models are linear and all random variables are Gaussian, this amounts to finding the covariance between $\boldsymbol{Y}$ and $\boldsymbol{U}$,

$$\begin{aligned} Cov[\boldsymbol{Y}, \boldsymbol{U}] &= Cov[\boldsymbol{X\beta} + \boldsymbol{ZU} + \boldsymbol{\epsilon}, \boldsymbol{U}] \\ &= \boldsymbol{Z}Cov[\boldsymbol{U}, \boldsymbol{U}] \\ &= \boldsymbol{Z\Psi}. \end{aligned} \tag{5.67}$$

Hence, the joint distribution of $\boldsymbol{Y}$ and $\boldsymbol{U}$ is a normal with mean value vector and variance-covariance

$$E\begin{bmatrix}\boldsymbol{Y}\\ \boldsymbol{U}\end{bmatrix} = \begin{bmatrix}\boldsymbol{X\beta}\\ \boldsymbol{0}\end{bmatrix}; \quad V\begin{bmatrix}\boldsymbol{Y}\\ \boldsymbol{U}\end{bmatrix} = \begin{bmatrix}\boldsymbol{Z\Psi Z}^T + \boldsymbol{\Sigma} & \boldsymbol{Z\Psi}\\ \boldsymbol{\Psi Z}^T & \boldsymbol{\Psi}\end{bmatrix}. \tag{5.68}$$

The conditional random effects may be found via the so-called hierarchical likelihood, and we will treat this later in the last Section 5.4.2 and Chapter 6; here, we use the projection theorem (see, Madsen (2008)).

Theorem 5.3 (Projections for normally distributed variables) *Let* $(\boldsymbol{Y}^T, \boldsymbol{X}^T)^T$ *be a normally distributed random variable with*

$$E\begin{bmatrix}\boldsymbol{Y}\\ \boldsymbol{X}\end{bmatrix} = \begin{bmatrix}\boldsymbol{\mu_Y}\\ \boldsymbol{\mu_X}\end{bmatrix} \quad \text{and} \quad V\begin{bmatrix}\boldsymbol{Y}\\ \boldsymbol{X}\end{bmatrix} = \begin{bmatrix}\boldsymbol{\Sigma_{YY}} & \boldsymbol{\Sigma_{YX}}\\ \boldsymbol{\Sigma_{XY}} & \boldsymbol{\Sigma_{XX}}\end{bmatrix}. \tag{5.69}$$

Then $\boldsymbol{Y}|\boldsymbol{X}$ *is normally distributed with mean*

$$E[\boldsymbol{Y}|\boldsymbol{X}] = \boldsymbol{\mu_Y} + \boldsymbol{\Sigma_{YX}}\boldsymbol{\Sigma_{XX}^{-1}}(\boldsymbol{X} - \boldsymbol{\mu_X}), \tag{5.70}$$

and variance

$$V[\boldsymbol{Y}|\boldsymbol{X}] = \boldsymbol{\Sigma_{YY}} - \boldsymbol{\Sigma_{YX}}\boldsymbol{\Sigma_{XX}^{-1}}\boldsymbol{\Sigma_{YX}^{T}}. \tag{5.71}$$

Furthermore, the error, $(\boldsymbol{Y} - E[\boldsymbol{Y}|\boldsymbol{X}])$ *and* $\boldsymbol{X}$ *are* independent.

We will not prove this theorem, but refer the interested reader to Jazwinski (1970). The theorem shows us that the conditional expectation is a corrected version of the unconditional expectation ($E[\boldsymbol{Y}] = \boldsymbol{\mu_Y}$); this correction will be large when the covariance between $\boldsymbol{Y}$ and $\boldsymbol{X}$ is strong (they are highly correlated) the variance of $\boldsymbol{X}$ is small (i.e. the information from $\boldsymbol{X}$ is large), and when the difference between the expected value (the model) and $\boldsymbol{X}$ is large.

Using the projection theorem, we can find the conditional mean and variance of the random effects

$$\begin{aligned} E[\boldsymbol{U}|\boldsymbol{Y}=y] =& 0+\boldsymbol{\Psi}\boldsymbol{Z}^T\boldsymbol{V}^{-1}(\boldsymbol{y}-\boldsymbol{X}\boldsymbol{\beta}) \\ =& \boldsymbol{\Psi}\boldsymbol{Z}^T(\boldsymbol{Z}\boldsymbol{\Psi}\boldsymbol{Z}^T+\boldsymbol{\Sigma})^{-1}(\boldsymbol{y}-\boldsymbol{X}\boldsymbol{\beta}) \\ =& (\boldsymbol{Z}^T\boldsymbol{\Sigma}^{-1}\boldsymbol{Z}+\boldsymbol{\Psi}^{-1})^{-1}\boldsymbol{Z}^T\boldsymbol{\Sigma}^{-1}(\boldsymbol{y}-\boldsymbol{X}\boldsymbol{\beta}). \end{aligned} \tag{5.72}$$

For the third line, see Petersen and Pedersen (2012), eq. (158), and the variance can be written as, again using Petersen and Pedersen (2012), eq. (158)

$$\begin{aligned} V[\boldsymbol{U}|\boldsymbol{Y}=y] =& \boldsymbol{\Psi}-\boldsymbol{\Psi}\boldsymbol{Z}^T\boldsymbol{V}^{-1}\boldsymbol{Z}\boldsymbol{\Psi} \\ =& \boldsymbol{\Psi}-\boldsymbol{\Psi}\boldsymbol{Z}^T(\boldsymbol{Z}\boldsymbol{\Psi}\boldsymbol{Z}^T+\boldsymbol{\Sigma})^{-1}\boldsymbol{Z}\boldsymbol{\Psi} \\ =& \boldsymbol{\Psi}-(\boldsymbol{\Psi}^{-1}+\boldsymbol{Z}^T\boldsymbol{\Sigma}^{-1}\boldsymbol{Z})^{-1}\boldsymbol{Z}^T\boldsymbol{\Sigma}^{-1}\boldsymbol{Z}\boldsymbol{\Psi} \\ =& (\boldsymbol{\Psi}^{-1}+\boldsymbol{Z}^T\boldsymbol{\Sigma}^{-1}\boldsymbol{Z})^{-1}((\boldsymbol{\Psi}^{-1}+\boldsymbol{Z}^T\boldsymbol{\Sigma}^{-1}\boldsymbol{Z})\boldsymbol{\Psi}-\boldsymbol{Z}^T\boldsymbol{\Sigma}^{-1}\boldsymbol{Z}\boldsymbol{\Psi}) \\ =& (\boldsymbol{\Psi}^{-1}+\boldsymbol{Z}^T\boldsymbol{\Sigma}^{-1}\boldsymbol{Z})^{-1}(\boldsymbol{I}+\boldsymbol{Z}^T\boldsymbol{\Sigma}^{-1}\boldsymbol{Z}\boldsymbol{\Psi}-\boldsymbol{Z}^T\boldsymbol{\Sigma}^{-1}\boldsymbol{Z}\boldsymbol{\Psi}) \\ =& (\boldsymbol{Z}^T\boldsymbol{\Sigma}^{-1}\boldsymbol{Z}^T+\boldsymbol{\Psi}^{-1})^{-1}. \end{aligned} \tag{5.73}$$

We will illustrate the results with a small example.

Example 5.11 (One–way random effect model) *Consider again the simple balanced random effect model*

$$Y_{ij}=\mu+U_i+\epsilon_{ij}; \quad U_i\sim N(0,\sigma_u^2) \quad \epsilon_{ij}N(0,\sigma^2). \tag{5.74}$$

With n observations from each of k groups, we have

$$\boldsymbol{Z}^T\boldsymbol{\Sigma}^{-1}\boldsymbol{Z}=\frac{n}{\sigma^2}\boldsymbol{I}_k; \quad \boldsymbol{\Psi}^{-1}=\frac{1}{\sigma_u^2}\boldsymbol{I}_k, \tag{5.75}$$

and therefore

$$\begin{aligned} (\boldsymbol{Z}^T\boldsymbol{\Sigma}^{-1}\boldsymbol{Z}+\boldsymbol{\Psi}^{-1})^{-1} =& \frac{\sigma^2\sigma_u^2}{n\sigma_u^2+\sigma^2}\boldsymbol{I}_k \\ =& \frac{\sigma^2}{n+\sigma^2/\sigma_u^2}\boldsymbol{I}_k. \end{aligned} \tag{5.76}$$

With a bit of algebra, we can write

$$\begin{aligned} E[U_i|\boldsymbol{Y}] =& \frac{1}{n+\sigma^2/\sigma_u^2}\sum_j(y_{ij}-\mu)=\frac{n}{n+\sigma^2/\sigma_u^2}(\bar{y}_i-\mu) \\ V[U_i|\boldsymbol{Y}] =& \frac{\sigma^2}{n+\sigma^2/\sigma_u^2}. \end{aligned} \tag{5.77}$$

Hence, when σ^2 is small or n is large, the estimated random effect is close to the observed difference and in this case the variance on the random effect is also small.

In the example below, we again turn to our clothing insulation example;

Example 5.12 (Clothing insulation level (continued)) *As shown in Example 3.2 on page 47, the variance in clothing insulation level can be partly explained by the outdoor air temperature and the sex of a person. However, the unexplained part of variance is still high.*

At the same time, one can hypothesise that the level of clothing insulation worn depends on the preferences of the individual. As the data includes several values from each individual, we can formulate a random effects model with an identification of an individual as random factor.

Combining the findings presented in Example 3.2, where we showed that the outdoor air temperature, `tOut`*, has a significant influence on the clothing insulation level, with the random effect of the individual leads to the formulation of a mixed effect model with* `tOut` *as fixed effects and the individual as random effect.*

Hence, the following model is tested

$$\boldsymbol{Y} = \boldsymbol{X\beta} + \boldsymbol{XU} + \epsilon, \tag{5.78}$$

or

$$\texttt{clo}_{ij} = \beta_0 + \texttt{tOut}_{ij}\beta_1 + U_{1,i} + \texttt{tOut}_{ij}U_{2,i} + \epsilon_{ij}, \tag{5.79}$$

where $\epsilon_{ij} \sim N(0, \sigma^2)$ *and iid,* $[U_{1,i}, U_{2,i}]^T \sim N(\mathbf{0}, \boldsymbol{\Psi}_i)$ *with* $\boldsymbol{\Psi}_i$ *a* 2×2 *variance-covariance matrix. The model is estimated in the R-code given below (parts of the summary is left out), and* `tOutCen` *is a centred version of the outside temperature;*

```
# define and fit model
modMEMEx12 <- lmer(clo ~ tOutCen + (1 + tOutCen | subjId),
                   data = dfMEMEx1)

## [...]
## Random effects:
##  Groups   Name        Variance Std.Dev. Corr
##  subjId   (Intercept) 0.013642 0.1168
##           tOutCen     0.000159 0.0126   0.11
##  Residual             0.007378 0.0859
## Number of obs: 803, groups:  subjId, 47
##
## Fixed effects:
##             Estimate Std. Error t value
## (Intercept)  0.55286    0.01757   31.47
## tOutCen     -0.00970    0.00214   -4.52
## [...]
```

From the R-output, we get the following model

$$\texttt{clo}_{ij} = 0.55 - \texttt{tOut}_{ij}^{cen}0.0097 + U_{1,i} + \texttt{tOut}_{ij}^{cen}U_{2,i} + \epsilon_{ij}, \tag{5.80}$$

where $\epsilon_{ij} \sim N(0, 0.086^2)$ *and iid, and* $[U_{1,i}, U_{2,i}]^T \sim N(\mathbf{0}, \mathbf{\Psi})$, *with*

$$\begin{aligned}\mathbf{\Psi} &= \begin{bmatrix} 0.117 & 0 \\ 0 & 0.0126 \end{bmatrix} \begin{bmatrix} 1 & 0.11 \\ 0.11 & 1 \end{bmatrix} \begin{bmatrix} 0.12 & 0 \\ 0 & 0.0126 \end{bmatrix} \\ &= \begin{bmatrix} 0.117^2 & 0.00016 \\ 0.00016 & 0.0126^2 \end{bmatrix}. \end{aligned} \tag{5.81}$$

The output shows that the addition of `tOut` *increases the explained variance and in order to demonstrate the difference between the results of the linear model presented in Example 3.2 and the mixed effect model fitted here, the random coefficients (i.e.* $\beta + E[\boldsymbol{U}|\boldsymbol{Y} = \boldsymbol{y}]$*) can be extracted by;*

```
betamodMEMEx12 <- coef(modMEMEx12)$subjId
```

Based on this, the following figure can be drawn:

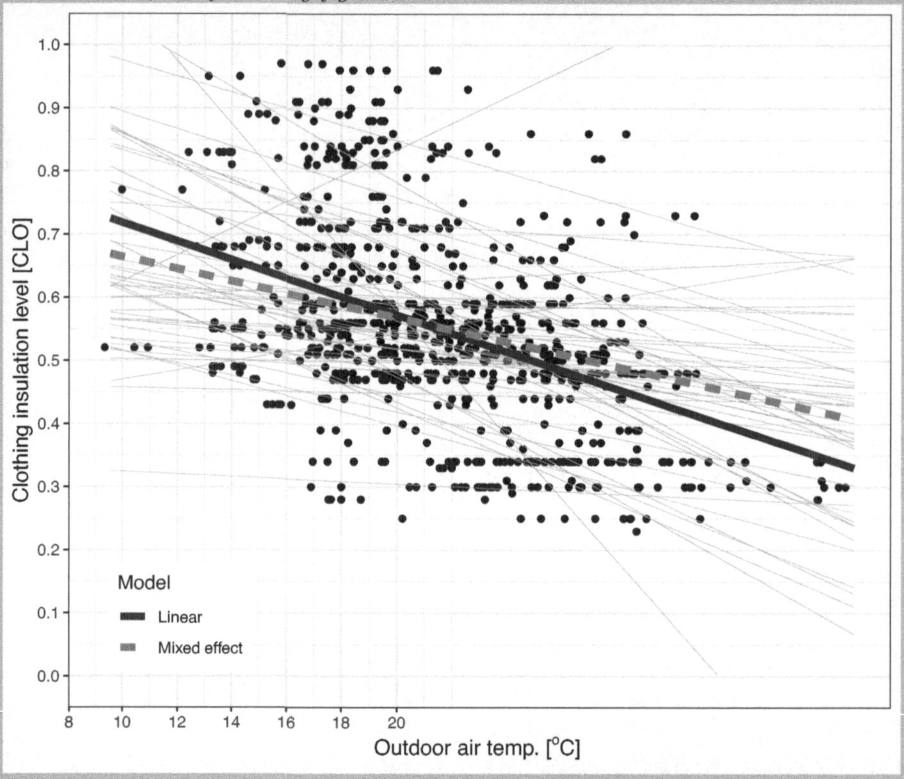

The figure demonstrates that the general linear model predicts a steeper decrease of the clothing insulation value with increasing `tOut` *compared to the mixed effect model. The thin grey lines highlight the variation in individuals clothing insulation level and individual variations in the extent that* `tOut` *affects the level of clothing worn (seen in the variation of steepness of the grey lines).*

5.4 More structures

We could of course imagine many different structures, for example that in addition to the variation between subjects, there is a variation between days within subjects; this means that we have a nested random effect (also called a split plot design). In the clothing example, this is truly nested (the variation between days works within the subject), but it is possible to have non-nested random effects, for example test score from exams with a random effect for the difficulty of the individual questions and one for the skill-fullness of the individual student. In any case, we may write the model as

$$\boldsymbol{Y} = \boldsymbol{X\beta} + \boldsymbol{Z}_1\boldsymbol{U}_1 + \boldsymbol{Z}_2\boldsymbol{U}_2 + \boldsymbol{\epsilon}, \tag{5.82}$$

where $\boldsymbol{U}_i$ follows a multivariate normal with variance-covariance $\boldsymbol{\Psi}_i$; such a model can also be written as

$$\boldsymbol{Y} = \boldsymbol{X\beta} + \boldsymbol{ZU} + \boldsymbol{\epsilon}, \tag{5.83}$$

where $\boldsymbol{Z} = [\boldsymbol{Z}_1 \quad \boldsymbol{Z}_2]$ and

$$V[\boldsymbol{Z}] = \begin{bmatrix} \boldsymbol{\Psi}_1 & \boldsymbol{0} \\ \boldsymbol{0} & \boldsymbol{\Psi}_2 \end{bmatrix}. \tag{5.84}$$

Example 5.13 *We again return to the example of clothing insulation level, and add a random effect of day within subject, that is*

$$\begin{aligned} \texttt{clo}_{ijl} =& \beta_0(\texttt{sex}_i) + \texttt{tOut}_{ijl}\beta_1(\texttt{sex}_i) + U_{1i} + \texttt{tOut}_{ijl}U_{2i} + V_{1ij} + \\ & \texttt{tOut}_{ijl}V_{2ij} + \epsilon_{ijl}, \end{aligned} \tag{5.85}$$

where i, j and l refer to subject i, on day j and observation l. The random variables U, V, and ϵ are all normal and independent of each other; further $cor(U_{1i}, U_{2i}) = \rho_u$, $cor(V_{1ij}, V_{2ij}) = \rho_V$, $cor(U_{1i}, U_{2i'}) = 0$ if $i \neq i'$ and $cor(V_{1ij}, V_{2i'j'}) = 0$ if $(i, j) \neq (i', j')$. The R code below tests which structure should be present;

```
modMEMEx16 <- lmer(clo ~ tOutCen * sex +
                       (1 + tOutCen | subjId)+
                       (1 | subjId: day), data = dfMEMEx1,
                   REML = TRUE)

modMEMEx17 <- lmer(clo ~ tOutCen * sex + (1 | subjId) +
                       (1 | subjId : day),
                   data = dfMEMEx1, REML = TRUE)

anova(modMEMEx15, modMEMEx16, modMEMEx17)[ ,c(1,4,6:8)]

## [...]
## modMEMEx17    7    958
## modMEMEx16    9    966 14.90  2    0.00058 ***
## modMEMEx15   11    968  4.22  2    0.12119
## ---
## [...]
```

The conclusion from the above is that we should choose the model

$$\texttt{clo}_{ijl} = \beta_0(\texttt{sex}_i) + \texttt{tOut}_{ijl}\beta_0(\texttt{sex}_i) + U_{1i} + \texttt{tOut}_{ijl}U_{2i} + V_{ij} + \epsilon_{ijl}, \tag{5.86}$$

that is each individual not only have their own overall sensitivity to the outdoor temperature, but also the overall level varies from day to day. Further, we can get the parameters by

```
summary(modMEMEx16)
## [...]
## Random effects:
##  Groups     Name        Variance  Std.Dev. Corr
##  subjId:day (Intercept) 0.0085370 0.09240
##  subjId     (Intercept) 0.0099303 0.09965
##             tOutCen     0.0000751 0.00867  0.25
##  Residual               0.0027880 0.05280
## Number of obs: 803, groups:  subjId:day, 136; subjId, 47
##
## Fixed effects:
##               Estimate Std. Error t value
## (Intercept)    0.58862    0.02361   24.93
## tOutCen       -0.00718    0.00222   -3.24
## sex2          -0.07863    0.03374   -2.33
## tOutCen:sex2  0.00491    0.00316    1.56
## [...]
```

Hence the model is

$$\begin{aligned} clo_{ijl} =& 0.589 - 0.079 I(sub_i = male) - 0.0072 \cdot tOut^{cen}_{ijl} + \\ & 0.0049 \cdot tOut^{cen}_{ijk} I(sub_i = male) + U_{1i} + \\ & \texttt{tOut}_{ijl}U_{2i} + V_{ij} + \epsilon_{ijl}, \end{aligned} \tag{5.87}$$

with $\boldsymbol{U}_i \sim N(\boldsymbol{0}, \boldsymbol{\Psi}_u)$, $V_{ij} \sim N(0, 0.092^2)$, $\epsilon_{ijl} \sim N(0.053^2)$, *and*

$$\begin{aligned} \boldsymbol{\Psi}_u &= \begin{bmatrix} 0.10 & 0 \\ 0 & 0.0087 \end{bmatrix} \begin{bmatrix} 1 & 0.25 \\ 0.25 & 1 \end{bmatrix} \begin{bmatrix} 0.1 & 0 \\ 0 & 0.0087 \end{bmatrix} \\ &= \begin{bmatrix} 0.10^2 & 0.0002 \\ 0.0002 & 0.0082^2 \end{bmatrix}. \end{aligned} \tag{5.88}$$

The interpretation is that males wear slightly less clothing insulation that females, and also that the variation between subjects is larger than the difference between sex. With increasing temperature, the insulation level decreases, and males are less sensitive to temperature than females, but again the difference between sex is smaller than the variation between subjects.

5.4.1 Correlation structures

We saw above that we can include more than one level of random effects; in the specific example, the variation over a specific day (for a specific subject id) is

$$\tilde{\boldsymbol{\epsilon}}_{ij} = \boldsymbol{V}_{ij}\mathbf{1} + \boldsymbol{\epsilon}_{ij}, \tag{5.89}$$

with $\boldsymbol{\epsilon}_{ij} = [\epsilon_{ij1}, ..., \epsilon_{ijn_{ij}}]$, where n_{ij} is the number of reported observations form subject i on day j, the variance-covariance structure of $\tilde{\boldsymbol{\epsilon}}$ has the compound symmetry as discussed earlier; this implies that the correlation between $\tilde{\boldsymbol{\epsilon}}_{ijl}$ and $\tilde{\boldsymbol{\epsilon}}_{ijk}$ is constant for all $l \neq j$. In many situations, it is however natural to hypothesise that the correlation depends on the distance between data-points. In our example, data is recorded during one day and it is natural measure distance in terms of time or equivalent, in our case, observation number within the day. A simple hypothesis in this case is that $\tilde{\boldsymbol{\epsilon}}_{ijk}$ follows an AR(1) process; we will discuss time–series models later in Chapter 8. Here it suffice to note that the AR(1) model is given by ($\phi \in (-1, 1)$)

$$X_k = \phi X_{k-1} + \lambda_k; \quad \lambda_k \sim N(0, \sigma_\lambda^2) \text{ and iid.} \tag{5.90}$$

This model implies that $Cor[X_k, X_l] = \phi^{|k-l|}$. There exist different R-packages that can handle this type of structure. We will use the package `glmmTMB` (Brooks et al., 2017), as it uses the same syntax as `lme4`, and further it generalise to include non-Gaussian models as we will see in Chapter 6.

Example 5.14 *As a starting point, the AR(1) structure is added to the last example; in order to use the build in ANOVA methods, the model is re-estimated using* `glmmTMB`*;*

```
modMEMEx16tmb <- glmmTMB(clo ~ tOutCen * sex +
                         (1 + tOutCen|subjId) +
                         (1|subjId:day), data = dfMEMEx1)

modMEMEx16AR1 <- glmmTMB(clo ~ tOutCen *as.numeric(sex) +
                         (1 + tOutCen|subjId) +
                         ar1(as.factor(obs.no) - 1 |
                             subjId:day), data = dfMEMEx1)

anova(modMEMEx16tmb, modMEMEx16AR1)[ ,c(1,4,6:8)]

##               Df logLik Chisq Chi Df Pr(>Chisq)
## modMEMEx16tmb  9    966
## modMEMEx16AR1 10   1029   125      1     <2e-16 ***
## ---
## [...]
```

As we can see, the likelihood increases with about 63 on 1 degree of freedom, which is significant on any reasonable level. There are some convergence problems (not shown in output); this can be traced back to correlation of the random effect being close to 1. Hence a possible fix is to reduce the model. Below two reductions are tested;

```
modMEMEx17AR1 <- glmmTMB(clo ~ tOutCen * sex + (1 | subjId) +
                         ar1(as.factor(obs.no) - 1 |
                             subjId : day), data=dfMEMEx1,
                     REML = FALSE)

modMEMEx18AR1 <- glmmTMB(clo ~ tOutCen + sex + (1 | subjId) +
                         ar1(as.factor(obs.no) - 1 |
                             subjId:day), data = dfMEMEx1,
                     REML = FALSE)


anova(modMEMEx16AR1, modMEMEx17AR1,
      modMEMEx18AR1)[ ,c(1,4,6:8)]

##                Df logLik Chisq Chi Df Pr(>Chisq)
## modMEMEx18AR1  7   1027
## modMEMEx17AR1  8   1028  3.42       1      0.065 .
## modMEMEx16AR1 10   1029  0.46       2      0.795
## ---
## [...]
```

We see that we would accept the simple model here, and hence the final model is

$$Y_{ijl} = \beta_0(sex_i) + \beta_1 tOut_{ijl}^{cen} + U_i + V_{ijl} + \epsilon_{ijl}, \tag{5.91}$$

with $U_i \sim N(0, \sigma_u^2)$ *iid,* $V_{ijl} \sim N(0, \sigma_v^2)$ *with* $cor(V_{ijl,ijk}, V_{ijk}) = \rho^{|l-k|}$*, and* $\epsilon_{ijl} \sim N(0, \sigma^2)$ *and iid. The specific REML-estimates are;*

```
## [...]
## Conditional model:
##  Groups     Name                Variance Std.Dev. Corr
##  subjId     (Intercept)         0.01015  0.1008
##  subjId:day as.factor(obs.no)1 0.01061  0.1030   0.90 (ar1)
##  Residual                       0.00079  0.0281
## Number of obs: 803, groups:  subjId, 47; subjId:day, 136
##
## Dispersion estimate for gaussian family (sigma^2): 0.00079
##
## Conditional model:
##             Estimate Std. Error z value Pr(>|z|)
## (Intercept)  0.58625    0.02338   25.07  < 2e-16 ***
## tOutCen     -0.00476    0.00117   -4.06 0.000048 ***
## sex2        -0.07785    0.03350   -2.32     0.02 *
## ---
## [...]
```

Hence, we see that on average clothing insulation level is slightly lower for males than females and that higher outdoor temperature results in lower insulation levels. The difference between sex is, however within one standard error for subjects. That is, the variation between subjects is larger than the difference between sex; also the variation between days is larger than the variation between subjects, the remaining standard error is less than 1/3 of the between day standard deviation. Lastly the correlation within the day is high (0.9), which implies the subject is reluctant to change their clothing insulation level during a day.

The model described above leads to a new correlation structure, where the correlation between observations taken on the same day is high if the observations are close in time and lower if these are further apart in time. Observations taken from the same subject, but on different days, are also correlated. Finally, observations on different subjects are not correlated (independent subjects). Figure 5.3 shows the resulting correlation matrix for one subject.

There are a number of different correlation structures that could be included, for example Gaussian spatial correlation, exponential decaying and others; we will not pursue this further here but refer the interested reader to Brooks et al. (2017).

5.4.2 Estimation of random effects

So far we have focused on the estimation of the parameters for the models, i.e. variances, correlation and, of course, fixed effects. It is sometimes of interest to estimate the random effects of a model; to this end, we formulate a so-called *hierarchical likelihood* (Lee and Nelder, 1996) by writing the joint density for observable as well as unobservable random quantities $(\boldsymbol{Y}, \boldsymbol{U})$,

$$f(\boldsymbol{y}, \boldsymbol{u}; \boldsymbol{\beta}, \boldsymbol{\psi}) = f_{Y|u}(\boldsymbol{y}; \boldsymbol{\beta}) f_U(\boldsymbol{u}; \boldsymbol{\psi}). \tag{5.92}$$

The conditional density and the density of the random effect have been discussed above and we therefore have the log-likelihood

$$l(\boldsymbol{\beta}, \boldsymbol{\psi}, \boldsymbol{u}) = -\frac{1}{2}\log|\boldsymbol{\Sigma}| - \frac{1}{2}(\boldsymbol{y} - \boldsymbol{X\beta} - \boldsymbol{Zu})^T \boldsymbol{\Sigma}^{-1}(\boldsymbol{y} - \boldsymbol{X\beta} - \boldsymbol{Zu}) - \frac{1}{2}\boldsymbol{u}^T \boldsymbol{\Psi}^{-1}\boldsymbol{u}, \tag{5.93}$$

and the derivative of the log-likelihood wrt. the random effects becomes

$$\frac{\partial}{\partial \boldsymbol{u}} l(\boldsymbol{\beta}, \boldsymbol{\psi}, \boldsymbol{u}) = \boldsymbol{Z}^T \boldsymbol{\Sigma}^{-1}(\boldsymbol{y} - \boldsymbol{X\beta} - \boldsymbol{Zu}) - \boldsymbol{\Psi}^{-1}\boldsymbol{u}, \tag{5.94}$$

setting the derivative equal tp 0 and solving for $\boldsymbol{u}$, we find that $\hat{\boldsymbol{u}}$ is the solution to

$$(\boldsymbol{Z}^T \boldsymbol{\Sigma}^{-1} \boldsymbol{Z} + \boldsymbol{\Psi}^{-1})\boldsymbol{u} = \boldsymbol{Z}^T \boldsymbol{\Sigma}^{-1}(\boldsymbol{y} - \boldsymbol{X\beta}), \tag{5.95}$$

or explicitly we get

$$\hat{\boldsymbol{u}} = (\boldsymbol{Z}^T \hat{\boldsymbol{\Sigma}}^{-1} \boldsymbol{Z} + \hat{\boldsymbol{\Psi}}^{-1})^{-1} \boldsymbol{Z}^T \hat{\boldsymbol{\Sigma}}^{-1}(\boldsymbol{y} - \boldsymbol{X}\hat{\boldsymbol{\beta}}). \tag{5.96}$$

Where we have stressed that the parameters to construct variance-covariance matrices are estimated.

In addition to the above, we can also assess the uncertainty of the random effect estimates; this is done through the second derivative of the log-likelihood (the information matrix) wrt. the random effect

$$\frac{\partial}{\partial \boldsymbol{u}} l(\boldsymbol{\beta}, \boldsymbol{\psi}, \boldsymbol{u}) = -\boldsymbol{Z}^T \boldsymbol{\Sigma}^{-1} \boldsymbol{Z} - \boldsymbol{\Psi}^{-1}, \tag{5.97}$$

and the Fisher's information wrt. $\boldsymbol{u}$ becomes

$$\boldsymbol{I}(\hat{\boldsymbol{u}}) = \boldsymbol{Z}^T\boldsymbol{\Sigma}^{-1}\boldsymbol{Z}, \boldsymbol{\Psi}^{-1} \tag{5.98}$$

and finally the inverse Fisher information can be used as a variance estimate for the random effects, and thereby allows for assessing the uncertainty.

Example 5.15 *In order to illustrate the estimation of random effects, we look at a slightly modified version of the last example, namely the following formulation*

$$Y_{ijk} = \boldsymbol{x}_{ijk}^T\boldsymbol{\beta} + U_i + V_{ij} + \epsilon_{ijk}, \tag{5.99}$$

where $\boldsymbol{x}_{ijk}^T$ *is one row in the design matrix,* $U_i \sim N(0, \sigma_u^2)$*,* $V_{ij} \sim N(0, \sigma_v^2)$ *both iid, and*

$$\epsilon_{ijk} = \phi\epsilon_{ij,k-1} + \lambda_k; \quad \lambda_k \sim N(0, \sigma_\lambda^2) \quad \text{and iid,} \tag{5.100}$$

The interpretation here is that every individual has a preference (U_i) *but also that this preference might change for different days* (V_{ij}) *and finally that there is a correlation between the clothing insulation at one time of reporting to the next (on the same day). In terms of reported clothing insulation level, this implies that* $\boldsymbol{\Sigma}$ *is constructed as a block-diagonal with diagonal elements*

$$\boldsymbol{\epsilon}_{ij} = \begin{bmatrix} \epsilon_{ij1} \\ \epsilon_{ij1} \\ \vdots \\ \epsilon_{ijn_{ij}} \end{bmatrix} \sim N_{n_{ij}}(\boldsymbol{0}, \boldsymbol{\Sigma}^{ij}), \tag{5.101}$$

where $\boldsymbol{\Sigma}^{ij}$ *is a matrix with elements* $\boldsymbol{\Sigma}_{lk}^{ij} = \sigma^2\phi^{|k-l|}$*. The parameters for the model are* $\boldsymbol{\theta} = [\boldsymbol{\beta}, \sigma_u^2, \sigma_v^2, \sigma^2, \phi]$*. This exact structure is not implemented through* `glmmTMB`*, but is implemented for the specific problem using general likelihood estimation. The mean value structure is as in Example 5.3, and the result of the estimation is (variance components given as standard deviations);*

```
round(theta,digits=4)

##  beta.0  beta.1  beta.m sigma.u sigma.v   sigma     phi
##  0.6677 -0.0045 -0.0795  0.0974  0.0832  0.0676  0.5873

logLik19

## [1] 1030
```

We see an increase in the likelihood, but also that the correlation coefficient is somewhat lower; however, the correlation matrix as presented in Figure 5.3 is almost identical.

The model given here can be formulated as

$$\boldsymbol{Y} = \boldsymbol{X\beta} + \boldsymbol{Zu} + \boldsymbol{\epsilon}; \quad \boldsymbol{\epsilon} \sim N(\boldsymbol{0}, \boldsymbol{\Sigma}), \tag{5.102}$$

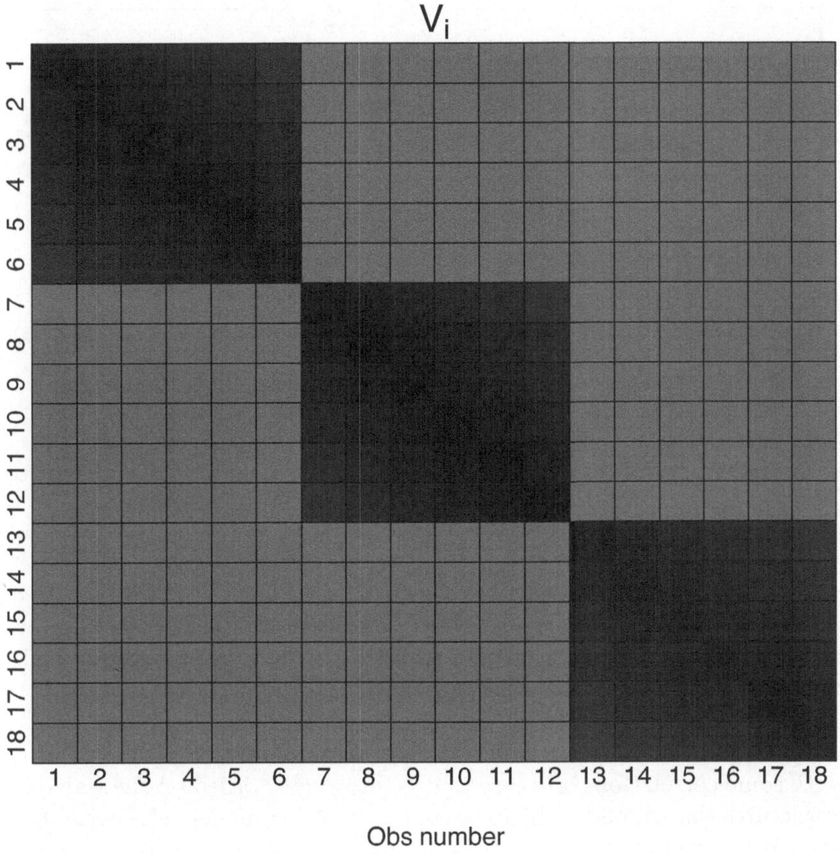

Figure 5.3: Plot of the correlation matrix for the random effect model presented in Example 5.14. The correlation matrix is for one subject that visits the lab 3 days and makes 6 records of clothing insulation level per day.

where $\boldsymbol{\Sigma}$ constructed from $\boldsymbol{\Sigma}^{ij}$ as described above and $\boldsymbol{Z}$ reflects the order of the random effect. Hence we can choose the structure we like and place correlations either on the residuals or between the random effects.

Figure 5.4 shows the random effects including two standard deviations for the random effect of subject and the random effect for days within subject. The uncertainty varies between subjects and days because the number of times each subject visits the lab differs and also the number of reported clothing insulation levels differs from day to day. We see that the variation between days is almost as large as the variation between subjects.

5.5 Concluding remarks

In this chapter, we have discussed linear mixed effect models that extend the general linear model by including normal random effects. This implies that we can describe random variation between subjects. The models also include nested variability, e.g. different persons have different preferences and the preferences may vary between days and even with a

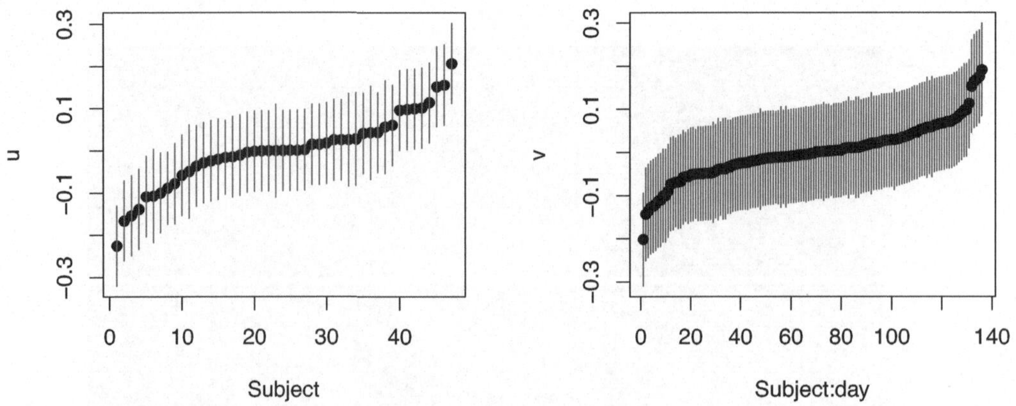

Figure 5.4: Visualisation of the random effects presented in Example 5.15. The random effects are sorted from lowest to highest and hence the x-axis does not refer to subject id. or subject/day id.

strong correlation of the within–day variability. Such models are useful for describing many types of occupant data where multiple observations on multiple subject are available.

The mixed effect models can be used to learn population parameters and at the same time estimate subject–specific parameters, which may be used for adapting the indoor environment for the general population or the individual user.

From a computational point of view, the presented models can be seen as normal linear models with some correlation structure that is introduced based on the way data is sampled and the structures that should be included in the resulting model. The normality assumption is of course restrictive in the same way as normality assumption is restrictive in the general linear model. In the next chapter, we will relax this assumption to deal with more general structures that include the generalized linear models with random effects.

6

Hierarchical models

In Chapter 5, we assumed that all random variables were Gaussian and that the models were linear. This implies that the likelihood can be calculated explicitly (though we may need a numerical optimiser to find the maximum likelihood estimate).

However, the distribution of outcome variables in the field of occupant's behaviour is often non-Gaussian. Therefore, in this chapter we will discuss non-linear and non-Gaussian random effect models. The main complication is, as we will see, that the likelihood can no longer be calculated explicitly, and the likelihood will hence be approximated. Some special cases where the likelihood can be calculated explicitly are treated in Section 6.4.

In general the models treated in the present chapter can be written as a first- and a second- stage model. The first–stage models is described by a conditional probability density or mass function

$$f_{\boldsymbol{Y}|\boldsymbol{U}=\boldsymbol{u}}(\boldsymbol{y};\boldsymbol{X},\boldsymbol{u},\boldsymbol{\beta}), \tag{6.1}$$

that is the conditional density of the observations depends on one more independent variables ($\boldsymbol{X}$), the random effects ($\boldsymbol{u}$) and the parameters ($\boldsymbol{\beta}$). The second–stage model is described by the probability density function

$$f_{\boldsymbol{U}}(\boldsymbol{u};\boldsymbol{\Phi}), \tag{6.2}$$

where $\boldsymbol{\Phi}$ presents the parameters for the random effects model, and the total set of parameters is denoted by $\boldsymbol{\theta} = (\boldsymbol{\beta}, \boldsymbol{\Phi})$.

In general, the likelihood is described by the marginal distribution of the observations

$$L(\boldsymbol{\theta}) = f_{\boldsymbol{Y}}(\boldsymbol{y};\boldsymbol{\theta}) = \int f_{\boldsymbol{Y}|\boldsymbol{U}=\boldsymbol{u}}(\boldsymbol{y};\boldsymbol{X},\boldsymbol{u},\boldsymbol{\beta}) f_{\boldsymbol{U}}(\boldsymbol{u};\boldsymbol{\Phi}) d\boldsymbol{u}. \tag{6.3}$$

The main complication is to solve the integral given in (6.3). An important example, of course, is represented by the linear mixed effect models covered in Chapter 5. The models treated in this chapter can be divided into 3 overall categories:

1. **Generalized linear mixed effect models (GLMM):** In this class of models, the first–stage model is a member of the exponential dispersion family (see Section 4.2), and $g(E[\boldsymbol{Y}|\boldsymbol{U}]) = \boldsymbol{X\beta} + \boldsymbol{ZU}$, and further $\boldsymbol{U} \sim N(\boldsymbol{0}, \boldsymbol{\Sigma})$. In addition there might be a dispersion parameter. This family of models is implemented in a number of different R-packages; we will use `glmmTMB` (Brooks et al., 2017) and `ordinal` (Christensen, 2022) here (see Section 6.2).

2. **General mixed effect models:** In principle, this class of models includes all structures of the first– and second– stage models, and the integral needs to be solved by some numerical procedure; we use the Laplace approximation (Wolfinger and Lin, 1997) (the Laplace approximation is treated in Section 6.1.2 and general mixed effect models are treated in Section 6.3).

DOI: 10.1201/9781003340812-6

3. **Hierarchical models with an explicit solution:** In some cases, the second–stage model can be chosen such that an explicit solution to the likelihood can be given. We demonstrate some examples in Section 6.4. Estimation in this class of models is much faster than using the approximations we present for General Mixed Effect Models.

The use of hierarchical models can be motivated by the examples that we have covered in Chapters 3-5, as illustrated below.

Example 6.1 (Window opening) *We have discussed modelling of window opening using aggregated data in Chapter 4 (e.g. Examples 4.1 and 4.8). In these examples, the probability of window opening is either modelled for a specific dwelling or the variability between dwellings is included in the general variability in the binomial model possibly with over-dispersion. It does, however, seem reasonable that those probabilities depend on the dwelling or the persons in the dwellings. Models with a dwelling-specific intercept can be modelled using a GLMM (Example 6.14), and examples where the timing of window opening is included as a random effect can be modelled using a general mixed effect model (Example 6.16), and finally special examples of the model structure allow for an explicit solution.*

Example 6.2 (Ceiling fan speed) *We examined an example of fan speed setting in Chapter 4 (Example 4.26). However, at that time, we were ignoring the variability between subjects. Such variability expresses the person–to–person (interpersonal) variability and might be important for predicting preferences of a random person. In Section 6.2.2 (Example 6.15), the ordinal regression model is expanded to include random effects.*

Example 6.3 (Clothing insulation, count data) *In Chapter 4, we modelled the number of times that a subject changed clothing insulation levels during a day using Poisson and binomial distributions (Example 4.19). In the present chapter, we analyse the data including interpersonal variability, using GLMM (Example 6.6), and also present models where the integral allows for an explicit solution (Examples 6.20 and 6.21).*

Example 6.4 (Clothing insulation, count data) *The clothing insulation level data has been analysed a number of times throughout the book, using a general linear model (Chapter 3), a generalized linear model with gamma-distributed outcomes (Chapter 4), and including subject–specific random effects, between–days variation (nested to subjects), and within day autocorrelation (Chapter 5). The results are combined in the present chapter, i.e. gamma-distributed outcomes including subject–specific random effects, between–days variation and within–day autocorrelation (Example 6.13).*

In addition we have seen that the variation between males and females is quite different and one could hypothesise that the variation in clothing insulation level varies greatly between subjects, which is examined in Example 6.22.

Before we discuss the three specific model classes introduced above, we will present the overall structure that all the classes rely on.

6.1 Formulation of hierarchical model

In this section, we will discuss the general formulation of hierarchical models. The model formulation is, as discussed above, given by the *first–stage* and *second–stage* models.

The *first–stage model* is

$$f_{Y|u}(\boldsymbol{y}; \boldsymbol{u}, \boldsymbol{\beta}), \tag{6.4}$$

and *second–stage model* is

$$f_U(\boldsymbol{u}; \boldsymbol{\Psi}), \tag{6.5}$$

which is a model for the random effects. The total set of parameters is $\boldsymbol{\theta} = (\boldsymbol{\beta}, \boldsymbol{\Psi})$. Let us first recall the Gaussian case.

Example 6.5 (Clothing insulation level) *In Chapter 5, we studied random effect models for clothing insulation level and the* first–stage *model is in this case*

$$\boldsymbol{Y}|\boldsymbol{U} = \boldsymbol{u} \sim N(\boldsymbol{X\beta} + \boldsymbol{Zu}, \boldsymbol{\Sigma}(\psi)) \tag{6.6}$$

that is, $f_{Y|u}(\boldsymbol{y}; \boldsymbol{u}, \boldsymbol{\beta}, \phi)$ *is the multivariate Gaussian density, and the* second–stage *model is*

$$\boldsymbol{U} \sim N(\boldsymbol{0}, \boldsymbol{\Sigma}_u(\boldsymbol{\Psi})) \tag{6.7}$$

i.e. $f_U(\boldsymbol{u}; \boldsymbol{\Psi})$ *is also the multivariate Gaussian density, and the total set of parameters is* $\boldsymbol{\theta} = (\boldsymbol{\beta}, \psi, \boldsymbol{\Psi})$.

Using the first– and second– stage models we can write the joint density of $\boldsymbol{Y}$ and $\boldsymbol{U}$ as

$$f_{Y,U}(\boldsymbol{y}, \boldsymbol{u}; \boldsymbol{\beta}, \psi, \boldsymbol{\Psi}) = f_{Y|u}(\boldsymbol{y}; \boldsymbol{u}, \boldsymbol{\beta}, \psi) f_U(\boldsymbol{u}; \boldsymbol{\Psi}). \tag{6.8}$$

This is the joint likelihood given both the observations and the random effect, which we write as

$$L(\boldsymbol{\beta}, \psi, \boldsymbol{\Psi}; \boldsymbol{u}, \boldsymbol{y}) = f_{Y|u}(\boldsymbol{y}; \boldsymbol{u}, \boldsymbol{\beta}, \psi) f_U(\boldsymbol{u}; \boldsymbol{\Psi}). \tag{6.9}$$

As we do not observe the random effects, we need to formulate the marginal likelihood. To this end, we need the marginal distribution of the observations, i.e.

$$f_Y(\boldsymbol{y}; \boldsymbol{\beta}, \psi\boldsymbol{\Psi}) = \int_{\mathbb{R}^q} f_{Y|u}(\boldsymbol{y}; \boldsymbol{u}, \boldsymbol{\beta}, \psi) f_U(\boldsymbol{u}; \boldsymbol{\Psi}) d\boldsymbol{u}, \tag{6.10}$$

where q is the dimension of the random effects. Written in terms of likelihood function, we have

$$L_M(\boldsymbol{\beta}, \psi, \boldsymbol{\Psi}; \boldsymbol{y}) = \int_{\mathbb{R}^q} L(\boldsymbol{\beta}, \psi, \boldsymbol{\Psi}; \boldsymbol{u}, \boldsymbol{y}) d\boldsymbol{u}. \tag{6.11}$$

This integral is the main complication for dealing with general random effect models. Except for special cases like the mixed effect models in Chapter 5, closed–form solutions for the integral are not available. If the random effects are high dimensional, it becomes very difficult to deal with this integral. In the first part of this chapter (Sections 6.2-6.3), we will focus on approximations using the Laplace approximation for solving the integral, while the last part of the chapter (Section 6.4) will focus on some special examples where explicit solutions for the integral are available.

Below we present an example of a non-Gaussian first–stage model and a Gaussian second–stage model.

Example 6.6 (Change in clothing insulation level) *In Example 4.19, we modelled the number of changes in clothing insulation level over 1 day. In that (and other examples), we saw that individuals actually have an effect and the possibility of overdispersion was discussed. Another way of modelling the dispersion is to include a random effect,*

$$Y_i|U_i \sim Binom(n_i, p_i), \tag{6.12}$$

with

$$\begin{aligned} p_i =& \frac{e^{\boldsymbol{x}_i^T\boldsymbol{\beta}+U_i}}{1+e^{\boldsymbol{x}_i^T\boldsymbol{\beta}+U_i}} \\ U_i \sim& N(0,\sigma^2). \end{aligned} \tag{6.13}$$

Hence the marginal likelihood is given by

$$\begin{aligned} L_M(\boldsymbol{\beta},\sigma^2;\boldsymbol{y}) =& \int_{\mathbb{R}^q} L(\boldsymbol{\beta},\sigma^2,\boldsymbol{u};\boldsymbol{y})d\boldsymbol{u} \\ =& \int_{\mathbb{R}^q} \prod_{i=1}^{q}\prod_{j=1}^{d_i} \binom{n_{i,j}}{y_{i,j}} p(\boldsymbol{x}_{i,j};\boldsymbol{\beta},u_i)^{y_{i,j}}(1-p(\boldsymbol{x}_{i,j};\boldsymbol{\beta},u_i))^{n_{i,j}-y_{i,j}} \times \\ & \frac{1}{\sqrt{2\pi\sigma^2}} e^{\frac{u_i^2}{\sigma^2}} du_1 \cdots du_q, \end{aligned} \tag{6.14}$$

where q is the number of subjects, d_i is the number of times, subject i visited the LOBSTER, i.e. the number of experimental days of that subject, $n_{i,j}$ is the number of observations made of subject i on day j, $y_{i,j}$ is the number of times the clothing insulation level changed between observations, $\boldsymbol{x}_{i,j}$ is the explanatory variable (e.g. sex) and u_i is the effect of subject i.

There is no explicit solution to the integral above, and we will therefore need numerical methods. Further the dimension of the integral is q; in our specific case, this is 47, which makes direct numerical solution of the integral challenging (or impossible).

One method to approximate the integral is to use the Laplace approximation which we will discuss below. At the same time, some simplifications can be obtained by considering the grouping structures, which we will discuss first.

6.1.1 Grouping structures and nested effects

The complexity of evaluating the marginal likelihood (6.11) is mostly related to the dimension of the integral (i.e. q), and hence large computational savings may be obtained if the structure of random effects implies that we can integrate over lower dimensions. For problems with only one level of grouping, the marginal likelihood can be simplified as

$$L_M(\boldsymbol{\beta},\boldsymbol{\Psi};\boldsymbol{y}) = \prod_{i=1}^{M}\int_{\mathbb{R}^{q_i}} f_{Y|u_i}(\boldsymbol{y};\boldsymbol{u}_i,\boldsymbol{\beta})f_{U_i}(\boldsymbol{u_i};\boldsymbol{\Psi})\,d\boldsymbol{u}_i. \tag{6.15}$$

The main purpose of the exercise is to reduce the dimension of the integral in (6.11), i.e. it is cheaper to solve M low-dimensional integrals than one high dimensional integral. Again we refer to the clothing insulation examples in Chapter 5. Here, the subjects were assumed independent and therefore the integral can be done for each subject rather than for all subjects at the same time.

Example 6.7 (Change in clothing insulation level) *In Example 6.6 above, the grouping structure is defined by subjects and the integral can be written as*

$$L_M(\boldsymbol{\beta}, \sigma^2; \boldsymbol{y}) = \prod_{i=1}^{q} \int_{\mathbb{R}} \prod_{j=1}^{d_i} \binom{n_{i,j}}{y_{i,j}} p(\boldsymbol{x}_{i,j}; \boldsymbol{\beta}, u_i)^{y_{i,j}} (1 - p(\boldsymbol{x}_{i,j}; \boldsymbol{\beta}, u_i))^{n_{i,j} - y_{i,j}} \times \\ \frac{1}{\sqrt{2\pi\sigma^2}} e^{\frac{u_i^2}{\sigma^2}} du_i. \tag{6.16}$$

Hence rather than one 47-dimensional integral, we have 47 one-dimensional integrals, which is much simpler to handle.

Considerations on grouping structures are relevant regardless of how the integral is solved, i.e. no matter if the Laplace approximation is used or if the integral is solved by another numerical method.

6.1.2 Laplace approximation

After our excurs to grouping structures, we will get the Laplace approximation. The Laplace approximation is used for approximating the integral in (6.11) or (6.15). To this end, let us consider the log-joint likelihood

$$l(\boldsymbol{\theta}, \boldsymbol{u}; \boldsymbol{y}) = \log f_{Y|u}(\boldsymbol{y}; \boldsymbol{u}, \boldsymbol{\beta}) + \log f_U(\boldsymbol{u}; \boldsymbol{\Psi}), \tag{6.17}$$

where $\boldsymbol{\theta} = (\boldsymbol{\beta}, \boldsymbol{\Psi})$. For the Laplace approximation, we will need the likelihood with respect to the random effects for fixed parameters, that is

$$l(\boldsymbol{u}; \boldsymbol{\theta}, \boldsymbol{y}) = \log f_{Y|u}(\boldsymbol{y}; \boldsymbol{u}, \boldsymbol{\beta}) + \log f_U(\boldsymbol{u}; \boldsymbol{\Psi}). \tag{6.18}$$

This is shown in the example below.

Example 6.8 (Change in clothing insulation level) *Using Example 6.7, we have (after applying the grouping structure in previous section)*

$$l(u_i; \boldsymbol{y}, \boldsymbol{\theta}) = \sum_{j=1}^{d_i} y_{ij} \log(p(\eta_{ij}, u_i)) + (n_{ij} - y_{ij}) \log(1 - p(\eta_{ij}, u_i)) - \frac{u_i^2}{\sigma^2} + \\ c(n_{ij}, y_{ij}, \sigma^2), \tag{6.19}$$

where $\eta_i = \boldsymbol{x}_{ij}^T \boldsymbol{\beta}$, $c(\cdot)$ *is a constant not depending on* u_i, *and*

$$p(\eta_{ij}, u_i) = \frac{e^{\eta_{ij} + u_i}}{1 + e^{\eta_{ij} + u_i}}, \tag{6.20}$$

In order to implement the Laplace approximation, we need to implement the negative log-likelihood function as an R-function;

```
nll.u <- function(u, eta, y, n, sigma){
    p <- exp(eta + u)/(1 + exp(eta + u))
    - sum(dbinom(y, size = n, prob = p, log =TRUE)) -
        dnorm(u, mean=0, sd = sigma, log = TRUE)
}
```

The idea is to approximate the joint log-likelihood by a second order Taylor approximation around the optimal $\boldsymbol{u}$ (given the parameters in $\boldsymbol{\theta}$), formally

$$\tilde{\boldsymbol{u}}(\boldsymbol{\theta}) = \arg\max_{\boldsymbol{u}} l(\boldsymbol{u}; \boldsymbol{y}, \boldsymbol{\theta}), \tag{6.21}$$

where we have emphasised that $\tilde{\boldsymbol{u}}$ depends on $\boldsymbol{\theta}$, but we will suppress this dependence in the following. The second–order Taylor expansion of the joint log-likelihood around $\tilde{\boldsymbol{u}}$ is given by (note that $\frac{\partial l}{\partial \boldsymbol{u}}|_{\boldsymbol{u}=\tilde{\boldsymbol{u}}} = \boldsymbol{0}$)

$$\ell(\boldsymbol{\theta}, \boldsymbol{u}; \boldsymbol{y}) \approx \ell(\boldsymbol{\theta}, \tilde{\boldsymbol{u}}, \boldsymbol{y}) - \frac{1}{2}(\boldsymbol{u} - \tilde{\boldsymbol{u}})^T \boldsymbol{H}(\tilde{\boldsymbol{u}})(\boldsymbol{u} - \tilde{\boldsymbol{u}}), \tag{6.22}$$

where $\boldsymbol{H}(\tilde{\boldsymbol{u}})$ is the Hessian matrix of the joint log-likelihood taken in $\tilde{\boldsymbol{u}}$.

Hence the marginal likelihood in (6.11) can be approximated by

$$\begin{aligned} L_{M,LA}(\boldsymbol{\beta}, \boldsymbol{\Psi}; \boldsymbol{y}) &= \int_{\mathbb{R}^q} \exp\left(\ell(\boldsymbol{\theta}, \tilde{\boldsymbol{u}}, \boldsymbol{y}) - \frac{1}{2}(\boldsymbol{u} - \tilde{\boldsymbol{u}})^T \boldsymbol{H}(\tilde{\boldsymbol{u}})(\boldsymbol{u} - \tilde{\boldsymbol{u}})\right) d\boldsymbol{u} \\ &= e^{\ell(\boldsymbol{\theta}, \tilde{\boldsymbol{u}}, \boldsymbol{y})} \int_{\mathbb{R}^q} \exp\left(-\frac{1}{2}(\boldsymbol{u} - \tilde{\boldsymbol{u}})^T \boldsymbol{H}(\tilde{\boldsymbol{u}})(\boldsymbol{u} - \tilde{\boldsymbol{u}})\right) d\boldsymbol{u}. \end{aligned} \tag{6.23}$$

The expression under the integral is recognised as the kernel of the multivariate Gaussian (with the variance-covariance matrix $\boldsymbol{H}(\tilde{\boldsymbol{u}})^{-1}$), and hence the Laplace approximation can be written as

$$L_{M,LA}(\boldsymbol{\beta}, \boldsymbol{\Psi}; \boldsymbol{y}) = e^{\ell(\boldsymbol{\theta}, \tilde{\boldsymbol{u}}, \boldsymbol{y})} \left| \frac{\boldsymbol{H}(\tilde{\boldsymbol{u}})^{-1}}{2\pi} \right|. \tag{6.24}$$

Using the approximations in (6.24) and in (6.11), the Laplace approximation of the marginal log-likelihood becomes

$$\ell_{M,LA}(\boldsymbol{\theta}, \boldsymbol{y}) = \ell(\boldsymbol{\theta}, \tilde{\boldsymbol{u}}, \boldsymbol{y}) - \tfrac{1}{2} \log \left| \frac{\boldsymbol{H}(\tilde{\boldsymbol{u}})}{2\pi} \right|. \tag{6.25}$$

We see that the integral in (6.15) is replaced by an optimisation and an evaluation of the Hessian matrix. It is important to note that regardless of the approximation above, it is important to use grouping structures if relevant, in particular if numerical methods are used for evaluation of the Hessian matrix and the optimisation.

The maximum likelihood estimate of the model parameters is given by

$$\hat{\boldsymbol{\theta}} = \arg\max_{\boldsymbol{\theta}} \ell_{M,LA}(\boldsymbol{\theta}, \boldsymbol{y}). \tag{6.26}$$

Estimation of random effects

Using the Laplace approximation, the random effects are estimated as part of the procedure itself and hence the estimated random effects are given by

$$\hat{\boldsymbol{u}} = \tilde{\boldsymbol{u}}(\hat{\boldsymbol{\theta}}). \tag{6.27}$$

Further, as we have seen, the variance estimate of the conditional random effect random effects is given by

$$\hat{V}[\boldsymbol{u}|\boldsymbol{Y} = \boldsymbol{y}] = \boldsymbol{H}_{\boldsymbol{y}}(\hat{\boldsymbol{u}})^{-1}, \tag{6.28}$$

which is also calculated directly in the Laplace approximation we will omit the subscript $\boldsymbol{y}$ in the developments of the models and take it as implicit. The example below demonstrates an example of an implementation of the Laplace likelihood in R.

Example 6.9 (Laplace log-likelihood) *Continuing Example 6.8, the R-function to calculate the Laplace approximation for the marginal negative log-likelihood is given below. Notice, that in this set up, the variance of the random effects is estimated in the log-domain.*

In the function given below, the first lines (before the `for`*-loop) initialise parameters. The main part of the algorithm is the* `for`*-loop over the subjects. Since subjects are independent, the optimisation can also be done independently, and hence we have in this case 47 one-dimensional optimisations, and also because of the independence between random effects, the Hessian matrix is a diagonal matrix and therefore*

$$|\boldsymbol{H}(\boldsymbol{u})| = \prod_{i=1}^{q} \boldsymbol{H}_{ii}(\boldsymbol{u}), \tag{6.29}$$

and hence the contributions can be treated individually.

When calling the function, one can choose to have only the negative log-*likelihood returned (*`full=FALSE`*); this is used when the function is used for finding the parameter estimates. When the parameters have been found, the random effects (and their variance) can be found by calling the function with* `full=TRUE`.

```
## Laplace app. negative log-likelihood
nll.la <- function(theta, X, y, n, subj.lev, subjId, full = FALSE){
    beta <- theta[1:dim(X)[2]] ## Fixed effect parameters
    ## Random effect parameters
    sigma <- exp(theta[-(1:dim(X)[2])]/2)
    eta <- as.numeric(X %*% beta) ## Linear predictor
    nll <- 0 ## Initalize negative marginal log-likelihood
    u <- numeric(length(subj.lev))
    ## For loop over random effects
    for(i in 1:length(subj.lev)){
        I <- subjId == subj.lev[i] ## Choose subject
        ## Optimise random effect for given parameters
        opt <- nlminb(0, nll.u, eta = eta[I], y = y[I],
                      n = n[I], sigma = sigma)
        ## Extract random effect and Hessian
        ## of random effect likelihood
        u[i] <- opt$par
        H[i] <- hessian(nll.u, u[i], eta = eta[I], y = y[I],
                        n = n[I], sigma = sigma)
        ## Calculate marginal negative log-likelihood
        nll <- nll + opt$objective + 0.5 * log(as.numeric(H[i])) -
            0.5 * log(2 * pi)
    }
    ## If full = TRUE return marginal likelihood and random effects
    ## including variance of random effects
    if(full){return(list(ll=-nll,u=u,Vu=1/H))}
    nll
}
```

Notice, in the above example, that the Laplace approximation requires an inner optimisation that replaces the calculation of the integral over the random effect. In the simple example above, it would actually be possible to numerically calculate the integral, but for more complex examples, this is not possible. Notice also that the grouping structure is used here by replacing the 47-dimensional integral with 47 one-dimensional optimisations (the `for`-loop).

The final step is the outer optimisation.

Example 6.10 (Estimate parameters) *The last step needed to complete Example 6.9 is to run the functions that we have now formulated. Therefore, we need the initial set of the inputs and the first step is to set up the design matrix for the example.*

```
## Design matrix
X <- cbind(1,  dfcount$sex=="male")
```

With the initial set up, we can now find the parameters in the model using an outer optimisation;

```
## Outer optimisation:
fit <- nlminb(c(0,0,0), nll.la, X = X, y = y, n = n,
              subj.lev = subj.lev, subjId=dfcount$subjId )

## Find standard errors and report results
H <- hessian(nll.la, fit$par,  X = X, y = y, n=n,
             subj.lev = subj.lev, subjId=dfcount$subjId)
se <- sqrt(diag(solve(H)))

##         Est lower upper glm-pars
## beta0 -1.79 -2.28 -1.29     -1.6
## beta1 -1.35 -2.17 -0.54     -1.2
## sigma  0.91  0.55  1.49       NA
```

The last column shows the results using a similar generalized linear model (see Example 4.19). We can observe that the mean value parameters changes; hence the random effect model has slightly different mean value parameters than the over-dispersion model. The main effect here is that the random effect is a normal random variable inside a non-linear function and that

$$E\left[\frac{e^{\beta_0(sex)+U}}{1+e^{\beta_0(sex)+U}}\right] \neq \frac{e^{\beta_0(sex)+E[U]}}{1+e^{\beta_0(sex)+E[U]}}. \tag{6.30}$$

This point is illustrated below, where the observed probabilities are compared with the generalized linear model and the result of the random effect model. The column `glmm.fix` *is the right-hand side of* (6.30) *while* `glmm.boot` *is the left-hand side of* (6.30) *found by simulation from the estimated normal distribution.*

```
##              obs glmm.fix glmm.boot  glm
## (Intercept) 0.17    0.144     0.176 0.17
## sexmale     0.06    0.041     0.058 0.06
```

We see that while the `glm` *gives the observed probabilities, the random effect model, using only the fixed effects, does not. However, if we bootstrap from the estimated model, the result is very close to the observed probabilities. This illustrates that the parameters from the generalized linear mixed effect model cannot directly be interpreted as probabilities in the population.*

As a final remark, we can compare the likelihood of the two methods;

```
##  glm.logLik glmm.logLik
##        -135        -129
```

We see that the difference in likelihood is about 6 on 1 degree of freedom, which is significant on any reasonable level (i.e. the p-value << 0.05). Implying that the GLMM gives a significantly better descriptions of the data, this means that the random effects (i.e. a model for individuals) are important for the modelling of these data.

6.1.3 Interpretation of parameters

Example 6.10 illustrates that care should be taken when interpreting the parameters from a GLMM. The key complication is that

$$E[g(U)] \neq g(E[U]), \tag{6.31}$$

and in our particular setting

$$E[g(\eta + U)] \neq g(\eta + E[U]). \tag{6.32}$$

Hence the linear predictor will have a different interpretation in the generalized linear mixed effect model compared to the generalized linear model, where, as we saw in Example 6.10, the parameters can be compared directly to averages in data.

Two-level hierarchical model

In summary, the joint log-likelihood function is

$$\ell(\boldsymbol{\theta}, \boldsymbol{u}, \boldsymbol{y}) = \ell(\boldsymbol{\beta}, \boldsymbol{\Psi}, \boldsymbol{u}, \boldsymbol{y}) = \log f_{Y|u}(\boldsymbol{y}; \boldsymbol{u}, \boldsymbol{\beta}) + \log f_U(\boldsymbol{u}; \boldsymbol{\Psi}), \tag{6.33}$$

and can be approximated by the Laplace approximation

$$\ell_{M,LA}(\boldsymbol{\theta}, \boldsymbol{y}) = \log f_{Y|u}(\boldsymbol{y}; \tilde{\boldsymbol{u}}, \boldsymbol{\beta}) + \log f_U(\tilde{\boldsymbol{u}}; \boldsymbol{\Psi}) - \frac{1}{2}\log\left|\frac{\boldsymbol{H}(\tilde{\boldsymbol{u}})}{2\pi}\right|. \tag{6.34}$$

The Laplace approximation works for much more general models than presented in the example above, e.g. Gaussian models with non-linear (in the parameters) mean-value functions as well as non-Gaussian models. We will consider some examples in Section 6.3.

6.1.4 Automatic differentiation

The Laplace approximation includes an "inner" optimisation in a potentially high dimensional space, and hence large improvements in terms of calculation speed can be obtained if we would have access to the derivative of the log-likelihood wrt. the random effects. Automatic differentiation is a process where the likelihood is broken down to a long chain of simple operations that can be differentiated by a computer program. This is implemented in e.g. `TMB`. In addition to the increase in speed, the results will also be more accurate if we have access to a function that calculates the derivative. The computational cost of finding the derivative is quite low in particular if the number of parameters is large, see Madsen and Thyregod (2011) for a discussion.

Example 6.11 (Normal-Poisson model) *To illustrate the principle, we again consider the model and data in Example 6.8, the derivative of $l(u_i; \boldsymbol{y}, \boldsymbol{\theta})$ is (we use the notation $p(\eta_{ij}, u_i) = p_{ij}$)*

$$\begin{aligned}\frac{\partial l(u_i; \boldsymbol{y}, \boldsymbol{\theta})}{\partial u_i} &= \sum_{j=1}^{d_i} \frac{y_{ij}}{p_{ij}} \frac{\partial p_{ij}}{\partial u_i} - \frac{n_{ij} - y_{ij}}{1 - p_{ij}} \frac{\partial p_{ij}}{\partial u_i} - \frac{u_i}{\sigma^2} \\ &= \sum_{j=1}^{d_i} \frac{\partial p_{ij}}{\partial u_i} \left(\frac{y_{ij}(1 - p_{ij}) - p_{ij}(n_{ij} - y_{ij})}{p_{ij}(1 - p_{ij})} \right) - \frac{u_i}{\sigma^2} \\ &= \sum_{j=1}^{d_i} \frac{\partial p_{ij}}{\partial u_i} \left(\frac{y_{ij} - p_{ij} n_{ij}}{p_{ij}(1 - p_{ij})} \right) - \frac{u_i}{\sigma^2},\end{aligned} \tag{6.35}$$

the derivative of p_{ij} can be calculated as

$$\begin{aligned}\frac{\partial p_{ij}}{\partial u_i} &= \frac{\partial}{\partial u_i} \frac{e^{\eta_{ij} + u_i}}{1 + e^{\eta_{ij} + u_i}} \\ &= \frac{e^{\eta_{ij} + u_i}}{1 + e^{\eta_{ij} + u_i}} - \frac{e^{2\eta_{ij} + 2u_i}}{(1 + e^{\eta_{ij} + u_i})^2} \\ &= p_{ij}(1 - p_{ij}),\end{aligned} \tag{6.36}$$

and hence

$$\frac{l(u_i; \boldsymbol{y}, \boldsymbol{\theta})}{\partial u_i} = \sum_{j=1}^{d_i} y_{ij} - p_{ij} n_{ij} - \frac{u_i}{\sigma^2}. \tag{6.37}$$

The Hessian with respect to u_i is given by

$$\frac{\partial^2 l(u_i; \boldsymbol{y}, \boldsymbol{\theta})}{\partial u_i^2} = -\sum_{j=1}^{d_i} p_{ij}(1 - p_{ij}) n_{ij} - \frac{1}{\sigma^2}. \tag{6.38}$$

The expressions in (6.37) *and* (6.38) *can be implemented in R by;*

```
## derivative of nll.u
dnll.u <- function(u, eta, y, n, sigma){
    p <- exp(eta + u)/(1 + exp(eta + u))
    - sum(y - n * p) +  u / sigma^2
}

## Hessian matrix of nll.u
Hnll.u <- function(u, eta, y, n, sigma){
    p <- exp(eta + u)/(1 + exp(eta + u))
    matrix(sum( p * (1 - p) * n) +  1 / sigma^2)
}
```

Further the R-function `nll.la` *(see Example 6.9, on page 151) needs to be modified as (only modified lines are shown).*

```
nll.la.Deriv <- function(theta,...)
## ...
        ## Optimise random effect for given parameters
        opt <- nlminb(0, nll.u, gradient = dnll.u,
                      hessian = Hnll.u, eta = eta[I], y = y[I],
                      n = n[I], sigma = sigma)
        ## Extract random effect and Hessian
        ## of random effect likelihood
        u[i] <- opt$par
        H[i] <-  Hnll.u(u[i], eta = eta[I], y = y[I],
                        n = n[I], sigma = sigma)
## ...
```

Hence we see that `nlminb` *takes the derivative and the Hessian matrix as input, and uses that in the inner optimisation. Optimisation using the two functions (*`nll.la` *and* `nll.la.Deriv`*) give almost identical output, but using* `nll.la.Deriv` *is about 2 times faster.*

As illustrated in the example above, it is in some cases straightforward to implement the first and second derivatives and by that get improved speed and accuracy. For more complicated structures, it might be quite hard to implement the derivatives and in such cases automatic differentiation is an effective way of getting improved speed.

In the R-package `TMB`, the Laplace approximation is implemented as part of the optimisation procedure. This step implies that we (as modellers) do not have to think about implementing the Laplace approximation. The R-package `TMB` also includes automatic differentiation of the likelihood. Hence we can focus on the formulation of the likelihood, i.e. the model. This formulation is done as `C++` code as shown, later on, in Section 6.3.

6.2 Generalized linear mixed effect models

We will refer to generalized linear models with Gaussian random effects added to the linear predictor as generalized linear mixed effect models (GLMMs). The structure in this class of models is very similar to the generalized mixed effect models, as discussed in Chapter 4, i.e.

$$\begin{aligned} \boldsymbol{\mu} &= g^{-1}(\boldsymbol{X\beta} + \boldsymbol{ZU}) \\ \boldsymbol{U} &\sim N(\boldsymbol{0}, \boldsymbol{\Psi}), \end{aligned} \tag{6.39}$$

and the density of $\boldsymbol{Y}$ being a member of the exponential family. This class of models is implemented in a number of different R-packages. Here we will use the package `glmmTMB`, which uses automatic differentiation and Laplace approximation in the optimisation as described above.

Example 6.12 (Change in clothing insulation level) *The model that we implemented in Example 6.6–6.10 can be estimated directly using* `glmmTMB` *as follows;*

```
## library(glmmTMB)
summary(fit.tmb.binom <- glmmTMB(cbind(clo, nobs - clo) ~ sex +
                                 (1 | subjId), family = binomial,
                             data = dfcount))

## [...]
## Conditional model:
##  Groups Name        Variance Std.Dev.
##  subjId (Intercept) 0.82     0.905
## Number of obs: 136, groups:  subjId, 47
##
## Conditional model:
##             Estimate Std. Error z value Pr(>|z|)
## (Intercept)   -1.786      0.254   -7.03  2.1e-12 ***
## sexmale       -1.354      0.416   -3.25   0.0011 **
## ---
## [...]
```

We see here that the syntax is the same as we saw in `lme4`. *We can also note that the effect of male is -1.354 while the standard deviation of the random effect is 0.905, and hence the variation between subjects is comparable to the difference between males and females. As discussed in Chapter 4 (Example 4.19), it is also reasonable to model these data as Poisson outcomes, which is implemented as*

```
fit.tmb.pois <- glmmTMB(clo ~ sex + offset(log(time)) + (1 | subjId),
                        family = poisson, data = dfcount)
```

Comparing the log-likelihood for the two models to the results in Chapter 4 reveals

```
## l.binom.glmm  l.pois.glmm  l.binom.glm   l.pois.glm
##         -129          -130          -135         -133
```

We see that a) the binomial-normal model (`l.binom.glmm`*) has a slightly higher likelihood than the normal-Poisson model (*`l.pois.glmm`*) but also that the difference is small, and b) the likelihood is in both cases larger than the likelihood for the fixed effect models (*`l.binom.glm` *and* `l.pois.glm`*).*

In Chapter 4, we discussed over-dispersion that includes more dispersion than would be expected from a model (e.g. binomial or Poisson). While over-dispersed models push the extra variation in the data to the parameter uncertainty and hence help us to give an honest assessment of the uncertainty, they do not explain where the extra uncertainty originates from. In a random effect model, the over-dispersion is explained by the structure in data (in the example above subjects). Referring to Example 6.12, Figure 6.1 shows the distribution of the random effect for males and females in the linear domain and in the probability domain. While we saw that the difference between males and females was significant, it is also clear that the distributions of males and females have a large overlap.

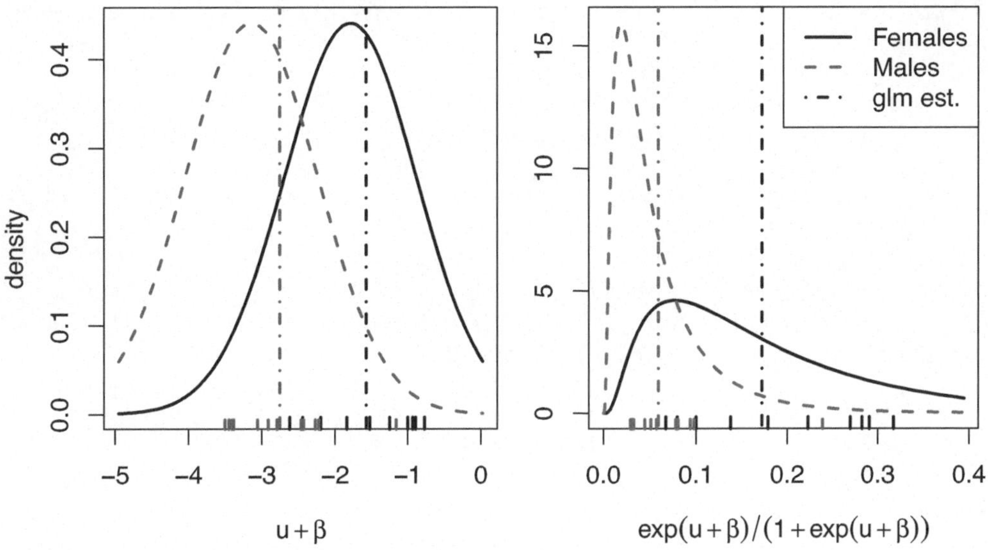

Figure 6.1: Distribution of linear predictor (left) and probability (right) in Example 6.12. Rugs show the estimated random effects.

6.2.1 More structures

In Chapter 5, we discussed both nested random effects and correlation structures related to the time between measurements. We can extend the model to a glmm by, for example

$$\eta_{ijk} = \boldsymbol{x}_{ijk}^T \boldsymbol{\beta} + u_i + v_{ijk}, \tag{6.40}$$

where $u_i \sim N(0, \sigma_u^2)$, and $\boldsymbol{v}_{ij\cdot} = [v_{ij1}, ..., v_{ijn_{ij}}] \sim N(\boldsymbol{0}, \boldsymbol{\Sigma}_v)$, with $\boldsymbol{\Sigma}_v$ defined by a correlation structure like $(\boldsymbol{\Sigma}_v)_{kl} = \rho^{|k-l|}$. This implies that the linear predictor for subject i can be written as

$$\boldsymbol{\eta}_{i\cdot\cdot} \sim N(\boldsymbol{X}_i \boldsymbol{\beta}, \boldsymbol{\Sigma}_i), \tag{6.41}$$

where $\boldsymbol{\Sigma}_i$ is given by $(n_j = 2)$

$$\boldsymbol{\Sigma}_i = \begin{bmatrix} \sigma_u^2 \boldsymbol{J}_{11} + \boldsymbol{\Sigma}_v^1 & \sigma_u^2 \boldsymbol{J}_{12} \\ \sigma_u^2 \boldsymbol{J}_{21} & \sigma_u^2 \boldsymbol{J}_{22} + \boldsymbol{\Sigma}_v^1 \end{bmatrix}, \tag{6.42}$$

and $\boldsymbol{J}_{ij}$ are matrices of ones with appropriate dimensions, and finally the expected value of the observations (first–stage model) is

$$E[\boldsymbol{Y}|U = u_i, \boldsymbol{V} = \boldsymbol{v_{ij\cdot}}] = g(\boldsymbol{\eta}_{i\cdot\cdot}). \tag{6.43}$$

The concept is illustrated using the clothing insulation example that we have used a number of times throughout Chapters 3–5.

Example 6.13 (Clothing insulation level) *In Chapter 3, we found that transformations of the dependent variable gave better models in terms of likelihood. In Chapter 4, we found that the gamma distribution, using the log-link, gave better results than most of the general linear models (except when applying quite complicated transformations). Finally, in Chapter 5, we saw that including subjects as random effects and* *within day autocorrelation increased the model performance. The results from Chapter 3 and 5 can be combined in the generalized mixed effect set up described here. Such a model is*

$$\begin{aligned}\log(\mu_{ijk}) =& \beta_0(\texttt{sex}_i) + \texttt{tOut}_{ijk}\beta_1(\texttt{sex}_i) + u_i + v_{ijk} \\ \texttt{clo}_{ijk} \sim& Gamma(\mu_{ijk}, 1/\lambda),\end{aligned} \tag{6.44}$$

where $u_i \sim N(0, \sigma_u^2)$ *is the subject effect* ($Cor[u_{i_1}, u_{i_2}] = 0$, *for* $i_1 \neq i_2$), *and* v_{ijk} *is effect of a measurement on subject* i *on day* j *at time* k, *i.e.* $V[v_{ijk}] = \sigma_v^2$, $cor[v_{i_1j_1k_1}, v_{i_2j_2k_2}] = 0$ *for* $(i_1, j_1) \neq (i_2, j_2)$ *and* $cor[v_{i_1j_1k_1}, v_{i_1j_1k_2}] = \rho^{|k_1-k_2|}$. *The model is implemented in the code below:*

```
modMEMEx18AR1Gam <- glmmTMB(clo ~ tOutCen * sex + (1 | subjId) +
                         ar1(as.factor(obs.no) -1 | subjId:day),
                      family = Gamma(log), data = dfMEMEx1,
                      disp = ~ sex)

## [...]
##
## Conditional model:
##  Groups     Name               Variance Std.Dev. Corr
##  subjId     (Intercept)        0.0311   0.176
##  subjId:day as.factor(obs.no)1 0.0346   0.186    0.99 (ar1)
## Number of obs: 803, groups:  subjId, 47; subjId:day, 136
##
## Conditional model:
##              Estimate Std. Error z value   Pr(>|z|)
## (Intercept)  -0.58029    0.04260  -13.62    < 2e-16 ***
## tOutCen      -0.01347    0.00273   -4.93 0.00000083 ***
## sex2         -0.11235    0.06078   -1.85      0.065 .
## tOutCen:sex2  0.01186    0.00288    4.11 0.00003911 ***
## ---
## Signif. codes:  0 '***' 0.001 '**' 0.01 '*' 0.05 '.' 0.1 ' ' 1
## [...]
```

This result can be compared to the results in Chapter 5, as the models are not nested; we cannot use the likelihood ratio test, and we therefore compare using AIC;

```
c(AIC(modMEMEx18AR1Gam),AIC(modMEMEx18AR1))

## [1] -2788 -2039
```

Hence, we see a large gain in AIC, implying that the gamma model provides a more adequate description of the data.

The example above together with examples throughout Chapters 3-5 show how considerations on distribution types (we have considered Gaussian, gamma and transformation) should be an integrated part of the modelling, and also how the structures in data imply a interdependence structure through the correlation structures in the random effects.

In Chapter 4, we looked at aggregated window opening data. We continue this example and include the variation introduced by different dwellings as random effects:

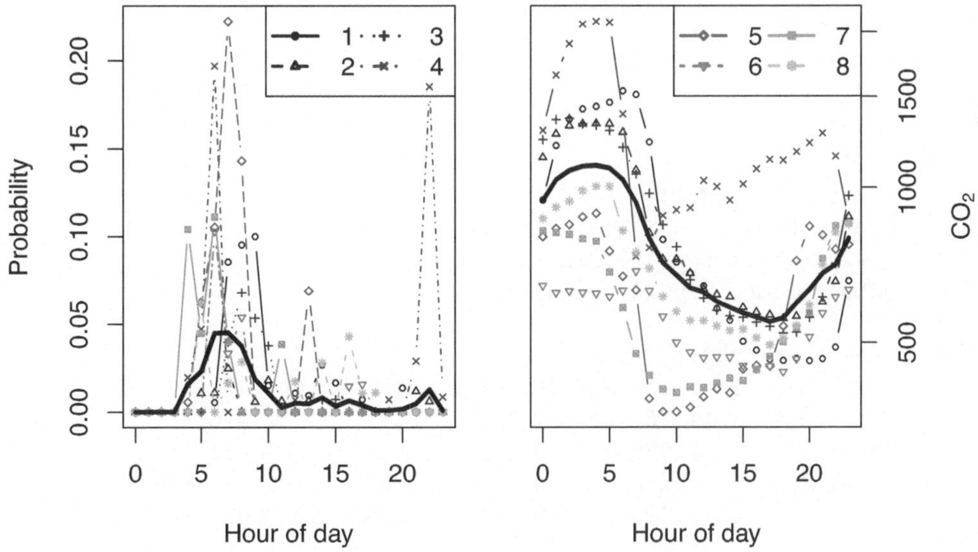

Figure 6.2: Observed probability of window opening in the 8 dwellings and CO_2 concentration as a function of hour of day.

Example 6.14 (Window opening) *We consider a subset of data from Andersen et al. (2013). The output here is the window opening action as a function of the explanatory variable. Here, we will only consider CO_2-concentration and hour of day in the bedroom of 8 dwellings and further restrict the analysis to the month of May. It is natural to consider the variable dwelling as a random effect. Most dwellings have a high probability of opening the window in the morning (Figure 6.2), but for dwelling 4, we observe an additional high probability in the evening. There is also a considerable variation in CO_2-concentration over a day.*

Hence a reasonable model could be

$$P(\textit{Window open in next 10 minutes}|\textit{dwelling}) = p(\boldsymbol{x}, u), \tag{6.45}$$

where

$$p(\boldsymbol{x}, u) = \frac{e^{\boldsymbol{x}^T\boldsymbol{\beta}+u}}{1 + e^{\boldsymbol{x}^t\boldsymbol{\beta}+u}}, \tag{6.46}$$

and $u \sim N(0, \sigma_u^2)$ is the dwelling effect. The explanatory variables are

$$\boldsymbol{x}^T = \left[1, \cos\left(\frac{2\pi H}{24}\right), \sin\left(\frac{2\pi H}{24}\right), log(CO_2)\right], \tag{6.47}$$

where H is hour of day.

The model can be estimated with `glmmTMB` *by* (`h=`$\frac{2\phi H}{24}$);

```
fit.window <- glmmTMB(y ~  I(cos(h)) + I(sin(h)) + I(lco2) +
                          ( 1 | dwel), family = binomial,
                      data = df.window)

summary(fit.window)

## [...]
## Conditional model:
##  Groups Name        Variance Std.Dev.
##  dwel   (Intercept) 0.372    0.61
## Number of obs: 25976, groups:  dwel, 8
##
## Conditional model:
##             Estimate Std. Error z value   Pr(>|z|)
## (Intercept)  -11.760      1.375   -8.55    < 2e-16 ***
## I(cos(h))     -1.086      0.132   -8.21    < 2e-16 ***
## I(sin(h))      1.285      0.135    9.52    < 2e-16 ***
## I(lco2)        0.995      0.201    4.94 0.00000079 ***
## ---
## [...]
```

The summary shows that there is a significant variation over the day and that increasing CO_2-concentration results in higher probability of opening the window. Furthermore, the standard deviation of the variation between dwellings is $\sigma_u = 0.61$. Testing if this is a significant contribution to the likelihood is done by;

```
fit.window.glm <- glmmTMB(y ~  I(cos(h)) + I(sin(h)) +  I(lco2),
                          family = binomial, data = df.window)

anova(fit.window.glm,fit.window)[ c(1,5:8)]

##                Df deviance Chisq Chi Df Pr(>Chisq)
## fit.window.glm  4     2523
## fit.window      5     2483  39.5      1    3.3e-10 ***
## [...]
```

The output shows a significant contribution to the likelihood. The random effects can be extracted by;

```
fit.window$sdr$par.random
```

The fixed effects are illustrated in the figures below. The top row shows the linear predictor as a function of hour of day (with $CO_2 = 800$), and as a function of $\log(CO_2)$ (at 8 in the morning). The bottom row shows the linear predictor and probability of a window opening as a function of the hour of the diurnal cycle with CO_2 concentration as given in Figure 6.2.

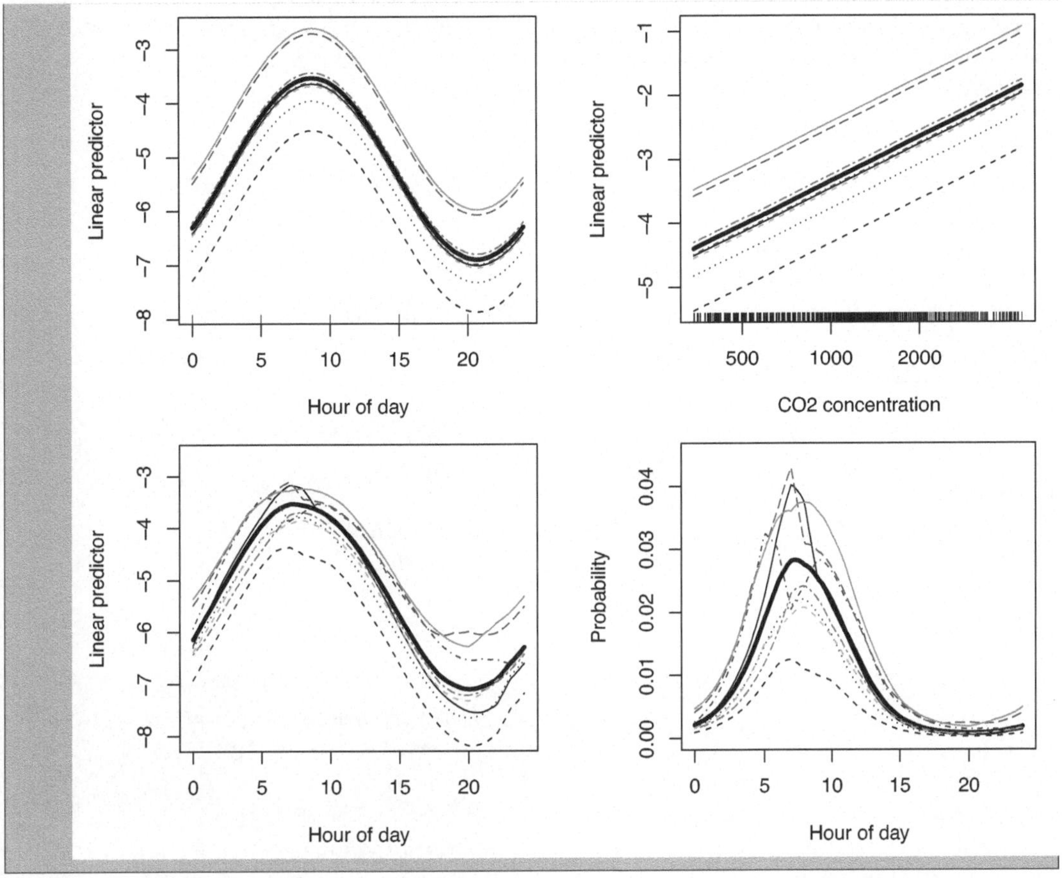

The examples above illustrate how simple random effects can be added to the exponential family. The method generalises to more complex examples, for example analysing different CO_2-sensitivity depending on the dwelling, but exclude nonlinear functions of random effects. Based on Figure 6.2, we might for example hypothesise that the timing of window opening depends on the dwelling. Such more general assumptions are treated in Section 6.3.

The usual principles for model building apply also here, e.g. keep it simple and in particular keep the random effect structure simple, and examine the possible fixed effects that can explain your data. For example, in the example above, we did not look at differences between work days and weekends, which might explain some of the variation in data.

6.2.2 Ordinal regression

In Example 4.28 on page 112, we studied the ceiling fan example using ordinal regression. In that example, we ignored the effect of different users. These may, however, be modelled in the same way as we have done so far by including a random effect attributed to subjects. The basic idea is the same: the random effect is added inside the link function. In the case of ordinal regression, this implies

$$\eta_{j,l}(\boldsymbol{x}) = \theta_j - \boldsymbol{x}^T\boldsymbol{\beta}_j - u_l; j = 1, ..., k-1, \tag{6.48}$$

where $u_l \sim N(0, \sigma_u^2)$. The estimation is carried out through the Laplace approximation as discussed in Section 6.1.

Example 6.15 (Ceiling fan speed (cont.)) *Coming back to the example of the ceiling fan speed, we can formulate the following cumulative link mixed models in R:*

```
## library(ordinal)
## Null model
m0S0 <- clmm(fanSpeed ~ 1 + (1 | subjId), data = dfCeilingFan)
## Including TSV
m0S1 <- clmm(fanSpeed ~ TSV + (1 | subjId), data = dfCeilingFan)
## Including random effect of TSV
m0S2 <- clmm(fanSpeed ~ TSV + (1 + TSV | subjId),
             data = dfCeilingFan)
anova(m0S0,m0S1,m0S2)

## Likelihood ratio tests of cumulative link models:
##      no.par AIC logLik LR.stat df Pr(>Chisq)
## m0S0      3 443   -218
## m0S1      5 425   -207    22.2  2   0.000015 ***
## m0S2     10 424   -202    10.8  5      0.055 .
## ---
## [...]
```

The likelihood table indicates that TSV *should be included as a fixed effect, but there is not sufficient statistical evidence of an random effect related to* TSV*.*

We may test if the random effect should be included by

```
m00S1 <- clm(fanSpeed ~ 1 + TSV, data=dfCeilingFan)
anova(m0S1,m00S1)

## [...]
##       no.par AIC logLik LR.stat df Pr(>Chisq)
## m00S1      4 434   -213
## m0S1       5 425   -207    11.9  1    0.00057 ***
## ---
## [...]
```

The output shows that there is a significant variation between subjects, and hence the random effect model is preferred.

We see that the model m0S1 *is the preferred one (comparing* m0S1 *and* m0S1 *suggests the inclusion of* TSV *as a fixed effect, comparing* m0S1 *and* m0S2 *suggests that the parameters for* TSV *should not be included as random effects and finally comparing* m00S1 *and* m0S1 *suggests that a random intercept should be included). Hence the model including TSV as a fixed effect and random intercepts should be preferred. The parameters can be extracted from the object by*

```
summary(m0S1)

## [...]
## Random effects:
##  Groups Name        Variance Std.Dev.
##  subjId (Intercept) 0.746    0.864
## Number of groups:  subjId 21
##
## Coefficients:
##      Estimate Std. Error z value  Pr(>|z|)
## TSV1    0.606      0.321    1.89     0.059 .
## TSV2    2.243      0.508    4.42 0.0000099 ***
## ---
##
## Threshold coefficients:
##     Estimate Std. Error z value
## 0|1    0.986      0.285    3.46
## 1|2    2.346      0.330    7.10
## [...]
```

The model output tells us that solely the coefficient for TSV2, i.e. a "warm" vote is significant different from 0.

The model coefficients can be used to predict the probability of a chosen fan stage depending on TSV, which is visualised below. Further, the difference between subject preferences of fan speed is visualised.

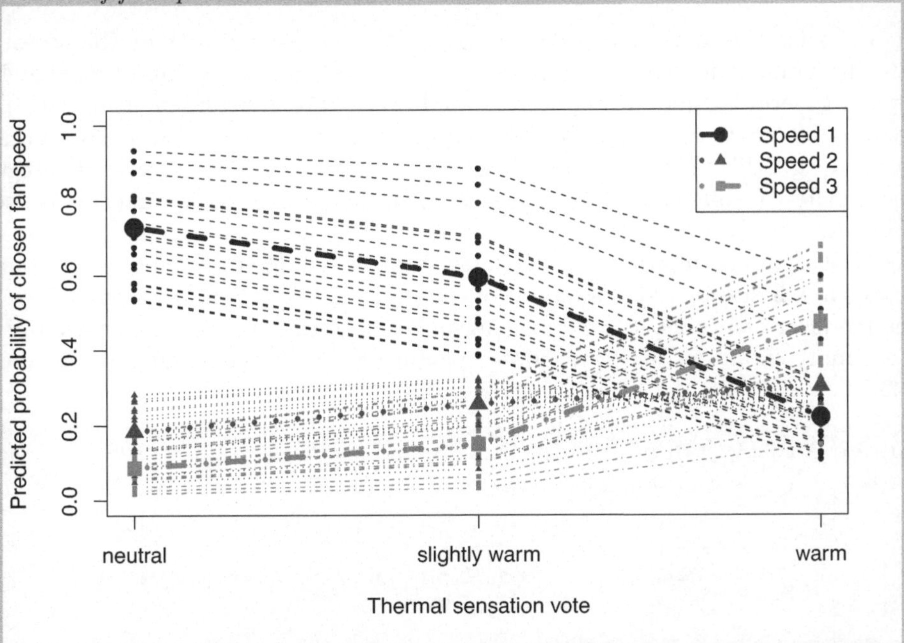

In order to test the second hypothesis, whether there is a moderating effect of the ceiling fan type on the relationship between ceiling fan speed and TSV, we can add the variable fan type to our previous model and apply the likelihood ratio test in order to

test its effect:

```
m1S <- clmm(fanSpeed ~ TSV * fanType + (1 |subjId),
            data=dfCeilingFan)
anova(m0S1,m1S)

## [...]
##       no.par AIC logLik LR.stat df Pr(>Chisq)
## m0S1       5 425   -207
## m1S        8 428   -206    2.84  3       0.42
## [...]
```

The likelihood ratio test indicates that the fan type does not affect the fan speed ($\chi^2 = 2.84$, compared to a χ^2-distribution with 3 degrees of freedom resulting in a p-value of 0.42*).*

The models we have discussed so far in this chapter are all implemented in standard software, implying that the methods can be implemented without programming the likelihood. In the following sections, we will cover some non-standard methods, which also require that the modeller programme the likelihood or at least part of it.

6.3 General mixed effect models

In some cases non-linear models might be more appropriate and include the use of a non-standard distribution for either the first– or second– stage model. One tool that handles this type of general likelihood functions is the library `TMB` (Kristensen et al., 2016), where the model is formulated in `C++` code. `TMB` can be used as a general tool for optimisation of any function e.g. a likelihood function. In the case of mixed effect models, the integral wrt. the random effect is difficult to evaluate even when the Laplace approximation is used, and the automatic differentiation, which is integrated in `TMB`, will speed up the calculations. `TMB` uses the Laplace approximation, and includes automatic differentiation of the second–stage model, which will ensure fast evaluation of the inner optimisation. This procedure implies that the modeller can concentrate on the formulation of the likelihood. The example below shows a simple extension of the window opening example that cannot be handled with `glmmTMB`[1].

Example 6.16 (Window opening) *We could hypothesise that the timing of window opening is dependent on the house where the window is situated. This could be formulated as*

$$P(\textit{Window open in next 10 minutes}|\textit{dwelling}) = p(\boldsymbol{x}, \boldsymbol{u}), \tag{6.49}$$

[1] For examples and further introduction, see `https://kaskr.github.io/adcomp/Introduction.html`

where

$$\begin{aligned} p(\boldsymbol{x}, u) =& \frac{e^{\eta(\boldsymbol{x},u)}}{1 + e^{\eta(\boldsymbol{x},u)}} \\ \eta(\boldsymbol{x}, u) =& \beta_0 + \beta_1 \cos\left(\frac{2\pi H}{24} - \beta_3 - u_2\right) + \beta_2 \log(CO_2) + u_1 \end{aligned} \tag{6.50}$$

with $u_1 \sim N(0, \sigma_1^2)$ *and* $u_2 \sim N(0, \sigma_2^2)$ *being the dwelling effect.* β_3 *is the typical time of day where windows are opened, but the model allows this time of day to be dependent on the dwelling (by* u_2*).*

The likelihood is declared in `C++`*, using the code below, we start by declaring the data matrices;*

```
#include <TMB.hpp>        // Links in the TMB libraries

template<class Type>
Type objective_function<Type>::operator() ()
{
  // Data
  DATA_MATRIX(Y);             // Observation matrix from R
  DATA_MATRIX(X);             // Regression matrix from R
  DATA_FACTOR(dwel);          // Data vector from R
```

Followed by declaring random effect parameters, and fixed effect parameters;

```
  // Random Effects Parameters
  PARAMETER_MATRIX(u);

  // Fixed Effects Parameters
  PARAMETER_VECTOR(pars);
```

Initialisation of parameters;

```
  // initialize variables
  int nobs = Y.col(0).size(), nsub = u.col(0).size();
  int j;
  Type mean_u1 = Type(0);
  Type mean_u2 = Type(0);
  Type p = Type(0);
  Type c;
  Type c2;
  Type f = 0;       // Declare neg. log. likelihood
```

and finally the likelihood is calculated in 2 loops; first over the random effects;

```
  // Random effect part of model
  for(int j=0; j < nsub; j++){
    f -= dnorm(u(j,0), mean_u1 , exp(pars(0)), true);
    f -= dnorm(u(j,1), mean_u2 , exp(pars(1)), true);
  }
```

and then over the observations,

```
  // Conditional model
  for(int i =0; i < nobs; i++){
    j  = dwel[i];
    c  = cos(X(i,0) - pars(5) - u(j,1));
    c2 = pars(2) + pars(3) * c + pars(4) * X(i,1)
    p  = invlogit(c2 + u(j,0));
    f -= dbinom(Y(i,0),Y(i,1),p, true);
  }
  return f;
}
```

Here, `f` *is the negative log-likelihood. The code is stored in the file* `window2.cpp`.

For estimating the parameters, the first step is to compile the model (i.e. the C++ *files). The compilation is done in R by;*

```
## library(TMB)
compile("cpp-files/window2.cpp")
dyn.load(dynlib("cpp-files/window2"))
```

The second step is to create lists containing the parameters and the data. Note that the parameters contain both the fixed effects parameters and the random effects and that they are separated by the likelihood setup above.

```
tmbpars = list(
  u    = cbind(rep(0,8),rep(0,8)),
  pars = c(0,0,-12,2.2,1,2)
)

tmbdata = list(
  Y    = as.matrix(df.window[,c("y","n")]),
  X    = as.matrix(df.window[,c("h","lco2")]),
  dwel = as.vector(df.window$dwel)
)
```

In the last step, the likelihood function is developed using the data and parameters as just declared, and the MLE is found using an optimiser. The object `obj` *includes the objective function (*`obj$fn`*), i.e. the likelihood, and the derivative of the objective function (*`obj$gr`*), i.e. the score function. In that step, we also declare that* $\mathbf{u}$ *is the random effects (by* `random = "u"`*);*

```
obj <- MakeADFun(data = tmbdata,
                 parameters = tmbpars,
                 random = "u",
                 DLL = "window2",
                 silent=FALSE)


opt3 <- nlminb(obj$par, obj$fn, obj$gr,
               lower=c(-Inf,-Inf,-Inf,0,-Inf,-Inf),
               upper=c(Inf,Inf,Inf,2*pi,Inf,Inf))
```

The random effects can be extracted by;

```
## Random effects
rap <- sdreport(obj,getJointPrecision = TRUE)
u1 <- rap$par.random[1:8]
u2 <- rap$par.random[9:16]
```

The figure below summarises the results. We see, for eaxmple that in most dwellings the window will mostly be open in the morning, but in one dwelling the window is mostly open late in the evening (around midnight). The top row of the figure shows the linear predictor for each dwelling (for fixed CO_2 and fixed hour of day, respectively) while the second row shows the linear predictor and the predicted probabilities as a function of the hour of day (using the average observed CO_2 per dwelling as input)

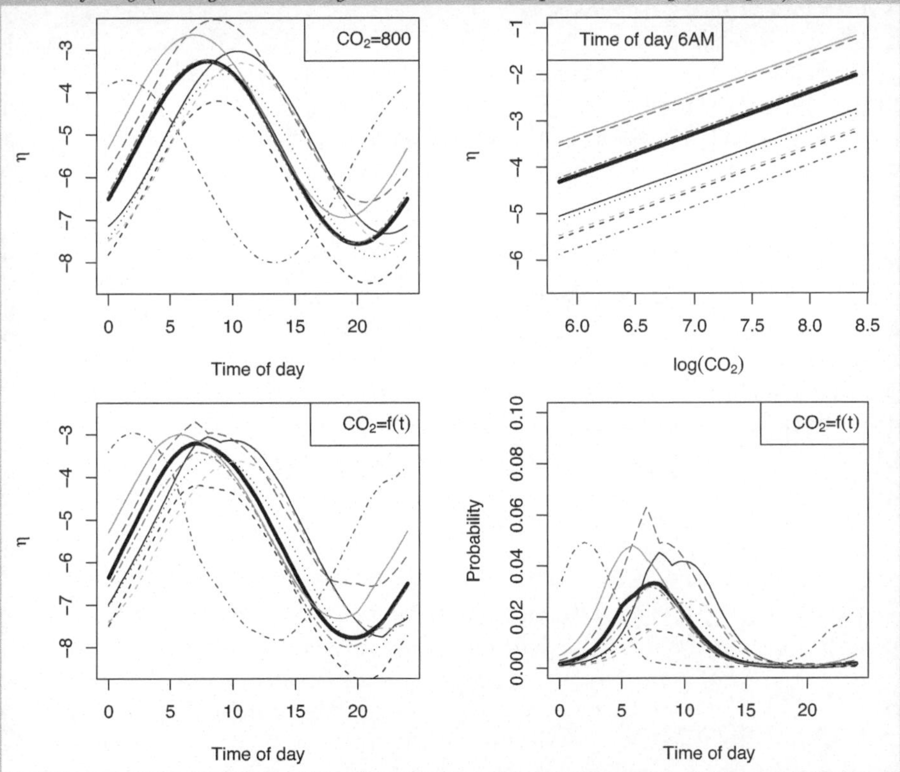

In the figure above, we see the "linear predictor", topleft: for fixed level of CO_2

(i.e. different random effect for each dwelling), top right: the linear predictor as a function CO_2 at 6 AM, bottom left: "linear predictor" with CO_2 concentration modelled as a function of time of day (as local linear interpolation). Bottom right: inverse link function of the linear predictor in the bottom left panel.

The example above illustrates how non-linear, but Gaussian, random effects can be included. However, in some cases, it is more reasonable to include non-Gaussian random effects. The decision regarding Gaussian or non-Gaussian random effects may be based directly from data or we might simply try different distribution assumptions. The example below illustrates how the latter can be implemented using TMB.

Example 6.17 *The model presented above gives good results. However, care must be taken when defining the initial values and it does not seem reasonable that u_2 is normally distributed. A better alternative is to use the von Mises distribution (or circular normal distribution), i.e.*

$$u_2 \sim VM(\mu, \kappa), \tag{6.51}$$

with pdf

$$f_{u_2}(u_2) = \frac{e^{\kappa \cos(u_2-\mu)}}{2\pi I_0(\kappa)}, \tag{6.52}$$

where $I_0(\cdot)$ is the modified Bessel function of order 0. The overall layout of the C++ file is the same, except for the lines defining the likelihood. The line treating the random effects are changed to

```
c2 = log(2*M_PI) + log(besselI(exp(lkappa), Type(0)));
for(int j=0; j < nsub; j++){
  f -= dnorm(u1[j], mean_ran1, exp(sigma_1), true);
  c1 = exp(lkappa) * cos(u2[j] - beta[3]);
  f -= c1 - c2;
}
```

and the lines related to the observation are changed to

```
for(int i =0; i < nobs; i++){
  j = dwel[i];
  c3 = beta[1] * cos(h[i] - u2[j]);
  c4 =  beta[0] + c3 + beta[2] * lco2[i] + u1[j];
  p  = invlogit(c4);
  f -= dbinom(y[i],n[i],p, true);
}
```

The distribution of the second-stage model for the two models is shown in the figure below (black line von Mises, grey line Gaussian random effects from Example 6.16) along with the estimated random effects. We see that the distributions are similar, but also that the Von Mises have the periodic behaviour that is expected here.

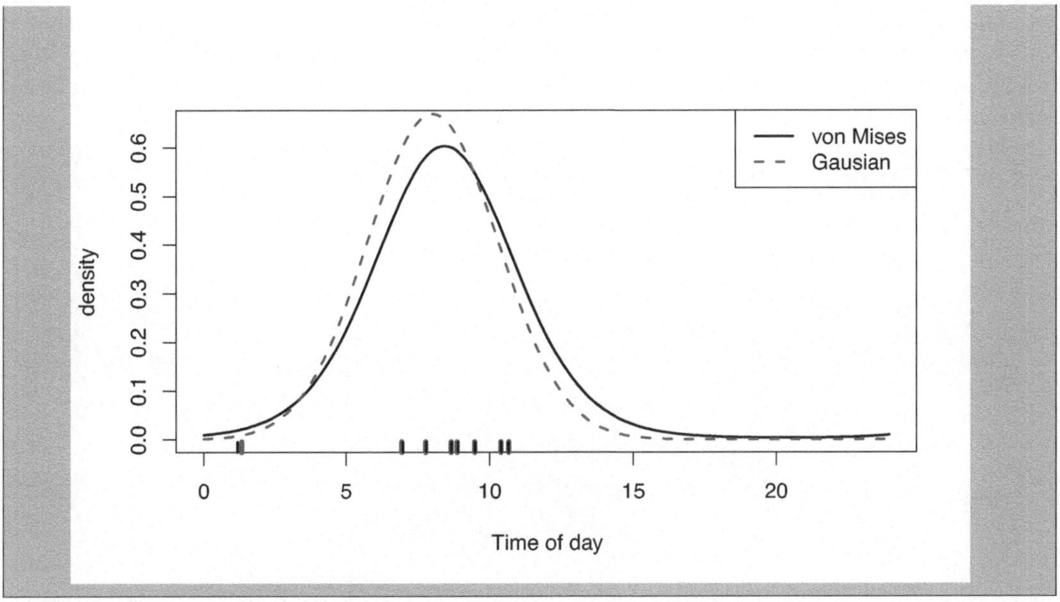

The examples above illustrate the flexibility of the TMB. As discussed earlier, the implementation relies on the Laplace approximation and is implying that the likelihood is not exact. The value of the integral may be checked by importance sampling, see, for example Madsen and Thyregod (2011).

6.4 Some models that allow explicit solutions

The challenging part of including random effects is finding the integral (6.3); so far, in this chapter, we have dealt with this challenge using the Laplace approximation for non-linear/non-Gaussian models. In Chapter 5, we considered linear Gaussian mixed effect models, implying an explicit solution to the integral in (6.3). The explicit solution is not restricted to the linear Gaussian case, but can be generalised to other distributions, using results from Baysian statistics, as we will discuss blow.

Inspired by Bayesian statistics, a conjugate prior is a parameter distribution such that the prior and the posterior belong to the same family. These are tractable from a computational point of view, and the theorem below gives the general recipe for finding the conjugate prior in the exponential family.

Theorem 6.1 (Standard conjugate distribution, exponential dispersion family) *Consider an exponential dispersion family* $\mathrm{ED}(\mu, V(\mu)/\lambda)$ *with density* (4.8) *for* $\theta \in \Omega$. *Let* $\mathcal{M} = \tau(\Omega)$ *denote the mean value space for this family. Let* $m \in \mathcal{M}$ *and consider*

$$g_\theta(\theta; m, \gamma) = \frac{1}{C(m,\gamma)} \exp\left(\frac{\theta m - \kappa(\theta)}{\gamma}\right) \tag{6.53}$$

with

$$C(m,\gamma) = \int_\Omega \exp\left(\frac{\theta m - \kappa(\theta)}{\gamma}\right) d\theta \tag{6.54}$$

for all (positive) values of γ *for which the integral converges.*

Then (6.53) *defines the conjugate prior density function of* θ.

Theorem 6.1 is constructive (i.e. the precise recipe for calculations is given) in that it gives the direct method for calculating the conjugate prior. However, for the concept to be really useful, we will need to be able to solve the integral in (6.54) explicitly (which is not guaranteed by Theorem 6.1).

Example 6.18 (Conjugate prior for Poisson distribution) *Consider the Poisson distribution with pdf (see Example 4.3)*

$$f_Y(y;\mu) = \frac{\mu^y e^{-\mu}}{y!} = \frac{e^{\log(\mu)y-\mu}}{y!} = c(y)e^{\theta y - e^\theta}, \tag{6.55}$$

with $\theta(\mu) = \log(\mu)$, $\kappa(\theta) = e^\theta = \mu$, *and* $c(y) = 1/y!$. *Following Theorem 6.53, the conjugate prior become*

$$\begin{aligned} g_\theta(\theta, m, \gamma) =& \frac{1}{C(m,\gamma)} \exp\left(\frac{\theta m - \kappa(\theta)}{\gamma}\right) \\ =& \frac{1}{C(m,\gamma)} \exp\left(\frac{\log(\mu) m - \mu}{\gamma}\right) \\ =& \frac{1}{C(m,\gamma)} \mu^{\frac{m}{\gamma}} \exp\left(-\frac{\mu}{\gamma}\right) \\ =& \frac{1}{C(m,\gamma)} \mu^{\frac{m+\gamma}{\gamma}-1} \exp\left(-\frac{\mu}{\gamma}\right), \end{aligned} \tag{6.56}$$

the part that depends on μ *is recognised as the kernel of a gamma distribution with shape parameter* $\frac{m+\gamma}{\gamma}$ *and scale parameter* γ *and hence*

$$C(m,\gamma) = \frac{1}{\Gamma\left(\frac{m+\gamma}{\gamma}\right)\left(\frac{m+\gamma}{\gamma}\right)^{\gamma}}, \tag{6.57}$$

And the conjugate prior for the Poisson distribution is the gamma distribution.

We will present the specific result for some distributions in the next sub-sections.

6.4.1 Gamma–Poisson model

The conjugated prior distributions can also form the basis for finding the marginal distributions for hierarchical models, as illustrated in the following theorem.

Theorem 6.2 (Compound Gamma-Poisson model) *Consider a hierarchical model for* Y *specified by*

$$\begin{aligned} Y|\gamma &\sim Pois(\mu\gamma) \\ \gamma &\sim G(1,\phi), \end{aligned} \tag{6.58}$$

i.e. $E[\gamma] = 1$ *and* $V[\gamma] = 1/\phi$. *Then the marginal distribution of* Y *is a* negative binomial distribution, $Y \sim \mathrm{NB}(\phi, \frac{\mu}{\phi+\mu})$. *The probability mass function for* Y *is*

$$P[Y = y] = \binom{y+\phi+1}{y}\left(\frac{\phi}{\phi+\mu}\right)^{\phi}\left(\frac{\mu}{\phi+\mu}\right)^{y}, \tag{6.59}$$

where we have used the convention

$$\binom{z}{y} = \frac{\Gamma(z+1)}{\Gamma(z+1-y)\, y!}, \tag{6.60}$$

for z *real and* y *integer values.*

Proof *The marginal distribution is*

$$f_Y(y) = \int f_{Y|\gamma}(y) f_\gamma(\gamma) d\gamma, \tag{6.61}$$

where

$$f_\gamma(\gamma) = \frac{\phi^\phi}{\Gamma(\phi)} \gamma^{\phi-1} e^{-\gamma\phi}. \tag{6.62}$$

Hence we can calculate the marginal distribution as

$$\begin{aligned} f_Y(y) &= \int \frac{(\mu\gamma)^y}{y!} e^{-\mu\gamma} \frac{\phi^\phi}{\Gamma(\phi)} \gamma^{\phi-1} e^{-\gamma\phi} d\gamma \\ &= \frac{\mu^y}{y!} \frac{\phi^\phi}{\Gamma(\phi)} \int \gamma^{\phi+y-1} e^{-\gamma(\phi+\mu)} d\gamma \\ &= \frac{\mu^y}{y!} \frac{\phi^\phi}{\Gamma(\phi)} \frac{\Gamma(\phi+y)}{(\phi+\mu)^{\phi+y}} \\ &= \frac{\Gamma(\phi+y)}{\Gamma(\phi) y!} \left(\frac{\phi}{\phi+\mu}\right)^\phi \left(\frac{\mu}{\phi+\mu}\right)^y \\ &= \binom{y+\phi+1}{y} \left(\frac{\phi}{\phi+\mu}\right)^\phi \left(\frac{\mu}{\phi+\mu}\right)^y, \end{aligned} \tag{6.63}$$

which was to be shown.

Note here that (using properties of the negative binomial distribution)

$$\begin{aligned} E[Y] &= \phi \frac{\mu}{\phi+\mu} \frac{\phi+\mu}{\phi} = \mu \\ V[Y] &= \phi \frac{\mu}{\phi+\mu} \left(\frac{\phi+\mu}{\phi}\right)^2 = \mu \left(\frac{\phi+\mu}{\phi}\right), \end{aligned} \tag{6.64}$$

Hence we see that the model introduces over-dispersion, and for ϕ large, the over-dispersion is small (i.e. $V[Y] \approx \mu$).

Inference on individual group means

It is sometimes of interest to find the conditional distribution of the random effects given the observations. The following theorem gives the result for the gamma-Poisson model.

Theorem 6.3 (Conditional distribution of γ) *Consider the hierarchical gamma-Poisson model in (6.58), and assume that a value $Y = y$ has been observed.*

Then the conditional distribution of γ for given $Y = y$ is a gamma distribution, i.e.

$$\gamma | Y = y \sim G(\phi + y, \mu + \phi), \tag{6.65}$$

with mean and variance

$$\begin{aligned} E[\gamma | Y = y] &= \frac{\phi+y}{\phi+\mu} \\ V[\gamma | Y = y] &= \frac{\phi+y}{(\phi+\mu)^2}. \end{aligned} \tag{6.66}$$

Proof *The conditional distribution of $f_{\gamma|Y}(\gamma)$ is given by*

$$\begin{aligned} f_{\gamma|Y}(\gamma) =& \frac{f_{\gamma,Y}(y,\gamma)}{f_Y(y)} = \frac{f_{Y|\gamma}(y)f_\gamma(\gamma)}{f_Y(y)} \\ =& \frac{(\mu\gamma)^y}{y!}e^{-\mu\gamma}\frac{\phi^\phi}{\Gamma(\phi)}\gamma^{\phi-1}e^{-\gamma\phi}\frac{\Gamma(\phi)y!}{\Gamma(\phi+y)}\left(\frac{\phi+\mu}{\phi}\right)^\phi\left(\frac{\mu+\phi}{\mu}\right)^y \\ =& \gamma^{\phi+y-1}e^{-\gamma(\phi+\mu)}\frac{1}{\Gamma(\phi+y)}(\phi+\mu)^{\phi+y}, \end{aligned} \tag{6.67}$$

which is recognised as the pdf of a gamma distributed random variable, with rate parameter $\mu+\phi$ and shape parameter $\phi+y$.

The conditional mean is close to 1 if either y is close to μ (we observe the expected value) or if ϕ is large (the variance of the random effect is small). To see what information is added consider

$$\frac{V[\gamma|Y=y]}{E[\gamma|Y=y]} = \frac{\phi+y}{(\phi+\mu)^2}\frac{\phi+\mu}{\phi+y} = \frac{1}{\phi+\mu} < \frac{1}{\phi} = \frac{V[\gamma]}{E[\gamma]}. \tag{6.68}$$

So the ratio between the mean and variance becomes smaller after observing the outcome of the Poisson distribution, and further the ratio does not depend on the observation, and the effect is minor when ϕ is large.

Note that the result we have presented here is related to individual observations and hence it immediately generalises to

$$\begin{aligned} Y_i|\gamma_i \sim& Pois(\mu_i\gamma_i) \\ \gamma_i \sim& G(1,\phi), \end{aligned} \tag{6.69}$$

where μ_i might be modelled as a regression type of problem, e.g. $\mu_i = e^{\boldsymbol{x}_i\boldsymbol{\beta}}$. The likelihood is in this case

$$L(\boldsymbol{\beta},\phi;\boldsymbol{y}) = \prod_i \binom{y_i+\phi+1}{y_i}\left(\frac{\phi}{\phi+\mu_i}\right)^\phi\left(\frac{\mu_i}{\phi+\mu_i}\right)^{y_i}. \tag{6.70}$$

We illustrate the results with a small example.

Example 6.19 *We again turn to the example of changes in clothing insulation levels (e.g. Example 6.12). Considering the examples we have analysed so far, a natural model would be*

$$\begin{aligned} Y_i \sim& Pois(\mu_i\gamma_i) \\ \gamma_i \sim& G(1,\phi), \end{aligned} \tag{6.71}$$

with $\mu_i = e^{\log(time)+\boldsymbol{x}_i\boldsymbol{\beta}}$, and $\boldsymbol{x}_i = [1 \quad 0]$ for females and $\boldsymbol{x}_i = [1 \quad 1]$ for males. The likelihood is implemented in the code-listing below, using the notation introduced above. In addition to the likelihood, the expected value and the variance of the conditional values of the random effects can also be calculated.

```
## Negative log-likelihood
nll.dnbinom <- function(theta, X, y, off.set, full = FALSE){
    mu <- exp(X %*% theta[1:dim(X)[2]] + off.set)
    phi <- exp(theta[dim(X)[2] + 1])
    ll <- sum(dnbinom(y, size =  phi, mu=mu, log = TRUE))
    if(full){
        E <- (phi + y) / (phi + mu)
        V <- E / (phi + mu)
        return(list(ll = ll, theta = theta, E = E, V = V))
    }
    - ll
}
```

In order to make a direct comparison with the normal-Poisson, model this is refitted with a similar random effect structure (one random effect per observation).

```
## Estimate bench-mark models
## Poisson
pois0 <- glmmTMB(clo ~ offset(log(time)) + sex, data = dfcount,
                 family=poisson)
## Normal-Poisson
pois.norm1 <- glmmTMB(clo ~ offset(log(time)) + sex + (1|subjId:day),
                      data=dfcount,family=poisson)
## Gamma-Poisson
X <- model.matrix(fit.tmb.pois)
theta <- c(0,0,0)
opt.dnbinom1 <- nlminb(theta, nll.dnbinom, X = X, y = dfcount$clo,
                       off.set = log(dfcount$time))
```

The parameters for the models are given below;

```
##               gamma-Pois   Pois norm-Pois
## (Intercept)         -2.1   -2.1      -2.2
## sexmale             -1.1   -1.1      -1.1
## log-Like.         -132.6 -133.1    -132.4
```

First of all, comparing the last rows (the log-likelihood values), we see that the improvement in the likelihood does not justify an extra parameter here, and hence this is probably not the right structure for the models. Secondly, we notice that the mean value parameters (intercept and the coefficient for males) are not effected when using the negative binomial model. In contrast, the mean value parameter of the intercept changes slightly when using the normal-Poisson model. The difference can be illustrated by average multiplicative random effect as shown below;

```
summ.dbinom1 <- nll.dnbinom(theta = opt.dnbinom1$par, X,
                            y = dfcount$clo,
                            off.set = log(dfcount$time),
                            full = TRUE)
c(mean(summ.dbinom1$E), mean(exp(pois.norm1$sdr$par.random)))-1

## [1] 9.1e-08 3.6e-02
```

The average for the random effects for the gamma-Poisson model is 1, while it is slightly higher for the normal-Poisson model. Finally the multiplicative random effects are compared in the figure below, where we see that these are generally close to each other.

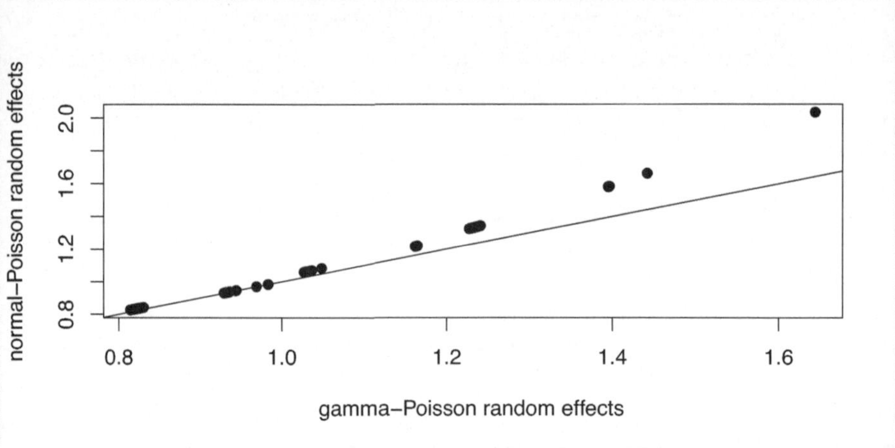

The introduction of gamma random effects did not give large improvements in the likelihood. On the other hand, we saw that random effects on subject level gave improvements in the likelihood when introduced in the normal-Poisson model (see Example 6.12). The example below generalises the result presented in Theorem 6.2.

Example 6.20 (Clothing count, gamma-Poisson model) *We start by writing the model*

$$Y_{ij}|U_i = u_i \sim Pois(\lambda_{ij}), \tag{6.72}$$

with

$$\lambda_{ij} = e^{x_{ij}\theta+v_i} = \tilde{\lambda}_{ij}u_i, \tag{6.73}$$

and

$$U_i \sim G(1,\phi). \tag{6.74}$$

In our particular case, we only consider sex as an explanatory variable and hence $x_{ij} = x_i$, with i referring to subject id. Hence also $\tilde{\lambda}_{ij} = \tilde{\lambda}_i$.

We can now write the joint density for subject i as

$$\begin{aligned} f(y_{i,1},...,y_{i,n_i},u_i) =& f_{Y|U}(y_{i,1},...,y_{i,n_i}|u_i)f_U(u_i) \\ =& f_U(u_i)\prod_{j=1}^{n_i} f_{Y|U}(y_{i,j}|U_i=u_i) \\ =& \frac{\phi^\phi}{\Gamma(\phi)}u_i^{\phi-1}e^{-u_i\phi}\prod_{j=1}^{n_i}(u_i\tilde{\lambda}_i)^{y_{ij}}\frac{e^{-u_i\tilde{\lambda}_i}}{y_{ji}!} \\ =& \frac{\phi^\phi u_i^{\phi+\sum_j y_{ij}-1}}{\Gamma(\phi)}e^{-u_i(\phi+\sum_j\tilde{\lambda}_i)}\frac{\tilde{\lambda}_i^{\sum_j y_{ij}}}{\prod_{j=1}^{n_i} y_{ji}!} \\ =& \frac{\phi^\phi u_i^{\tilde{\alpha}_i-1}}{\Gamma(\phi)}e^{-u_i/\tilde{\beta}_i}\frac{\tilde{\lambda}_i^{\sum_j y_{ij}}}{\prod_{j=1}^{n_i} y_{ji}!}, \end{aligned} \tag{6.75}$$

with

$$1/\tilde{\beta}_i = \phi + \sum_j \tilde{\lambda}_i = \phi + n_i\tilde{\lambda}_i, \tag{6.76}$$

or

$$\tilde{\beta}_i = \frac{1}{\phi + n_i\tilde{\lambda}_i}, \tag{6.77}$$

and

$$\tilde{\alpha}_i = \phi + \sum_j y_{ij}. \tag{6.78}$$

Hence the marginal density for observations on one subject can be written as

$$\begin{aligned} f(y_{i,1},...,y_{i,n_i}) =& \frac{\phi^\phi}{\Gamma(\phi)}\frac{\tilde{\lambda}_i^{\sum_j y_{ij}}}{\prod_{j=1}^{n_i} y_{ji}!}\int_0^\infty u_i^{\tilde{\alpha}_i-1}e^{-u_i/\tilde{\beta}_i}du_i \\ =& \frac{\phi^\phi}{\Gamma(1/\beta)}\frac{\tilde{\lambda}_i^{\sum_j y_{ij}}}{\prod_{j=1}^{n_i} y_{ji}!}\tilde{\beta}_i^{\tilde{\alpha}_i}\Gamma(\tilde{\alpha}_i) \\ =& \frac{\phi^\phi}{\Gamma(\phi)}\frac{\tilde{\lambda}_i^{n_i\bar{y}_i}}{\prod_{j=1}^{n_i} y_{ji}!}\tilde{\beta}_i^{\tilde{\alpha}_i}\Gamma(\tilde{\alpha}_i) \\ =& \frac{\phi^\phi}{\Gamma(\phi)}\frac{\tilde{\lambda}_i^{n_i\bar{y}_i}}{\prod_{j=1}^{n_i} y_{ji}!}\left(\frac{1}{\phi+n_i\tilde{\lambda}_i}\right)^{\phi+n_i\bar{y}_i}\Gamma\left(\phi+\sum_j y_{ij}\right). \end{aligned} \tag{6.79}$$

Which is the likelihood function, i.e. $L(\theta,\phi) = f(y_{i,1},...,y_{i,n_i})$.

Conditioning on observations from subject i ($\boldsymbol{Y}_i = [Y_{1,i},...,Y_{n_i,i}]$), *we can write the condition distribution as*

$$\begin{aligned}
g_{u_i|\boldsymbol{Y}_i}(u_i|\boldsymbol{Y}_i=\boldsymbol{y}_i) =& \frac{f_{u_i,\boldsymbol{Y}_i}(u_i,\boldsymbol{y}_i)}{f_Y(\boldsymbol{y}_i)} \\
=& \frac{\phi^\phi u_i^{\tilde{\alpha}_i-1}}{\Gamma(\phi)} e^{-u_i/\tilde{\beta}_i} \frac{\tilde{\lambda}_i^{\sum_j y_{ij}}}{\prod_{j=1}^{n_i} y_{ji}!} \times \\
& \frac{\Gamma(\phi)}{\phi^\phi} \frac{\prod_{j=1}^{n_i} y_{ji}!}{\tilde{\lambda}_i^{n_i \bar{y}_i}} \tilde{\beta}_i^{-\tilde{\alpha}_i} \frac{1}{\Gamma(\tilde{\alpha}_i)} \\
=& u_i^{\tilde{\alpha}_i-1} e^{-u_i/\tilde{\beta}_i} \tilde{\beta}_i^{-\tilde{\alpha}_i} \frac{1}{\Gamma(\tilde{\alpha}_i)} \\
=& \frac{1}{\tilde{\beta}_i^{\tilde{\alpha}_i}\Gamma(\tilde{\alpha}_i)} u_i^{\tilde{\alpha}_i-1} e^{-u_i/\tilde{\beta}_i}.
\end{aligned} \tag{6.80}$$

Hence

$$U_i|\boldsymbol{Y}_i=\boldsymbol{y}_i \sim G(\tilde{\alpha}_i,\tilde{\beta}_i) = G\left(\phi+n_i\bar{y}_i, \frac{1}{\phi+n_i\tilde{\lambda}_i}\right), \tag{6.81}$$

implying that the mean and variance conditional on the observations are

$$\begin{aligned}
E[U_i|\boldsymbol{Y}_i] &= (\phi+n_i\bar{y}_i)\left(\frac{1}{\phi+\beta n_i\tilde{\lambda}_i}\right) = \frac{\phi+n_i\bar{y}_i}{\phi+n_i\tilde{\lambda}_i} \\
V[U_i|\boldsymbol{Y}_i] &= (\phi+n_i\bar{y}_i)\left(\frac{1}{\phi+n_i\tilde{\lambda}_i}\right)^2 = \frac{\phi+\beta n_i\bar{y}_i}{(\phi+n_i\tilde{\lambda}_i)^2}
\end{aligned} \tag{6.82}$$

Hence, we see that when $\bar{y}_i$ *is close to the prediction* $(\tilde{\lambda}_i)$*, the posterior mean is close to 1 and also that the ratio between the mean and the variance is* $E[U_i|\boldsymbol{Y}_i]/V[U_i|\boldsymbol{Y}_i] = \phi+n_i\tilde{\lambda}_i > \phi = E[U_i]/V[U_i]$*, i.e. the precision increase as* n_i *increase.*

The likelihood is implemented in R by

```
nll.hir <- function(theta,dat,X){
    lambda <- exp(X %*% theta[1:2] + log(dat$time))
    beta <- exp(-theta[3])
    alpha<-1 / beta
    l <- 0
    lev.sub <- levels(dat$subjId)
    for(i in 1:length(lev.sub)){
        yi <- dat$clo[dat$subjId == lev.sub[i]]
        lambdai <- lambda[dat$subjId == lev.sub[i]]
        l <- l  - (lgamma(alpha + sum(yi)) - lgamma(alpha) -
                   sum(lfactorial(yi)) + sum(yi * log(lambdai)) -
                   alpha * log(1 + beta * sum(lambdai)) +
                   sum(yi) * log(beta) -
                   sum(yi) * log(1 + beta * sum(lambdai)))
    }
    l
}
```

Estimation is done using `nlminb`*, and the estimates are compared with the Poison-normal model below;*

```
##              gamma-Pois norm-Pois   Pois
## Intercept          -2.1     -2.24   -2.1
## sex                -1.1     -1.13   -1.1
## Disp                0.7     -0.41     NA
## log-Lik          -129.7   -129.75 -133.1
```

The log-likelihood of the two random effect models is very similar while the fixed effect model has a lower likelihood. The mean value parameters are comparable (first two rows), while the parameter for the random effect (third row) is very different. This is because these parameters have different interpretations. For the normal-Poisson model, it is the logarithm of the standard deviation of $U_i \sim N(0, \sigma_u^2)$*, with* $Y_i|U_i = u_i \sim Pois(\lambda_i e^{u_i})$*. While in the gamma-Poisson model it is the dispersion of the random effect, i.e.* $V_i \sim G(1, \phi)$ *with* $Y_i|V_i = v_i \sim Pois(\lambda_i v_i)$*. So we have*

$$E[e^{U_i}] = e^{\sigma^2/2} \quad ; \quad V[e^{U_i}] = (e^{\sigma^2} - 1)e^{\sigma^2} \tag{6.83a}$$

$$E[V_i] = 1 \quad ; \quad V[V_i] = \phi. \tag{6.83b}$$

The random effects mean and standard deviation are compared in the table below.

```
##    Pois-Norm Pois-Gam
## E       1.25     1.00
## sd      0.93     0.71
```

We see that the expected value for e^{U_i} *is higher than 1, which is also the reason for the smaller mean value parameters. The estimated random effects are compared in the figures below.*

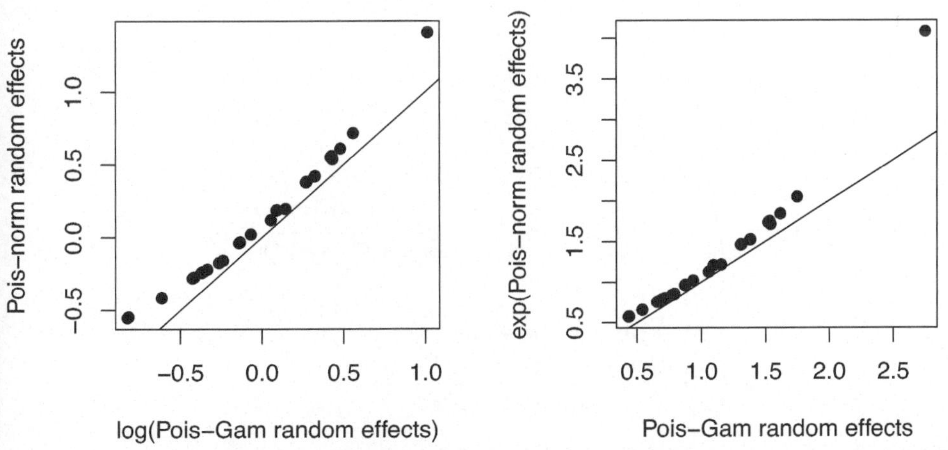

In general we see very good agreement between estimated random effects for the two models. The exception is one extreme point for both models. This is a male subject that

has two changes of clothing insulation level on each of his 3 visits (i.e. much higher than the average for males (which is 0.29)).

6.4.2 Hierarchical beta-binomial model

Another hierarchical model that allows for an explicit solution is the case of a binomial distribution in the first stage and a beta-distribution in the second stage. In order to keep the interpretation of the model and to enable the inclusion of regression type models, the beta distribution is parametrised by the mean and precision (Ferrari and Cribari-Neto, 2004; Grün and Leisch, 2012), i.e.

$$f_{Beta}(x) = \frac{x^{\mu\phi-1}(1-x)^{(1-\mu)\phi-1}}{B(\mu\phi,(1-\mu)\phi)}, \tag{6.84}$$

where $B(\alpha,\beta) = \frac{\Gamma(\alpha)\Gamma(\beta)}{\Gamma(\alpha+\beta)}$, and we write

$$X \sim Beta(\mu,\phi), \tag{6.85}$$

In this formulation, $E[X] = \mu$ and $V[X] = \frac{\mu(1-\mu)}{1+\phi}$. The concrete result is given below.

Theorem 6.4 *Consider the generalized one-way random effects model for* $Y_1, Y_2, \ldots, Y_k$ *given by*

$$\begin{aligned} Y_i|p_i &\sim Bin(n_i,p_i) \\ p_i &\sim \text{Beta}(\mu_i,\phi), \end{aligned} \tag{6.86}$$

that is the conditional distribution of Y_i *given* p_i *is a binomial distribution, and the distribution of the mean value* p_i *is a beta distribution. Then the marginal distribution of* Y_i *is a* Polya distribution *or beta-binomial distribution with probability mass function*

$$P[Y=y] = g_Y(y) = \binom{n}{y}\frac{B(y+\mu\phi, n-y+(1-\mu)\phi)}{B(\mu\phi,(1-\mu)\phi)}, \tag{6.87}$$

for $y = 0, 1, 2, \ldots, n$. *The mean and variance of* Y *are*

$$\begin{aligned} E[Y] &= n\mu \\ V[Y] &= n\mu(1-\mu)\frac{\phi+n}{\phi+1} = n^2\mu(1-\mu)\frac{\phi/n+1}{\phi+1}. \end{aligned} \tag{6.88}$$

Proof

$$\begin{aligned} g_Y(y) &= \int_0^1 \binom{n}{y} p^y(1-p)^{n-y}\frac{p^{\mu\phi-1}(1-p)^{(1-\mu)\phi-1}}{B(\mu\phi,(1-\mu)\phi)}dp \\ &= \binom{n}{y}\frac{1}{B(\mu\phi,(1-\mu)\phi)}\int_0^1 p^{y+\mu\phi-1}(1-p)^{n-y+(1-\mu)\phi-1}dp \\ &= \binom{n}{y}\frac{B(y+\mu\phi,n-y+(1-\mu)\phi)}{B(\mu\phi,(1-\mu)\phi)}. \end{aligned} \tag{6.89}$$

If we consider our observation as the fraction of success ($\hat{p} = \frac{Y}{n}$), then we get

$$\begin{aligned} E[\hat{p}] &= \mu \\ V[\hat{p}] &= \mu(1-\mu)\frac{\phi/n+1}{\phi+1}. \end{aligned} \tag{6.90}$$

For ϕ large ($\phi >> n$), we recover the well-known variance of the estimator for the probability ($V[\hat{p}] = \frac{\mu(1-\mu)}{n}$), and for $n >> \phi$, we get $V[\hat{p}] = \frac{\mu(1-\mu)}{\phi+1}$ or the variance from the beta-distribution. Just as we saw for the gamma-Poisson model, we could also consider the conditional distribution of p after observing y.

Theorem 6.5 (Conditional distribution of p for given y) *Consider the hierarchical beta-binomial model in (6.86) and assume that a value $Y = y$ has been observed. Then the conditional distribution of p is a beta distribution*

$$p|Y = y \sim \text{Beta}(\mu\phi + y, (1-\mu)\phi + n - y), \tag{6.91}$$

with mean and variance

$$\begin{aligned} E[p|Y = y] =& \tilde{\mu} = \frac{\mu\phi + y}{\phi + n} \\ V[p|Y = y] =& \frac{(\phi\mu + y)(\phi(1-\mu) + n - y)}{(\phi + n)^2(\phi + n + 1)}. \end{aligned} \tag{6.92}$$

The proof of Theorem 6.5 is done by directly inserting the joint and marginal distribution.

Proof

$$\begin{aligned} f_{p|Y}(p) =& \binom{n}{y} p^y (1-p)^{n-y} \frac{p^{\mu\phi-1}(1-p)^{(1-\mu)\phi-1}}{B(\mu\phi, (1-\mu)\phi)} \times \\ & \left(\binom{n}{y} \frac{B(y+\mu\phi, n-y+(1-\mu)\phi)}{B(\mu\phi, (1-\mu)\phi)}\right)^{-1} \\ =& \frac{p^{y+\mu\phi-1}(1-p)^{n-y+(1-\mu)\phi-1}}{B(y+\mu\phi, n-y+(1-\mu)\phi)} \\ =& \frac{p^{\tilde{\mu}\tilde{\phi}-1}(1-p)^{(1-\tilde{\mu})\tilde{\phi}-1}}{B(\tilde{\mu}\tilde{\phi}, (1-\tilde{\mu})\tilde{\phi})}, \end{aligned} \tag{6.93}$$

which is the pdf of a beta-distribution with parameters $\tilde{\mu} = \frac{\phi\mu + y}{\phi + n}$ and $\tilde{\phi} = \phi + n$, with mean

$$E[p|Y = y] = \tilde{\mu} = \frac{\mu\phi + y}{\phi + n}, \tag{6.94}$$

and variance

$$\begin{aligned} V[p|Y = y] =& \frac{\tilde{\mu}(1-\tilde{\mu})}{\tilde{\phi} + 1} \\ =& \frac{\frac{\phi\mu + y}{\phi + n} \frac{\phi(1-\mu) + n - y}{\phi + n}}{\phi + n + 1} \\ =& \frac{(\phi\mu + y)(\phi(1-\mu) + n - y)}{(\phi + n)^2(\phi + n + 1)}. \end{aligned} \tag{6.95}$$

Even though it may not be immediately clear from the expression of the conditional mean, and variance, we can write those as a function of weights on the prior and the observed probability. Define $\hat{p} = \frac{y}{n}$ and

$$w = \frac{1}{1 + \frac{\phi}{n}}. \tag{6.96}$$

Note that for $\phi >> n$, w is approximately 1 and for $n >> \phi$, w is approximately 0 (and then $1 - w$ is approximately 1). The conditional mean can be written as

$$E[p|Y = y] = w\mu + (1 - w)\hat{p}, \tag{6.97}$$

which shows that for large n, the conditional mean is approximately the observed probability.

Example 6.21 (Clothing) *Using the result above, we may formulate the model for number of changes in clothing insulation level*

$$\begin{aligned} Y_{ij}|p_{ij} &\sim Binom(p_{ij}, n_{ij}) \\ p_{ij} &\sim Beta(\mu_i, \phi), \end{aligned} \tag{6.98}$$

where Y_{ij} and n_{ij} are the number of changes in clothing insulation level and the number of recorded insulation levels for person i on day j, respectively. In addition, the probability of changing clothing insulation level on a particular day is p_{ij}. μ_i depends on the particular person, for example if it is a male or a female

$$\mu_i = \frac{e^{\beta(\texttt{sex}_i)}}{1 + e^{\beta(\texttt{sex}_i)}}. \tag{6.99}$$

Using Theorem 6.4, the likelihood is formulated by

$$L(\boldsymbol{\theta}; \boldsymbol{y}, \boldsymbol{n}) = \prod_{i=1}^{N} \binom{n_i}{y_i} \frac{B(y_i + \mu_i\phi, n_i - y_i + (1 - \mu_i)\phi)}{B(\mu_i\phi, (1 - \mu_i)\phi)}, \tag{6.100}$$

where μ_i is given by (6.99), and the log-likelihood is implemented below;

```
## Likelihood compound beta-binomial
ldCBb <- function(y,mu,phi,n){
    lchoose(n, y) + lbeta(y + mu * phi, n - y +(1 - mu) * phi) -
        lbeta(mu * phi, (1 - mu) * phi)
}
```

and combined in a regression by

```
CBinBetaReg <- function(pars,y,n,X){
  p <- dim(X)[2]
  mu <- 1 / (1 + exp( -(X %*% pars[1:p])))
  phi <- exp(pars[p+1])
  -sum(ldCBb(y, mu, phi, n))
}
```

As usual, the parameters are estimated by optimising the negative log-likelihood, and the results are compared below;

```
##                 Beta-binom Binom Gamma-Pois  Pois
## (Intercept)           -1.6  -1.6       -2.1  -2.1
## sexmale               -1.2  -1.2       -1.1  -1.1
## log-phi                2.4    NA        0.7    NA
## log-likelihood       -76.7 -84.5      -78.5 -81.9
```

The parameters cannot really be compared across distributions (i.e. Poisson and binomial), but the likelihood can and there is not a big difference between the likelihood for the two models. Further, we see that the mean value parameters do not change at all between fixed effect and random effect models.

Predictions from the models can be compared with the observations. In order to do such comparison in a coherent way, only subjects who have 15 records of changes in clothing insulation level are used for the comparison; the tables are constructed separately for males and females;

```
tabF.R

##      Obs Binom Pois Beta-Binom Gamma-Pois
## 0-1    7   5.4  6.0        8.1        9.6
## 2-3    7  11.2 10.3        7.4        4.9
## >3     8   5.5  5.7        6.5        7.4

tabM.R

##      Obs Binom Pois Beta-Binom Gamma-Pois
## 0     11  6.40 6.63        9.1        8.9
## 1-2    4  8.71 8.42        5.2        5.5
## >2     1  0.89 0.96        1.7        1.7
```

We see that overall the random effect models have better fits to the data, but also that the difference between beta-binomial and gamma-Poisson models is small.

From the discussion here, we have seen several possibilities for modelling outcomes: ranging from generalized linear mixed effect models, which are included in standard functions, to the extremely flexible setup using TMB, where basically any set up is allowed. However, the latter comes with the price of a more complex implementation of the model. Finally, we have examined some models that allow explicit solutions and a more clear interpretation of the model parameters, but with the price of less flexibility and an even more complex implementation task.

We close this chapter with a complex non-standard example of models that also allows for an explicit solution.

6.4.3 Normal distributions with random variance

In line with the results in Chapter 5, we may assume that the variance of observations depends on the subject in question, that is

$$Y_i|\sigma_i^2 \sim N(\mu, \sigma_i^2), \tag{6.101}$$

where the distribution of σ_i^2 is chosen in an appropriate way (i.e. allowing an explicit calculation of the likelihood). It turns out that the appropriate (in sense of the conjugated prior) distribution is related to the gamma distribution and the likelihood is a t-distribution. The standard Student t-distribution is given by (the parameter ν is usually called the degrees of freedom)

$$f_0(x;\nu) = \frac{\Gamma\left(\frac{\nu+1}{2}\right)}{\sqrt{\nu\pi}\,\Gamma\left(\frac{\nu}{2}\right)}\left(1+\frac{x^2}{\nu}\right)^{-\frac{\nu+1}{2}}. \tag{6.102}$$

The t-distribution is a member of the location-scale family, implying that the distribution can be written as

$$f(y;\mu,\sigma^2)=\frac{1}{\sigma}f_0\left(\frac{y-\mu}{\sigma}\right). \tag{6.103}$$

The family includes also e.g. the normal distribution. The main result (given below) of this section relies on the t-distribution being written as a member of the location-scale family.

Theorem 6.6 *Consider a generalized one-way random effects model specified by*

$$\begin{aligned} Y_i|\gamma_i^2 &\sim N(\mu,\sigma_0^2/\gamma_i)\\ \gamma_i &\sim G(1,\phi), \end{aligned} \tag{6.104}$$

where σ_i^2 are mutually independent for $i=1,\ldots,k$.

The marginal density of Y_i under this model is

$$f_Y(y)=\frac{1}{\sigma_0}f_0\left(\frac{y-\mu}{\sigma_0};2\phi\right), \tag{6.105}$$

that is a t-distribution with location parameter μ, scale parameter σ_0 and degrees of freedom 2ϕ.

The proof of this theorem follows the same line as we have seen in previous examples with the gamma-Poisson and the beta-binomial models.

Proof *The marginal distribution for Y (we suppress the index i for now) is*

$$\begin{aligned} f_Y(y)=&\int\frac{1}{\sqrt{2\pi\sigma_0^2/\gamma}}e^{-\frac{\gamma}{2\sigma_0^2}(y_i-\mu)^2}\gamma^{\phi-1}\frac{\phi^\phi}{\Gamma(\phi)}e^{-\gamma\phi}d\gamma\\ =&\frac{1}{\sqrt{2\pi\sigma_0^2}}\frac{\phi^\phi}{\Gamma(\phi)}\int\gamma^{1/2+\phi-1}e^{-\gamma\left(\frac{1}{2\sigma_0^2}(y_i-\mu)^2+\phi\right)}d\gamma\\ =&\frac{1}{\sigma_0}\frac{1}{\sqrt{2\pi}}\frac{\phi^\phi}{\Gamma(\phi)}\Gamma\left(\frac{1+2\phi}{2}\right)\left(\phi\left(\frac{1}{2\phi\sigma_0^2}(y_i-\mu)^2+1\right)\right)^{-\frac{1+2\phi}{2}}\\ =&\frac{1}{\sigma_0}\frac{1}{\sqrt{2\phi\pi}}\frac{1}{\Gamma(\phi)}\Gamma\left(\frac{1+2\phi}{2}\right)\left(\frac{1}{2\phi\sigma_0^2}(y_i-\mu)^2+1\right)^{-\frac{1+2\phi}{2}}\\ =&\frac{1}{\sigma_0}f_0\left(\frac{y-\mu}{\sigma_0};2\phi\right), \end{aligned} \tag{6.106}$$

which is (6.105)*, and hence the proof is complete.*

The construction above applies when we estimate one variance per observation. But the formulation gives a direct relation between the observation y and the parameters (μ,σ_0, and ϕ). Unlike for the beta-binomial distribution, the model above generalises, and for example

$$\begin{aligned} Y_{ij}|\sigma_i^2 &\sim N(\mu_{ij},\sigma_{ij}^2/\gamma_i)\\ \gamma_i &\sim G(1,\phi), \end{aligned} \tag{6.107}$$

where μ_{ij} and σ_{ij} could potentially be a function of some regressors.

Another generalisation of the model is

$$\begin{aligned} \mathbf{Y}_i|\gamma_i &\sim N(\boldsymbol{\mu}_i,\boldsymbol{\Sigma}_i/\gamma_i)\\ \gamma_i &\sim G(1,\phi), \end{aligned} \tag{6.108}$$

i.e. there is a within-subject variance-covariance (e.g. the solution to a mixed effect model), but the variance-covariance matrix is scaled by γ_i.

In this case, we get

$$f_Y(\boldsymbol{y}_i) = \int \frac{\gamma_i^{N_i/2}}{(2\pi)^{N_i/2}\sqrt{|\boldsymbol{\Sigma}_i|}} e^{-\frac{\gamma_i}{2}(\boldsymbol{\mu}_i-\boldsymbol{y}_i)^T\boldsymbol{\Sigma}_i^{-1}(\boldsymbol{\mu}_i-\boldsymbol{y}_i)}\gamma_i^{\phi-1}\frac{\phi^\phi}{\Gamma(\phi)}e^{-\gamma_i\phi}d\gamma_i. \tag{6.109}$$

In order to simplify the notation, we put

$$r_i^2 = (\boldsymbol{\mu}_i - \boldsymbol{y}_i)^T\boldsymbol{\Sigma}_i^{-1}(\boldsymbol{\mu}_i - \boldsymbol{y}_i), \tag{6.110}$$

and the marginal distribution becomes

$$\begin{aligned} f_Y(\boldsymbol{y}_i) =& \frac{\phi^\phi}{(2\pi)^{N_i/2}\sqrt{|\boldsymbol{\Sigma}_i|}\Gamma(\phi)}\int \gamma_i^{\phi+N_i/2-1}e^{-\gamma_i(\phi+\frac{r_i^2}{2})}d\gamma_i \\ =& \frac{\phi^\phi}{(2\pi)^{N_i/2}\sqrt{|\boldsymbol{\Sigma}_i|}\Gamma(\phi)}\left(\phi+\frac{r_i^2}{2}\right)^{-(\phi+N_i/2)}\Gamma(\phi+N_i/2) \\ =& \frac{\Gamma(\phi+N_i/2)\phi^\phi}{(2\pi)^{N_i/2}\sqrt{|\boldsymbol{\Sigma}_i|}\Gamma(\phi)}\left(\phi+\frac{r_i^2}{2}\right)^{-(\phi+N_i/2)} \\ =& \frac{\Gamma(\phi+N_i/2)}{(2\pi)^{N_i/2}\sqrt{|\boldsymbol{\Sigma}_i|}\Gamma(\phi)\phi^{N_i/2}}\left(1+\frac{r_i^2}{2\phi}\right)^{-(\phi+N_i/2)} \\ =& \frac{\Gamma\left(\frac{2\phi+N_i}{2}\right)}{(\pi)^{N_i/2}\sqrt{|\boldsymbol{\Sigma}_i|}\Gamma(\phi)(2\phi)^{N_i/2}}\left(1+\frac{r_i^2}{2\phi}\right)^{-\frac{\phi+N_i}{2}}. \end{aligned} \tag{6.111}$$

The multivariate t-distribution is given by

$$f_t(\boldsymbol{y};\boldsymbol{\mu},\boldsymbol{\Sigma},\nu) = \frac{\Gamma\left(\frac{\nu+n}{2}\right)}{\Gamma(\nu/2)\nu^{n/2}\pi^{n/2}\sqrt{|\boldsymbol{\Sigma}|}}\left(1+\frac{1}{\nu}(\boldsymbol{y}-\boldsymbol{\mu})^T\boldsymbol{\Sigma}^{-1}(\boldsymbol{y}-\boldsymbol{\mu})\right)^{-\frac{\nu+n}{2}}, \tag{6.112}$$

and hence for our model, we have the marginal distribution (and hence the likelihood)

$$\boldsymbol{y}_i \sim t_\nu(\boldsymbol{\mu}_i, \boldsymbol{\Sigma}_i), \tag{6.113}$$

with $\nu = 2\phi$. As illustrated in Example 6.22 below, this implies that we may combine the results from Chapter 5 with the random variance model presented (6.108).

As we have seen previously, it may be of interest to find the posterior distribution of the random effects; these are

$$\begin{aligned} \gamma_i|\boldsymbol{Y}_i =& \frac{f_{Y|\gamma}(\boldsymbol{Y})f_\gamma(\gamma_i)}{f_y(\boldsymbol{Y})} \\ =& \frac{\gamma^{N_i/2}e^{-\gamma_i\frac{r_i^2}{2}}}{(2\pi)^{N_i/2}\sqrt{|\boldsymbol{\Sigma}_i|}}\frac{\phi^\phi\gamma_i^{\phi-1}e^{-\gamma_i\phi}}{\Gamma(\phi)}\frac{(\pi)^{N_i/2}\sqrt{|\boldsymbol{\Sigma}_i|}\Gamma(\phi)(2\phi)^{N_i/2}}{\Gamma\left(\frac{2\phi+N_i}{2}\right)}\left(1+\frac{r_i^2}{2\phi}\right)^{\frac{\phi+N_i}{2}} \\ =& \gamma_i^{N_i/2+\phi-1}e^{-\gamma_i\left(\phi-\frac{r_i^2}{2}\right)}\frac{\phi^{\phi+N_i/2}}{\Gamma\left(\frac{2\phi+N_i}{2}\right)}\left(1+\frac{r_i^2}{2\phi}\right)^{\frac{\phi+N_i}{2}} \\ =& \frac{\gamma_i^{\frac{\phi+N_i}{2}-1}}{\Gamma\left(\frac{2\phi+N_i}{2}\right)}\left(\phi+\frac{r_i^2}{2}\right)^{\frac{\phi+N_i}{2}}e^{-\gamma_i\left(\phi-\frac{r_i^2}{2}\right)}, \end{aligned} \tag{6.114}$$

which is the pdf of a gamma distribution with shape parameter $\frac{2\phi+N_i}{2}$ and rate parameter $\phi + \frac{r_i^2}{2}$ and the posterior mean and variance are given by

$$\begin{aligned} E[\gamma_i|\mathbf{Y}_i] =& \frac{2\phi + N_i}{2}\left(\phi + \frac{r_i^2}{2}\right)^{-1} \\ V[\gamma_i|\mathbf{Y}_i] =& \frac{2\phi + N_i}{2}\left(\phi + \frac{r_i^2}{2}\right)^{-2}. \end{aligned} \tag{6.115}$$

In the example below, the formulation is used to extend the models for clothing insulation level that were developed in Chapter 5.

Example 6.22 (Clothing insulation – random variance) *As an example, we consider the clothing insulation example that we have looked at many times throughout the book. Here, we consider a simplified set up where only the average clothing insulation level of each day is considered. Hence, we consider the model*

$$\begin{aligned} Y_{ij} =& \mu_{ij} + u_i + \epsilon_{ij} \\ u_i|\gamma_i \sim& N(0, s_i^2/\gamma_i) \\ \epsilon_{ij}|\gamma_i \sim& N(0, \sigma_i^2/\gamma_i) \\ \gamma_i \sim& G(1, \phi), \end{aligned} \tag{6.116}$$

where

$$\begin{aligned} \mu_{ij} =& \beta_0 + \beta_1(\texttt{sex}_i) + \beta_2\texttt{tOut}_{ij} \\ s_i^2 =& s_0^2\alpha(\texttt{sex}_i) \\ \sigma_i^2 =& \sigma_0^2\alpha(\texttt{sex}_i). \end{aligned} \tag{6.117}$$

Using the results from Chapter 5, the model can be written as

$$\begin{aligned} \mathbf{Y}_i|\gamma_i \sim& N(\boldsymbol{\mu}_i, \boldsymbol{\Sigma}_i/\gamma_i) \\ \gamma_i \sim& G(1, \phi). \end{aligned} \tag{6.118}$$

The covariance matrix is implemented by

$$\boldsymbol{\Sigma}_i = \sigma_i^2 \boldsymbol{R}, \tag{6.119}$$

where $\boldsymbol{R}_{ii} = 1$ *and* $\boldsymbol{R}_{ij} = \rho_u$ $(i \neq j)$. *As implemented in the function below;*

```
make.sig <- function(rho,ns,sig.s){
    Sigma <- matrix(rho, ncol = ns, nrow = ns)
    diag(Sigma) <- 1
    Sigma <- Sigma * sig.s
}
```

For reporting posterior mean and variance, the following small code is used

```
report.fun <- function(M, y, mu , Sigma, phi, ns, sex, subj, s, sig.s){
    ri <- as.numeric((y - mu) %*% solve(Sigma) %*% (y - mu))
    M[s, c("Egam", "Vgam", "sex", "subj","sig.s")] <-
        c((ns / 2 + phi)/(ri / 2 + phi),
        (ns / 2 + phi) / (ri / 2 + phi)^2, sex, subj, sig.s)
    M
}
```

And finally the negative log-likelihood is implemented below (`dmvt` *is part of the R-package* `mvtnorm` *(Genz et al., 2021; Genz and Bretz, 2009))*

```
nll.t <- function(theta, y, X, X.sig, subj, full = FALSE){
    ## Parameters
    n.par <- c(mean = dim(X)[2], var = dim(X.sig)[2])
    mu <- X %*% theta[1:n.par[1]]
    sig.s <- exp(X.sig %*% theta[1:n.par[2] + n.par[1]])
    rho.u <- 2 / (1 + exp(-theta[sum(n.par) + 1])) - 1
    phi <- exp(theta[sum(n.par) + 2])
    ny <- 2 * phi
    subj.lev <- unique(subj)
    n.subj <- length(subj.lev)
    ll <- vec <- numeric(n.subj)
    ## Initialize dataframe for reporting
    M <- data.frame(sex = vec, subj = subj.lev,
                      Egam = vec, Vgam = vec, sig.s=vec)
    for(s in 1:n.subj){
        ## Calculate likelihood
        I <- subj == subj[s]
        ns <- sum(I)
        Sigma <- make.sig(rho.u,ns,sig.s[s])
        ll[s] <- dmvt(y[I] - mu[I], sigma = Sigma, df = ny,
                     log = TRUE)
        ## For reporting posterior
        M <- report.fun(M, y[I], mu[I] , Sigma, phi, ns,
                       sex[I][1], subj[s], s, sig.s[s])
    }
    if(full){return(list(ll = ll, M = M, rho.u = rho.u,
                       sigma=sqrt(sigma), phi = phi))}
    -sum(ll)
}
```

Estimation is carried out as usual by `nlminb`*;*

```
## library(mvtnorm)
dat.agrr <- aggregate(cbind(clo, sex, tOutCen) ~ subjId + day,
                      mean, data = dfMEMEx1)
dat.agrr$subjId <- as.numeric(factor(dat.agrr$subjId))
sex <- as.numeric(dat.agrr$sex)-1
X <- cbind(1,sex,dat.agrr$tOut)
X.sig <- cbind(1,sex)
y <- dat.agrr$clo
theta <- rep(0,7)

opt.tdist <- nlminb(theta, nll.t, lower = -10, upper = 10,
                    y = y, X = X, X.sig =X.sig,
                    subj = dat.agrr$subj)
```

The parameters and the Wald confidence interval for the parameters are given in the table below;

```
##        Estimate  lower    upper
## beta0     0.595  0.544   0.6462
## beta1    -0.083 -0.141  -0.0250
## beta2    -0.012 -0.018  -0.0051
## sigma     0.137  0.108   0.1750
## alpha     0.546  0.369   0.8067
## rho.u     0.741  0.629   0.8287
## phi       4.012  0.644  25.0015
```

Finally we compare the estimated random effects between males and females

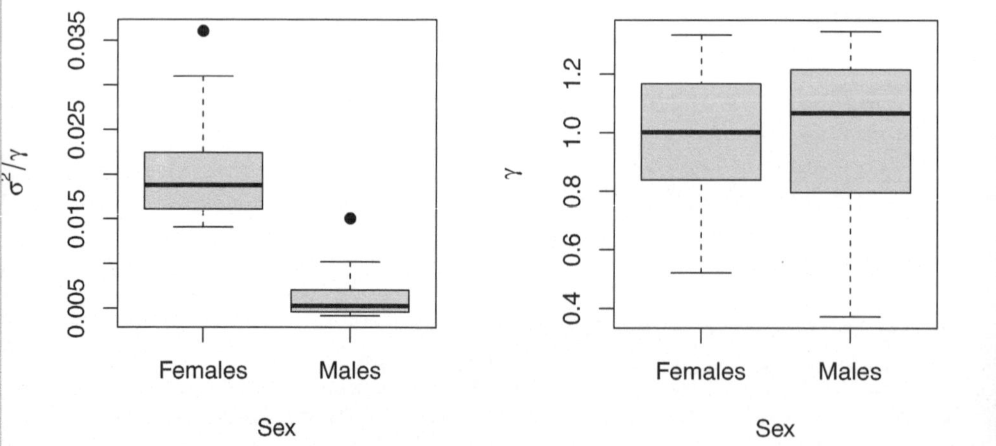

Hence, we see that the variances are very different for males and females, but that with the model setup we have here the random effects for males and females, are similar. Of course further structures could be included in the model, and further analysis of the results could also be conducted. We will, however, leave the hierarchical models with this example.

Table 6.1: Overview of models covered in Chapter 6

	Second–stage model	Exact Likelihood	Case–specific coding needed
GLMM (Section 6.2)	Normal	No	No
General Mixed effect model (Section 6.3)	Any	No	Yes
Explicit solution (Section 6.4)	Conjugate prior	Yes	Yes

6.5 Concluding remarks

This chapter covers a wide range of models that are all extensions of the models covered in Chapter 5, either by relaxing the distribution assumptions or by relaxing the linearity assumption. Table 6.1 gives an overview of the models covered in this chapter; the models also imply increasing complexity in the implementation of the models; generalized linear mixed effect models are easy to apply since these are implemented in standard software, we use `glmmTMB`, but `lme4` also include generalized linear mixed effect models (`glmer`). The benefit is that we can choose from a more general class of models.

A more general class of models than the generalized linear mixed effect model is included through the introduction of the Laplace approximation, that is handled using `TMB`, which includes estimation in very general classes of non-linear and non-Gaussian models, the models are computationally demanding and the Laplace approximation may not be accurate, the accuracy may be checked by importance sampling, see Madsen and Thyregod (2011).

The last class of models that are presented here is based on conjugate priors, where the second–stage model is chosen in such a way that the Laplace approximation of the likelihood is not needed. This gives a huge advantage in terms of calculation speed, but requires case–specific derivations and implementation. Lee and Nelder (1996) and Lee and Nelder (2001) present a compromise between the two approaches, where the second–stage distributions and link function are chosen in a proper way to set up iterative estimation algorithm, the method is implemented in the R-package `hglm` (Ronnegard et al., 2010).

Regardless of the way hierarchical models are implemented and optimised, they offer flexibility in terms of distribution assumption that is useful in cases where the normal distribution is not the proper choice.

Part II

Time–Dependent Observations, Directly Observed States

7

Survival analysis

7.1 Introduction

In many cases, occupant models are designed to make predictions of an occupant action (e.g. window opening) or the state of a building component with which occupants frequently interact (e.g. window position). These predictions are often made by looking at different environmental or categorical variables – e.g. predicting the likelihood of a window opening action as a function of the indoor and outdoor temperature. However, some types of occupant behaviour show no or very weak correlations with any environmental variables. For example, occupants' tendency to switch off their lights does not change with the daylight availability (Boyce et al., 2006). Similarly, surveys on window blinds use behaviour showed that occupants' blinds opening patterns show only a weak correlation with the ambient solar irradiance (Haldi and Robinson, 2010; Reinhart, 2004). In these cases, predicting the likelihood of an occupant action (or an occupancy event) by looking at a set of predictors becomes challenging.

Survival models in the context of occupant modelling predict the time between two consecutive occupant actions (or occupancy events) (Parys et al., 2011). In other words, they predict the lifetime of a certain occupant–controlled system's state or an occupancy state. For example, when an occupant departs for a break, a survival model can predict the duration of the absence period. In the occupant modelling literature, survival models have been used to predict the next arrival and departure time (Wang et al., 2005), or the next blinds opening (Gunay et al., 2017) and window closing times (Haldi, 2010). Because survival models do not require any environmental variables as inputs, they are well-suited for modelling aforementioned occupant behaviour types (Yan et al., 2015).

The survival models are built upon the distribution of the life-time observations of an occupancy state (or an occupant–controlled system's state). These empirical distributions are typically idealised as a continuous probability density function $f(t)$ – most commonly as a Weibull distribution. The cumulative distribution function $F(t) = \Pr(\text{Event} < t)$ provides the probability that the event has occurred at time t. The complement of the cumulative distribution function $F(t)$ gives the probability of not observing a state change before duration t – that is, the probability that the event of interest has not occurred by duration t. The complement of the cumulative distribution function $F(t)$ is known as the survival function $S(t)$.

Example 7.1 (Survival blinds use model) *The first example in this chapter is intended to illustrate the development process of a survival occupant behaviour model using a blinds use data-set from an academic office building in Ottawa, Canada. It is well documented in the literature that after a blind's closing action, it may take days or even weeks until a reopening action is observed (O'Brien et al., 2013). In line with this, eight months' worth of data gathered from eight private office spaces in an academic office*

DOI: 10.1201/9781003340812-7

building in Ottawa, Canada, contained only 72 blinds opening and 72 blinds closing actions. The survival model will be developed upon the time intervals (in days) during which a given window blind remained closed – $D_i = t_{opening,i+1} - t_{closing,i}$ *where* D *is the duration during which a given window blind remained closed and subscript* i *is the index number of the blinds opening and closing observations.*

The histogram in Figure 7.1 presents the distribution of the individual observations for duration D*. Using the R-function* `fitdistr`*, we trained a Weibull density function. The shape and scale parameters and their standard errors of the density function were determined as follows:* $\mu_{shape} = 0.62$, $\sigma_{shape} = 0.05$, $\mu_{scale} = 5.71$, $\sigma_{scale} = 1.16$.

To acquire the survival function, we will need to calculate the complement of the cumulative density function shown in the previous figure. Figure 7.2 presents the complement of the empirical cumulative density function and its approximation as a Weibull distribution. Our survival model confirms the previous observations reported in the literature. Once the occupants close their window blinds, it takes more than one–week for them to reopen in 30% of the cases (marked by horizontal grey line in the figure). The significance of this behaviour pattern for building simulation is that occupants' tardiness in blinds control leads to the underutilisation of daylight – i.e. , excess use of electric lighting. An advantage of survival models against commonly used discrete-time Markov models is that the occupant-driven events can be simulated as continuous time processes. For example, during the simulation of our blinds opening model, a uniform random number r *(between zero and one) is generated. Let's assume that* r *at* $t = 0$ *is 0.3. By looking at the survival function input that yields 0.3, one can predict that the next blinds opening instance will be in about ten days. Note that this random sampling procedure is also known as inverse transform sampling. Instead in a discrete-time Markov blinds use model, a simulated occupant makes decisions at each time-step about whether or not to change the blinds positions. Thus, survival models alleviate the building simulation from the burden of adjusting the temporal resolution of their building models to accommodate an occupant model.*

Example 7.2 (Survival occupancy model) *In this second example, a survival occupancy model is presented. The survival occupancy model in this example predicts the duration of intermediate vacancies (e.g. coffee and lunch breaks). To develop the model, one year worth of occupancy data records from 16 private offices in Ottawa, Canada,*

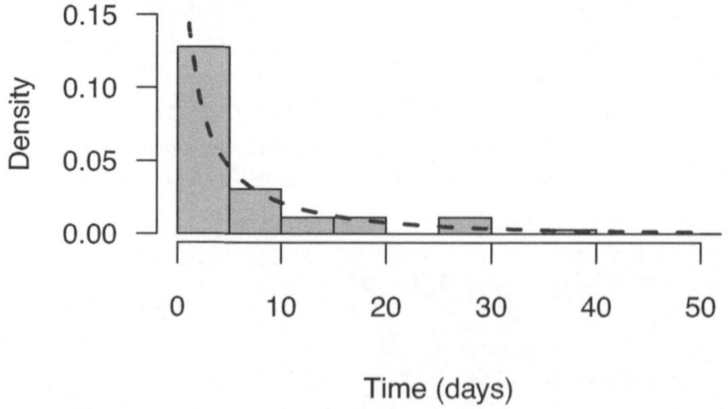

Figure 7.1: Distribution of the duration between consecutive blinds use actions. The line represents the Weibull fit.

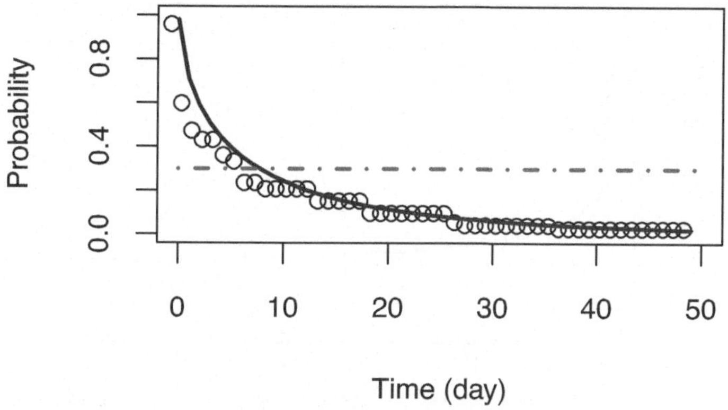

Figure 7.2: Survival function for the blinds position. The scatter plot shows the empirical and the solid line shows the Weibull fit.

were used. The offices were used by full-time faculty members. The uninterrupted vacancy periods were extracted from these data records to build the survival model. Figure 7.3 presents the complement of the empirical cumulative distribution function and its Weibull approximation. Results indicate that intermediate vacancy periods exceed 1.5 hrs in more than 30% of the observed cases. A survival vacancy model can be coupled with a Markov occupancy model such that once a simulated occupant decides to take an intermediate break, a survival vacancy model can predict the duration of this absence period. The ability to generate the number and duration of vacancy periods realistically is important because numerous occupant behaviour models (e.g. Pigg et al. (1996); Haldi (2010); Reinhart (2004); Gunay et al. (2016)) input the state of occupancy and vacancy as a predictor. For example, Pigg et al. (1996)'s light switch off model inputs the duration of absence to predict the likelihood of a switch off event at departure. Similarly, Haldi (2010)'s window and blinds use models, state of occupancy (e.g. just arrived, during intermediate occupancy) is an input.

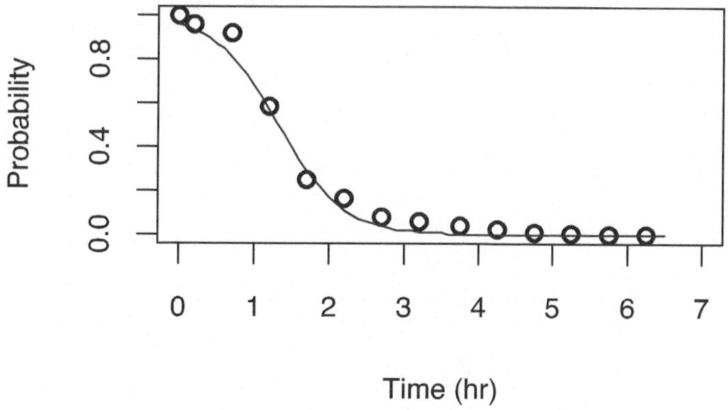

Figure 7.3: A survival model predicting the duration of intermediate vacancy periods in an office building.

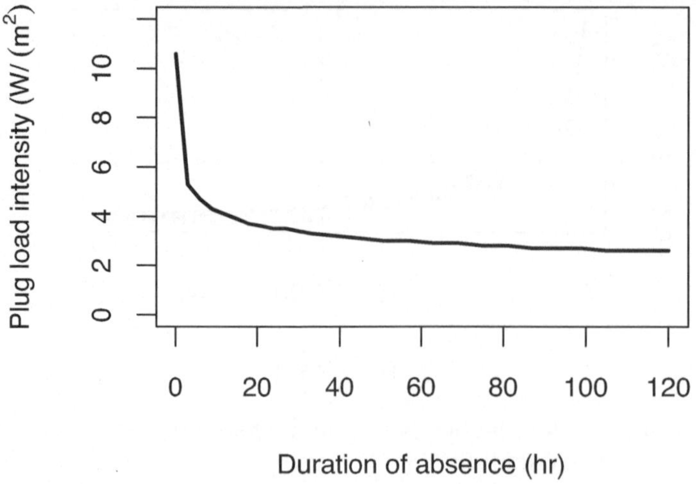

Figure 7.4: Plug load intensity as a function of the duration of absence.

Example 7.3 (Light switch of and plug load) *Two more behaviour model examples that can be used in tandem with our earlier survival vacancy model will be used. The first example is from Gunay et al. (2016). The model predicts the plug-in office equipment load intensity in vacancy periods (see Figure 7.4). The second example is from Gunay et al. (2017). The model predicts the probability of a light switch off action at departure (see Figure 7.5). Note that both models require realistic vacancy period predictions during a simulation.*

With these motivating examples presented above we will present the more formal treatment of survival data in the following sections. Starting with the visualisation of the survival function.

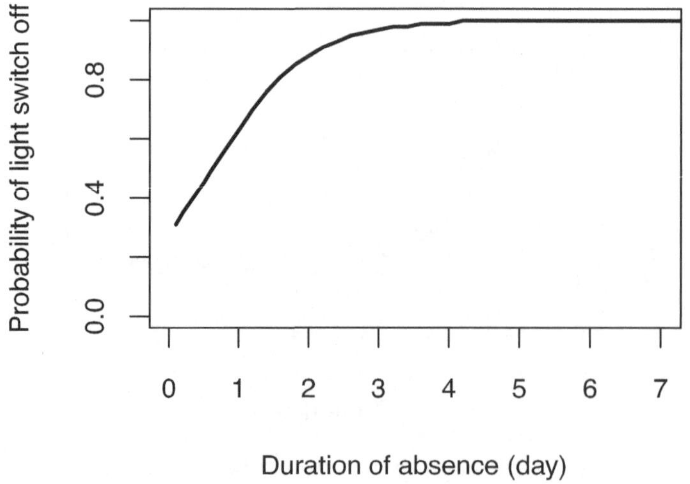

Figure 7.5: Light switch off probability at departure as a function of the duration of absence.

7.2 Survival data – the Kaplan–Meier estimator

In this section, we discuss survival data, and some non-parametric test in data, but the main discussion in this chapter is on parametric survival functions as indicated in the previous section.

In survival analysis, the objective is to describe the survival function, i.e. the probability of surviving beyond some time t; formally, we write

$$S(t) = P(T > t), \tag{7.1}$$

where T is a random variable, t is a fixed time and S is the survival function to be estimated.

7.2.1 Censoring

Survival data is often censored in some way, that is the time of the event is not measured, but we know that the time of the event is after some time t, this is called right censoring. Clearly we can imagine cases where we do not know the starting time of the experiment (called left censoring), however right censoring is the most common type of censoring; and we will only consider right censoring here. This implies that the observations consist of a vector of survival times

$$\boldsymbol{t} = \{t_1, ..., t_n\}, \tag{7.2}$$

and a vector containing censoring information

$$\boldsymbol{c} = \{c_1, ..., c_N\}. \tag{7.3}$$

7.3 Introduction to parametric models

In terms of the information from measured survival times, both censored and not censored points contain information about the survival function, but it is also clear that the information from the two different type of measurements are different. If we just briefly consider the likelihood of a set of survival times using some parametric distribution (e.g. the exponential distribution), then the likelihood can be written as

$$L(\boldsymbol{\theta}) = \prod_{i=1}^{N} f(t_i; \boldsymbol{\theta})^{1-c_i} S(t_i; \boldsymbol{\theta})^{c_i}, \tag{7.4}$$

where f is the density, t_i is the measured survival times and c_i is an indicator of censoring. The principle is illustrated in the example below.

Example 7.4 (Exponential distribution) *Assume that we have a sequence of survival time $t_1, ..., t_n$, and censoring information $c_1, ..., c_n$. Further, we assume that the survival times are exponentially distributed, i.e.*

$$f(t) = \frac{1}{\beta} e^{-t/\beta}, \tag{7.5}$$

the survival function is

$$S(t) = 1 - F(t) = 1 - (1 - e^{-t/\beta}) = e^{-t/\beta} \tag{7.6}$$

and the likelihood is therefore

$$\begin{aligned} L(\beta) &= \prod_{i=1}^{n} \left(\frac{1}{\beta}e^{-t_i/\beta}\right)^{1-c_i} \left(e^{-t_i/\beta}\right)^{c_i} \\ &= \prod_{i=1}^{n} \left(\frac{1}{\beta}\right)^{1-c_i} e^{-t_i/\beta}. \end{aligned} \tag{7.7}$$

Now let $n_c = \sum c_i$*, i.e. the number of censored points. The likelihood then reduces to*

$$\begin{aligned} L(\beta) &= \frac{1}{\beta^{n-n_c}} e^{-1/\beta \sum_{i=1}^{n} t_i} \\ &= \frac{1}{\beta^{n-n_c}} e^{-n/\beta \bar{t}}, \end{aligned} \tag{7.8}$$

and the log-likelihood is

$$l(\beta) = -(n - n_c)\log(\beta) - \frac{n}{\beta}\bar{t}, \tag{7.9}$$

with the derivative

$$l'(\beta) = -\frac{n - n_c}{\beta} + \frac{n}{\beta^2}\bar{t} \tag{7.10}$$

and the MLE

$$\hat{\beta} = \frac{n}{n - n_c}\bar{t}. \tag{7.11}$$

Hence in the case of no censored data ($n_c = 0$)*, the estimate is equal to the average observed survival time, but if censored data is present, the average is adjusted accordingly (i.e.* $\hat{\beta} > \bar{t}$*).*

Before we move on to the parametric models as introduced in the example above, we will discuss non-parametric methods for survival analysis. We start by considering an exploratory method for visualisation of the survival function.

7.4 The Kaplan–Meier estimator

In this section, we discuss one way of visualising the survival function, namely the Kaplan–Meier estimator for the survival function.

The basic idea of the Kaplan–Meier estimator is to estimate the survival function with as few assumptions as possible. To this end, all measured time points are ordered, i.e.

$$0 < t_{(1)} < t_{(2)} < .... < t_{(n)}. \tag{7.12}$$

At every recorded time a number of events happen (usually either 0 or 1), i.e. we may have some leaving the study without an event (censoring). We may denote the number of events at time t_i by $d(t_{(i)})$ (with d for death). Also at every time-point we have a record of the number at risk $R(t_{(i)})$. We can now estimate the conditional probability of surviving beyond time $t_{(i)}$ given that you survived until time $t_{(i-1)}$, using the number at risk and the number experiencing an event, we get

$$\hat{P}(T > t_{(i)}|T > t_{(i-1)}) = 1 - \frac{d(t_{(i)})}{R(t_{(i)})}, \tag{7.13}$$

where $\frac{d(t_{(i)})}{R(t_{(i)})}$ is the fraction of those at risk (at time $t_{(i)}$) that experience an event. In general, we can write the survival function as (for $t > s$)

$$S(t) = P(T > t) = P(T > t|T > s)P(T > s). \tag{7.14}$$

Now, the basic idea is to use the above formula recursively for $(t_{(i)} < t < t_{(i+1)})$, i.e.

$$S(t) = P(T > t_{(i)}|T > t_{(i-1)})P(T > t_{(i-1)}|T > t_{(i-1)}) \cdots P(T > t_{(1)}|T > 0). \tag{7.15}$$

Now, the Kaplan-Meier estimator for the survival function is

$$\begin{aligned} \hat{S}(t) &= \prod_{t \leq t_{(i)}} \hat{P}(T > t_{(i)}|T > t_{(i-1)})\hat{P}(T > t_{(i-1)}|T > t_{(i-1)}) \\ &= \prod_{t \leq t_{(i)}} \left(1 - \frac{d(t_{(i)})}{R(t_{(i)})}\right); \quad t \in [t_{(i)}, t_{(i+1)}). \end{aligned} \tag{7.16}$$

In this formulation, there is no parametric model for the survival function, and hence we do not rely on strong distribution assumptions. The principle is illustrated in the example below.

Example 7.5 (Small example of KM-estimate) *We consider a small sample of a larger data set, which will be used throughout this chapter. Time indicates how long the room is empty (no occupant) and censor indicate if data is censored (only weekdays and periods between 5am and 9pm, are considered and if a room is empty at 9pm the observation is marked as censored).*

```
##
dat.ex

##      time censor
## 1   2.00      1
## 2   2.75      0
## 3   1.00      0
## 4   2.25      0
## 5   2.00      0
## 6   0.25      1
## 7   3.00      0
## 8   2.00      0
## 9   0.50      0
## 10 2.75      1
## 11 3.75      0
## 12 1.00      0
## 13 0.25      0
## 14 3.25      1
## 15 3.00      0
```

As indicated above, we have 15 observations in our small example. From that, we can produce a table of all time-points, where there is an event, and calculate the estimated survival function;

```
survex.tab

##       t d c  R    S
## 1 0.25 1 1 15 0.93
## 2 0.50 1 0 13 0.86
## 3 1.00 2 0 12 0.72
## 4 2.00 2 1 10 0.57
## 5 2.25 1 0  7 0.49
## 6 2.75 1 1  6 0.41
## 7 3.00 2 0  4 0.21
## 8 3.25 0 1  2 0.21
## 9 3.75 1 0  1 0.00
```

Here, `t` *is the time from start of measuring (or duration before an event),* `d` *is the number of subjects that have an event at time* `t`*,* `c` *is the number of censored subjects at time* `t`*,* `R` *is the number at risk at time* `t` *and* `S` *is the estimated survival function at time* `t`*. From the example above, we see that at time 0.25 (i.e. the room has been empty for 15 minutes before some occupants enter the room), there are two events: 1 where occupants enter the room and one censored (i.e. the room is still occupied at 9pm). Hence just before time 0.25, we have $R(0.25) = 15$, therefore $S(0.25) = 1 - \frac{1}{15}$ and at time $t = 0.5$, we have $R(0.5) = 15 - 2 = 13$ and $S(0.5) = 1 - \frac{1}{13}$, and so on.*

The calculations given above give an estimate of the survival function, but it is of course of interest to evaluate the uncertainty of the estimator in order to find standard errors. This is treated in the next section.

7.4.1 Standard error and confidence interval of the Kaplan–Meier estimate

In order to find the standard error of the estimated survival function we consider the log-survival function. The log-survival function can be written as

$$\log \hat{S}(t) = \sum_{t_{(i)}<t} \log \hat{p}_i, \tag{7.17}$$

with $\hat{p}_i = \frac{R_i - d_i}{R_i}$ the mean and variance of $\hat{p}_i$ is given by (using the binomial distribution)

$$\begin{aligned} E[\hat{p}_i] &= p_i \\ V[\hat{p}_i] &= \frac{p_i(1-p_i)}{R_i}. \end{aligned} \tag{7.18}$$

Using the error propagation formula (first–order Taylor), the variance of the logarithm of a random variable can be approximated by

$$V[\log X] \approx \frac{V[X]}{E[X]^2}, \tag{7.19}$$

and an estimate of the variance of $\log(\hat{p}_i)$ is (using the mean and variance of estimated probabilities in the Binomial distribution)

$$\hat{V}[\log \hat{p}_i] = \frac{\hat{p}_i(1-\hat{p}_i)}{R_i}\frac{1}{\hat{p}_i^2} = \frac{d_i}{R_i(R_i-d_i)}, \tag{7.20}$$

and hence the variance of the log survival function can be written as (using independence)

$$\hat{V}[\log \hat{S}(t)] = \sum_{t_{(i)}<t} \hat{V}[\log \hat{p}_i] = \sum_{t_{(i)}<t} \frac{d_i}{R_i(R_i-d_i)}, \tag{7.21}$$

the final step for getting a variance estimate of the survival function is to use the inverse transformation (the exponential), and note that (again using error propagation)

$$V[e^X] \approx V[X]e^{2E[X]}. \tag{7.22}$$

Applying (7.22) to the log-survival function, we arrive at Greenwood's formula

$$\hat{V}[\hat{S}(t)] \approx \hat{S}(t)^2 \sum_{t_{(i)}<t} \frac{d_i}{R_i(R_i-d_i)}, \tag{7.23}$$

and the standard error of the survival function can be approximated by

$$SE(\hat{S}(t)) = \hat{S}(t)\sqrt{\sum_{t_{(i)}<t} \frac{d_i}{R_i(R_i-d_i)}}. \tag{7.24}$$

From this standard error, we could calculate a confidence interval directly, as $CI(t) = \hat{S}(t) \pm z_{1-\alpha/2}SE(\hat{S}(t))$; however such a confidence interval will usually extend above 1 or below 0 for some values of the estimated survival functions. The confidence intervals are therefore usually calculated in the another domain and then transformed back to the original domain. The R-package `survival` has implemented a number of different transformation (log, $\log - \log$, logit, and arcsin as well as the identity). Here, we will present the calculations for one of them namely the $\log - \log$ transformation. For this, we need (again using the error propagation method)

$$\hat{V}\left[\log\left(-\log \hat{S}(t)\right)\right] = \frac{1}{\log\left(\hat{S}(t)\right)^2} \sum_{t_{(i)}<t} \frac{d_i}{R_i(R_i-d_i)}, \tag{7.25}$$

and an $1-\alpha$ confidence interval in the $\log - \log$ domain is defined by

$$CI\left[\log\left(-\log \hat{S}(t)\right)\right] = \log\left(-\log \hat{S}(t)\right) \pm z_{1-\alpha/2}\frac{1}{\log \hat{S}(t)}\sqrt{\sum_{t_{(i)}<t} \frac{d_i}{R_i(R_i-d_i)}}, \tag{7.26}$$

and an $1-\alpha$ confidence intervalfor the survival curve is

$$CI\left[\hat{S}(t)\right] = e^{-e^{CI\left[\log(-\log \hat{S}(t))\right]}}. \tag{7.27}$$

Example 7.6 (Small example of KM-estimate - CI) *Continuing Example 7.5, confidence intervals for the survival function can now be calculated as*

```
attach(survex.tab)
se.loglog <- sqrt(1 / log(S)^2 * cumsum(d / (R * (R - d))))
lower.loglog <- log(-log(S)) - qnorm(0.975) * se.loglog
upper.loglog <- log(-log(S)) + qnorm(0.975) * se.loglog

survex.tab$lower <- exp(-exp(upper.loglog))
survex.tab$upper <- exp(-exp(lower.loglog))

survex.tab

##      t d c  R    S lower upper
## 1 0.25 1 1 15 0.93 0.613  0.99
## 2 0.50 1 0 13 0.86 0.550  0.96
## 3 1.00 2 0 12 0.72 0.411  0.88
## 4 2.00 2 1 10 0.57 0.287  0.78
## 5 2.25 1 0  7 0.49 0.218  0.72
## 6 2.75 1 1  6 0.41 0.158  0.65
## 7 3.00 2 0  4 0.21 0.036  0.47
## 8 3.25 0 1  2 0.21 0.036  0.47
## 9 3.75 1 0  1 0.00   NaN   NaN
```

Figure 7.6 shows the estimated survival function including a 95% confidence interval. The calculations above can all be done using the R-package `survival` *(Therneau et al., 2022) by*

```
## library(survival)
Surv.Ex <- survfit(Surv(time,censor == 0) ~ 1,
                   data = dat.ex, conf.type = "log-log")
```

Often, it is of interest to find confidence intervals for key numbers, for example a confidence interval for the median (or any other quantile). Such an interval can be seen in Figure 7.6 by the intersection between the horizontal dotted line and the confidence interval lines, and can also be extracted by

```
quantile(Surv.Ex)

## $quantile
##  25  50  75
## 1.0 2.2 3.0
##
## $lower
##   25   50   75
## 0.25 1.00 2.25
##
## $upper
##  25  50  75
## 2.2 3.0  NA
```

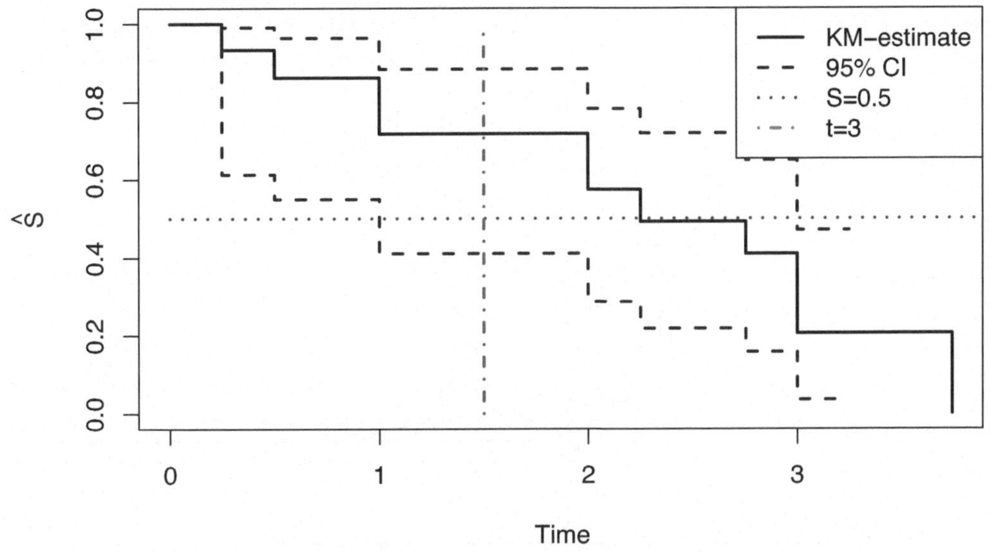

Figure 7.6: Kaplan–Meier estimate of the survival function for the data in Example 7.5.

So we see that the confidence interval for the median is $(1, 3)$, and the left end point of the confidence interval for the upper quartile is 2.25, while the right end point is undefined (which happens because of the low number of data points). We could of course also ask what is the confidence interval of the survival function after 1.5 hours (indicated by the vertical line in Figure 7.6). This is approximately $(0.4, 0.9)$, and the exact numbers can be seen in the table in Example 7.5 (as $(0.41, 0.88)$).

Example 7.7 *We continue Example 7.5, but consider all data from two rooms. The survival functions for 2 rooms can be estimated and plotted by;*

```
## KM-plot
## library(survival)
Surv.ByRoom <- survfit(Surv(time,censor == 0) ~ room,
                       data = dat.surv.room, conf.type = "log-log")

plot(Surv.ByRoom, conf.int = TRUE, xlab = "Tine [Hours]",
     col = gray(c(0, 0.5)), ylab=expression(hat(S)),lwd=2)

## Median
abline(h = 0.5, col = gray(0.25), lwd = 2,lty=3)
## Time t=5
abline(v = 5, col = gray(0.5), lwd = 2,lty = 4)
legend("topright", col = gray(c(0,0.5)),lty=c(1,1),
       c("Room 5203","Room 4206"), lwd = 2)
```

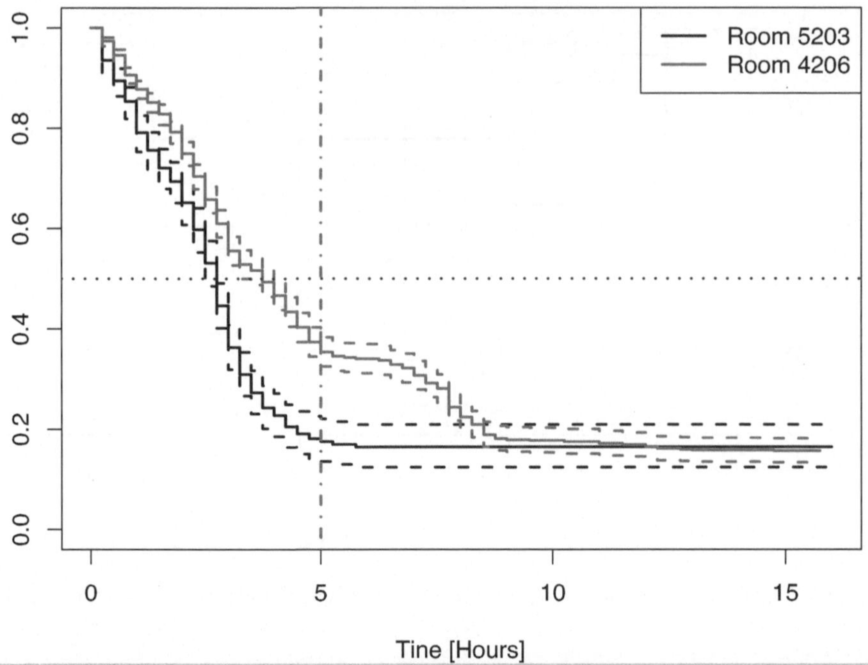

The vertical and horizontial lines are "guide to the eye", that help us to directly see the median time to event, and the survival functions at time $t = 5$. From the figure, it is clear that there are a lot of censored data (none of the curves reaches 0), and further it is clear that the survival curves are different. A summary of the data can be obtained by

```
Surv.ByRoom

## [...]
##
##                n events median 0.95LCL 0.95UCL
## room=4206  491    353   2.75    2.50    2.75
## room=5203 1334    901   3.75    3.25    4.00
```

so for room 4206, there are n=491 observations, 353 events and hence 491-353=140 censored points. For both rooms, the median and 95% confidence interval for the median are also reported (horizontal line in the figure above).

7.5 Comparing survival functions

Given two survival functions, a natural question is, if those functions are different, e.g. does the time between a window open and close differ between living room and bed room? In doing such a comparison, we should take into account all survival times; this implies that at every time point we can construct a table of events in the two groups,

Event/Group	1	0	Total
Event	d_{1i}	d_{0i}	d_i
Not event	$R_{1i} - d_{1i}$	$R_{0i} - d_{0i}$	$R_i - d_i$
At risk	R_{1i}	R_{0i}	R_i

Where the groups are for example two different rooms. The table refers to specific times, so one table for each survival time. If the totals ($R_{1,i}$, $R_{0,i}$, d_i and R_i) are fixed, then d_{1i} follows a hypergeometric distribution

$$f(d_{1,i}; R_{1,i}, R_{0,i}, d_i, R_i) = \frac{\binom{R_{1,i}}{d_{1,i}}\binom{R_i - R_{1,i}}{d_i - d_{1,i}}}{\binom{R_i}{d_i}} = \frac{\binom{R_{1,i}}{d_{1,i}}\binom{R_{2,i}}{d_{2,i}}}{\binom{R_i}{d_i}}, \tag{7.28}$$

with mean and variance

$$E[d_{1,i}] = \frac{R_{1,i}d_i}{R_i}; \quad V[d_{1,i}] = \frac{R_{1,i}R_{2,i}d_i(R_i - d_i)}{R_i^2(R_i - 1)}, \tag{7.29}$$

inserting the observed numbers, we get estimates of the expectation ($\hat{e}_{1,i}$) and variance ($\hat{v}_{1,i}$) at each survival time. Note that in most cases, $d_{1,i}$ will be equal 1 or 0. We can now construct a random variable

$$X = \sum_i d_{1,i}, \tag{7.30}$$

with expectation and variance (assuming independence between survival times)

$$E[X] = \sum_i E[d_{1,i}]; \quad V[X] = \sum_i V[d_{1,i}], \tag{7.31}$$

and hence the standardised version

$$Z = \frac{X - E[X]}{\sqrt{V[X]}}, \tag{7.32}$$

will asymptotically follow a standard normal distribution, if the null hypothesis of no difference between rooms is true, and hence $Q = Z^2 \sim \chi^2(1)$. To summarise, under the null hypothesis that there is no difference between the survival functions, the observed test statistics

$$Q = \frac{\left(\sum_i d_{1,i} - \hat{e}_{1,i}\right)^2}{\sum_i \hat{v}_{1,i}}, \tag{7.33}$$

should be compared with a χ^2-distribution with 1 degree of freedom; such test is called a log-rank test. The denominator is sometimes replaced by a weighted version of the variance estimator to give the Harrington-Fleming test statistic

$$Q_\rho = \frac{\left(\sum_i w_i(d_{1,i} - \hat{e}_{1,i})\right)^2}{\sum_i w_i^2 \hat{v}_{1,i}}, \tag{7.34}$$

w_i is found as

$$w_i = \hat{S}(t_{(i)})^\rho, \tag{7.35}$$

where $\hat{S}(t_{(i)})$ is the survival function ignoring differences between groups and $\rho \in [0, 1]$ define the weights. For $\rho = 0$, it reduces to the log-rank test(eq. (7.33)); for $\rho > 0$, the test is more likely to detect earlier survival times. The tests can be generalised to multi-group comparison, using the multivariate central hypergeometric distribution, the principles are the same as described above, but the variance is replaced by a variance covariance matrix and the differences are given as vectors, see Hosmer et al. (2008) for details.

Example 7.8 *Continuing Example 7.7, we can test if there is a significant difference between the rooms using the log-rank test with*

```
survdiff(Surv(time, censor == 0) ~ room, data = dat.surv.room)

## [...]
##                N Observed Expected (O-E)^2/E (O-E)^2/V
## room=4206  491      353      258     35.28      48.5
## room=5203 1334      901      996      9.12      48.5
##
##  Chisq= 48.5  on 1 degrees of freedom, p= 3e-12
## [...]
```

Which supports the conclusion we would make from the figure in Example 7.7, i.e. there is a difference and the p-value is very small ($3 \cdot 10^{-12}$).

The method presented in this and the previous section does not rely on distribution assumptions, and is useful for getting a first look at data, and comparing survival functions without using strong assumptions on the parametric model. For predictions, it is often useful to have a parametric model, for which some examples are presented in the next section.

7.6 Parametric survival models

The analyses so far in this chapter have focused on methods with no parametric distribution assumption on the survival function. This simplification has the advantage of very few assumptions. The drawback is that it is difficult to include many explanatory variables, and the methods covered so far do not support continuous input directly. It will therefore be difficult to make predictions based for example on the tme of day.

In this section, we focus on parametric survival models, where the parametric distribution is estimated based on the available data. As we will discuss below, the models are usually formulated for the so-called hazard function. Before we define that, we will discuss the general likelihood setup. Now assume that the data (survival times) is generated by some distribution with probability density function $f(t; \boldsymbol{\theta}, \mathbf{x}_i)$, from the *pdf* the survival function can be calculated as

$$S(t; \boldsymbol{\theta}, \mathbf{x}_i) = 1 - F(t; \boldsymbol{\theta}, \mathbf{x_i}) = 1 - \int_0^t f(u; \boldsymbol{\theta}, \mathbf{x}_i) du, \tag{7.36}$$

and the likelihood can be written as

$$L(\boldsymbol{\theta}) = \prod_{i=1}^{n} f(t_i, \boldsymbol{\theta}\mathbf{x}_i)^{d_i} S(t_i, \boldsymbol{\theta}, \mathbf{x}_i)^{1-d_i}, \tag{7.37}$$

where $\boldsymbol{\theta}$ is the vector of parameters to be estimated.

The modelling and estimation is similar to the formulation in Chapter 4, except that we usually need to consider censoring as part of the estimation (i.e. the survival function enters the likelihood). Further, the distributions that are used in survival are those that have been proven useful for that type of data.

Example 7.9 (Exponential model) *We already saw the likelihood and the MLE for the exponential survival model in Example 7.4. The usual implementation in R is given by*

$$f(t) = \frac{1}{\mu} e^{-t/\mu}, \tag{7.38}$$

μ is usually estimated in the log-domain, i.e. $\mu = e^{\beta}$ (hence $\theta = \beta$), and the log-likelihood function is

$$l(\beta) = (n - n_c)\beta - ne^{-\beta}\bar{t}, \tag{7.39}$$

with first and second derivatives

$$\begin{aligned} l'(\beta) =& (n - n_c) + ne^{-\beta}\bar{t} \\ l''(\beta) =& - ne^{-\beta}\bar{t}. \end{aligned} \tag{7.40}$$

Hence the MLE and standard error (using the inverse Hessian of the likelihood) for β is

$$\begin{aligned} \hat{\beta} =& \log\left(\frac{n}{n - n_c}\right) + \log(\bar{t}) \\ SE(\hat{\beta}) =& \sqrt{-l''(\hat{\beta})} = \left(ne^{-\hat{\beta}}\bar{t}\right)^{-1/2} \\ =& \sqrt{\frac{1}{n - n_c}}, \end{aligned} \tag{7.41}$$

that is the standard error is related to the number of non-censored datapoints.

If we consider the data from Example 7.5 we get the MLE and standard error

```
n <- dim(dat.ex)[1]
nc <- sum(dat.ex$censor)
c(MLE = log(n / (n - nc)) + log(mean(dat.ex$time)),
  SE = sqrt(1 / (n - nc)))

## MLE   SE
## 0.99 0.30
```

Using the R-package `survreg`*, we get the result*

```
summary(survreg(Surv(time, censor == 0) ~ 1, data = dat.ex,
                dist = "exponential"))

##
## Call:
## survreg(formula = Surv(time, censor == 0) ~ 1, data = dat.ex,
##     dist = "exponential")
##             Value Std. Error   z       p
## (Intercept) 0.995      0.302 3.3 0.00097
##
## Scale fixed at 1
##
## Exponential distribution
## Loglik(model)= -22   Loglik(intercept only)= -22
## Number of Newton-Raphson Iterations: 4
## n= 15
```

Which in addition to the MLE and standard error includes the Wald test statistics and the associated p-value for the null-hypothesis ($\beta = 0$), which is not really a relevant null-hypothesis in this case, but the standard error can be used to find confidence intervals in the usual way.

Likelihood ratio tests and standard errors can be calculated in the usual way, based on the likelihood and Hessian of the likelihood as we saw in Chapter 2.

7.6.1 The hazard function

Usually the survival function is not modelled directly, but rather a model for the so-called hazard function is instead formulated. The hazard function, $h(t)$, is the instantaneous rate of failure at time t, formally

$$h(t) = \lim_{\Delta t \to 0} \frac{Pr(t \leq T < t + \Delta t | T \geq t)}{\Delta t}. \tag{7.42}$$

The hazard function can take values between 0 and ∞, i.e. no risk at all or certain failure at time t. The hazard function is related to the survival function and with the pdf as

$$\begin{aligned} h(t) &= \lim_{\Delta t \to 0} \frac{Pr(t \leq T < t + \Delta t | T \geq t)}{\Delta t} \\ &= \lim_{\Delta t \to 0} \frac{1}{\Delta t} \frac{Pr(t \leq T < t + \Delta t)}{Pr(T \geq t)} \\ &= \frac{f(t)}{S(t)}, \end{aligned} \tag{7.43}$$

and further as

$$f(t) = \frac{dF(t)}{dt}, \tag{7.44}$$

we have

$$h(t) = -\frac{d \log S(t)}{dt}. \tag{7.45}$$

Finally, the accumulated Hazard functionis defined as

$$H(t) = \int_0^t h(u)du = -\int_0^t \frac{d\log S(u)}{du}du = -\log S(t) \tag{7.46}$$

Example 7.10 (Exponential) *We continue Example 7.9, and we can write the hazard function as (*$\log(S) = -t/\beta$*)*

$$\begin{aligned} h(t) =& \frac{d \log S(t)}{dt} \\ =& \frac{d}{dt}\left(\frac{t}{\beta}\right) \\ =& \frac{1}{\beta}. \end{aligned} \tag{7.47}$$

Hence the hazard function for the exponential model is constant, which implies that the accumulated hazard is a linear function of time and the survival function is an exponential decay.

7.6.2 Some specific failure time models

We have already treated the exponential distribution throughout the examples in this chapter, even though the choice is basically free when choosing the distribution for survival data, some distributions are more common (and useful) than others, e.g. survival times are positive and also usually right strewed. In this chapter, we will treat the following distributions for survival time:

- Exponential
- Weibull
- Log-logistic

Some features of the presented models are given in Table 7.1. For regression, it applies that (using the log-link)

$$\mu(\boldsymbol{x}_i) = e^{\boldsymbol{x}_i^T \boldsymbol{\beta}}. \tag{7.48}$$

From the table, we see that it is only in the case of the exponential distribution that the mean is equal μ, and also that the hazard function is only constant in the case of the exponential model, while for the log-logistic distribution μ equal the median, and μ is neither equal the mean or the median in the case of the Weibull distribution.

The models presented here and more generally survival models are often argued and written in the form, see e.g. Hosmer et al. (2008)

$$T_i = \mu\epsilon = e^{\boldsymbol{x}_i \boldsymbol{\beta}} \epsilon_i, \tag{7.49}$$

or

$$\log(T_i) = \boldsymbol{x}_i \boldsymbol{\beta} + \epsilon_i^*, \tag{7.50}$$

where ϵ_i follows some distribution (the choice of which leads to the distribution of T), depending on one parameter (σ). This formulation stresses the separation between regression

Table 7.1: Features of the presented distributions, for the parametric survival the regression is done wrt. $\mu(\boldsymbol{x}) = e^{\boldsymbol{x}^T\boldsymbol{\beta}}$.

Distribution	$f(t)$	$h(t)$	$S(t)$	$E[T]$	Median
Exponential	$\frac{1}{\mu}e^{-t/\mu}$	$\frac{1}{\mu}$	$e^{-t/\mu}$	μ	$\mu \log 2$
Weibull	$\frac{1}{\sigma\mu}\left(\frac{t}{\mu}\right)^{\frac{1}{\sigma}-1} e^{-(\frac{t}{\mu})^{\frac{1}{\sigma}}}$	$\frac{t^{1/\sigma-1}}{\sigma\mu^{1/\sigma}}$	$e^{-(\frac{t}{\mu})^{1/\sigma}}$	$\mu\Gamma(1+\sigma)$	$\mu(\log 2)^{\sigma}$
Log-logistic	$\frac{1}{\sigma\mu}\frac{(t/\mu)^{1/\sigma-1}}{(1+(t/\mu)^{1/\sigma})^2}$	$\frac{1}{\sigma t}\frac{(t/\mu)^{1/\sigma}}{1+(\frac{t}{\mu})^{\frac{1}{\sigma}}}$	$\frac{1}{1+(\frac{t}{\mu})^{\frac{1}{\sigma}}}$	$\frac{\mu\sigma\pi}{\sin(\pi\sigma)}$ †	μ

† For $\sigma < 1$ otherwise undefined

part μ (that is not stochastic) and the stochastic part (ϵ_i) that does not depend on the regressors. Either way, the resulting parametric distributions are the ones given in Table 7.1.

The log-likelihood can be written in terms of the hazard and the accumulated hazard function as

$$l(\theta, \sigma) = \sum_i d_i \log h(t_i, \theta, \sigma, \boldsymbol{x}_i) - H(t_i, \theta, \sigma, \boldsymbol{x}_i), \tag{7.51}$$

where d_i is the event indicator. We start by a small example:

Example 7.11 (Comparing models) *Continuing Example 7.9 (but using all data from 1 room), we start by comparing the different models for the survival time for vacancy in one room, the models are estimated by (*`status` *indicate event or censoring);*

```
fit0.wei <- survreg(Surv(time,status) ~ 1,
                    data = dat.surv.1room, dist = "weibull")
fit0.llogis<-survreg(Surv(time,status) ~ 1,
                     data = dat.surv.1room,
                     dist = "loglogistic")
```

The exponential and the Weibull models can be compared by likelihood ratio test (setting $\sigma = 1$ in the Weibull model give the exponential model), using the built-in ANOVA method.

```
anova(fit0.exp, fit0.wei)

##   Terms Resid. Df -2*LL Test Df Deviance Pr(>Chi)
## 1     1       856  2836      NA       NA       NA
## 2     1       855  2794    =  1       42  7.7e-11
```

As the p-value is extremely small, we would reject the null hypothesis and therefore prefer the Weibull model. The Weibull model and the log-logistic models are not nested and hence we cannot use a likelihood ratio test to compare those. Therefore we use AIC to compare the models;

```
AIC(fit0.wei, fit0.llogis)

##              df  AIC
## fit0.wei      2 2798
## fit0.llogis   2 2646
```

As the log-logistic models have a much smaller AIC, we would prefer the log-logistic survival function, for the given data.

The figure below compares the hazard function and the survival function for the 3 models, the hazard function (left panel) is constant for the exponential (meaning that the chance of occupants entering the room is constant independent of how long it has been empty). For the Weibull, the hazard is large in the beginning and monotonously decreasing with time (implying that the chance of occupants entering the room is large in the first minutes of it being empty). Finally for the log-logistic, the hazard is small in the beginning and increasing until a maximum at around 1 hour, and then decreasing (implying the rate of occupants arriving to (occupying) the room is greatest after approximately 1 hour). The survival functions are compared with the Kaplan-Meier plot in the right panel (the small +, means that the point is censored). From this figure, the difference between the survival functions seems small. The most important learning from the figure is that we can compare the models with data directly (i.e. compare the KM-plot and the parametric survival functions). This is possible here because we have simple models without covariates. We will get back to how this is done when we have more covariates in Section 7.8.

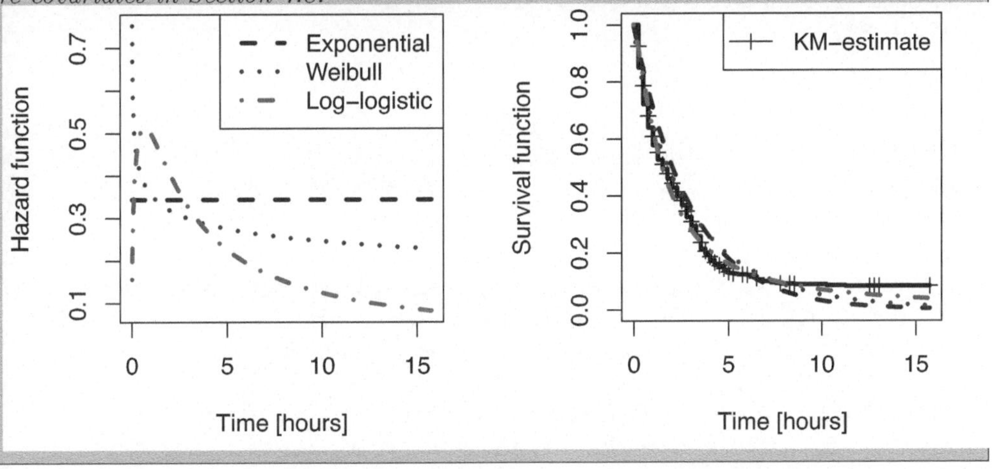

We see that the hazard functions are quite different, while it is difficult to see the difference in the resulting survival curves. As indicated above, more regressors can be included in the model, as illustrated in the example below.

Example 7.12 (More regressors) *Other regressors can be included in the usual way, and models can be compared in the usual way using analysis of deviance tables (*anova*). As an example we could include hour of day, which is also included in the data-set (this variable indicates the time of the day, when the room is emptied (or 5am for the first time of the day));*

```
fit0.llogish <- survreg(Surv(time,status) ~ hour, data = dat.surv.1room,
                        dist = "loglogistic")
```

The significance can be tested directly using Anova (i.e. likelihood ratio test);

```
anova(fit0.llogis,fit0.llogish)

##   Terms Resid. Df -2*LL Test Df Deviance Pr(>Chi)
## 1     1       855  2642       NA       NA       NA
## 2  hour       854  2572    =  1       70    5e-17
```

and a summary of the model is given below;

```
summary(fit0.llogish)

## [...]
##                Value Std. Error      z      p
## (Intercept)  1.33275    0.11222  11.88 <2e-16
## hour        -0.07704    0.00863  -8.92 <2e-16
## Log(scale)  -0.37959    0.03173 -11.96 <2e-16
##
## Scale= 0.684
##
## Log logistic distribution
## Loglik(model)= -1286   Loglik(intercept only)= -1321
##  Chisq= 70 on 1 degrees of freedom, p= 5e-17
## Number of Newton-Raphson Iterations: 4
## n= 857
## [...]
```

We see in the negative value of the coefficient for "hour" that the survival time is shorter when meeting starts later in the day (note no data during night time).

Example 7.13 (Compare rooms by parametric models) *We saw in Example 7.8 that a non-parametric test showed that the survival functions differ between the rooms. Such a hypothesis can also be tested using parametric models, for example by*

```
fit1.llogis <- survreg(Surv(time, censor) ~ factor(room),
                       data = dat.surv.room, dist = "loglogistic")
anova(fit1.llogis)

## [...]
## Terms added sequentially (first to last)
##              Df Deviance Resid. Df -2*LL Pr(>Chi)
## NULL                          1823  3929
## factor(room)  1     15.9      1822  3913 0.000065 ***
## [...]
```

Hence, there is a difference and the survival time is longer for room 5203, as can be seen from the coefficients below ($0.29 > 0$)*.*

```
coef(fit1.llogis)

##      (Intercept) factor(room)5203
##             1.79             0.29
```

In the above, we have considered some survival regression models using parametric survival models. The analysis focussed on comparing models, and while this is an important step, it does not say anything about goodness of fit, i.e. if the model is reasonable at all. For such analysis, we should consider the residual, and just as we saw in the case of generalized linear models in Chapter 4, it is not reasonable to assume that the response residuals are iid. and residuals that are related to the chosen distribution should therefore be calculated. We will consider this in Section 7.8, but first we will consider some basic key numbers that are often considered for survival models.

7.7 Features of survival functions

As we saw in the last section, the shape of the hazard function, and hence the survival function, can be determined by the chosen distribution, and often the distribution is characterized by one or a few key numbers. Some key numbers that are often used are

- Time ratio (TR): the ratio between the median time to event for different values of μ
- Hazard ratio: the ratio between hazard functions for different values of μ
- Odds ratio: the ratio between odds for different values of μ.

These 3 features are presented for the models treated here in Table 7.2, in cases where the feature is not constant the definition is given.

The time ratio tells us how much longer it takes (in median) before an event occurs in group 1 relative to group 2. We are treating linear models here and therefore the time ratio will be

$$\frac{\mu_2}{\mu_1} = e^{(\boldsymbol{x}_2^T - \boldsymbol{x}_1^T)\boldsymbol{\beta}}. \tag{7.52}$$

In cases where we compare two groups e.g. males and females, we could have $\boldsymbol{x}_1^T = (1, 0)$ for males and $\boldsymbol{x}_2^T = (1, 1)$ for females, with $\boldsymbol{\beta} = (\beta_0, \beta_1)^T$ and in this case we have

$$TR = e^{\beta_1}, \tag{7.53}$$

and in this case it implies that for $\beta_1 > 0$, females "live" longer, (or in our example stay longer out of the room), than males, further confidence intervals can be calculated directly as

$$CI_{TR} = e^{\beta_1 \pm z_{1-\alpha/2} SE(\beta_1)}, \tag{7.54}$$

where $SE(\beta_1)$ is the usual standard error and $z_{1-\alpha/2}$ the appropriate quantile in the standard normal distribution. In case of continuous regressors, we could answer questions like; Given everything else equal what would an increase of Δ in x_i imply?

In this case, we have

$$TR = e^{\beta_i \Delta}, \tag{7.55}$$

Table 7.2: Time ratio, hazard ratio, and odds ratio the presented distributions, for the parametric survival the regression is done wrt. $\mu(\boldsymbol{x}) = e^{\boldsymbol{x}^T\boldsymbol{\beta}}$.

Distribution	$TR(\mu_1, \mu_2)$	$HR(\mu_1, \mu_2)$	$OR(\mu_1, \mu_2)$
Exponential	$\frac{\mu_2}{\mu_1}$	$\frac{\mu_1}{\mu_2}$	$\frac{S(t,\mu_1)(1-S(t,\mu_2))}{S(t,\mu_2)(1-S(t,\mu_1))}$
Weibull	$\frac{\mu_2}{\mu_1}$	$\left(\frac{\mu_1}{\mu_2}\right)^{1/\sigma}$	$\frac{S(t,\mu_1)(1-S(t,\mu_2))}{S(t,\mu_2)(1-S(t,\mu_1))}$
Log-logistic	$\frac{\mu_2}{\mu_1}$	$\frac{h(t;\mu_1)}{h(t;\mu_2)}$	$\left(\frac{\mu_1}{\mu_2}\right)^{\sigma}$

with confidence interval

$$CI_{TR} = e^{\beta_i\Delta \pm \Delta z_{1-\alpha/2} SE(\beta_1)}. \tag{7.56}$$

We consider this in the example below.

Example 7.14 *To exemplify above aspects, we consider the models derived in the previous section. Referring to Example 7.12, the time ratio of survival when a meeting is started 1 hour later in the day is*

```
exp(coef(fit0.llogish)[2])

## hour
## 0.93

exp(confint(fit0.llogish)[2, ])

##  2.5 % 97.5 %
##   0.91   0.94
```

The time before the end of the occupied period will be approximately 0.93 times shorter when starting 1 hour later in the day and the confidence interval for that ratio is $[0.91, 0.94]$.

The odds ratio is calculated by

```
exp(coef(fit0.llogish)[2] * fit0.llogish$scale)

## hour
## 0.95
```

That is the odds of termination is 0.95 when the meeting starts 1 hour later. In this case, getting the confidence intervals is a bit more tricky as the odds ratio depends on two parameters.

Finally the hazard ratio is not constant in time and hence we would need to plot the ratio as a function of how long the occupation lasted.

7.8 Diagnostic plot

The purpose of diagnostic plots or residual analysis may be to check the assumptions of the model, e.g. the distributional assumptions made in the model set up, or to consider if other structures should be included, this may be other regressors or structures related to groups (e.g. random effect type of structures).

A number of different residuals are available for model checking in survival data. The simplest one is the so-called Cox-Snell residuals defined by, see e.g. Dobson and Barnett (2008),

$$r_i^{cs} = H(t_i; \boldsymbol{x}_i, \hat{\boldsymbol{\theta}}) = -\log(S(t_i; \boldsymbol{x}_i, \hat{\boldsymbol{\theta}})); \text{ if } t_i \text{ is not censored} \tag{7.57}$$

where H is the cumulative hazard function, S is the survival function and t_i is the observed survival time. If the model assumptions are true, then the Cox-Snell residual will follow an exponential distribution with mean equal 1. If there are censored data some constant is usually added to the residual. Dobson and Barnett (2008) suggest adding 1 or log(2) to censored observations.

It is not apparent from the general form for the Cox-Snell residuals given in (7.57), how the Cox-Snell residuals can be interpreted as residuals. To exemplify this point, consider the Cox-Snell residuals for the exponential model. In this case we have (see Table 7.1)

$$\begin{aligned} r_i^{CS,exp} &= -\log(S(t_i; \boldsymbol{x}_i, \hat{\boldsymbol{\theta}})) \\ &= \frac{t_i}{\hat{\mu}_i} \\ &= e^{\log(t_i) - \boldsymbol{x}_i^T \hat{\boldsymbol{\beta}}}, \end{aligned} \tag{7.58}$$

with $\mu_i = e^{\boldsymbol{x}_i^T \hat{\boldsymbol{\beta}}}$. If the assumption of the model is correct, then $t_i \sim Exp(\mu_i)$ and therefore $\frac{t_i}{\mu_i} \sim Exp(1)$ and hence the Cox–Snell residuals can be compared to the standard exponential distribution. So, the main point here is that the Cox–Snell residuals are free of the regressors, and hence they can all be compared to the same distribution. The Cox Snell residuals can now be compared directly to quantiles of the standard exponential distribution in a quantile-quantile plot similar to the one used for the general linear model, i.e. plot the pair

$$[q_{\rho_i}, r_{(i)}^{CS,exp}], \tag{7.59}$$

where $\rho_i = \frac{i-a}{n+1-2a}$ with $a \in [0,1]$ (a common choice is $a = 0.5$), and q_τ is the τ-quantile of the standard exponential distribution.

For the Weibull and log-logistic models similar arguments on transformation towards the standard exponential distribution hold, e.g. if

$$W \sim Weibull(\lambda, k), \text{ then, } \left(\frac{W}{\lambda}\right)^k \sim Exp(1). \tag{7.60}$$

Using the survival function given in Table 7.1, the Cox-Snell residuals can be written as

$$r_i^{CS,Wei} = \left(\frac{t_i}{\hat{\mu}_i}\right)^{1/\sigma}, \tag{7.61}$$

and hence since $t_i \sim Weibull(\mu_i, 1/\sigma)$ the Cox-Snell residuals will follow a standard exponential, given that the assumed Weibull distribution is correct. A similar transformation argument holds for the log-logistic distribution.

Example 7.15 *Referring to Examples 7.11 and 7.12, we can compare the Cox-Snell residuals, for the developed 1-room models. The result is shown in the figure below. The figure clearly shows that the log-logistic models have a better fit than the other two, and further that including time of day for start of the occupation of the room (Log-logistic2) also improves the fit.*

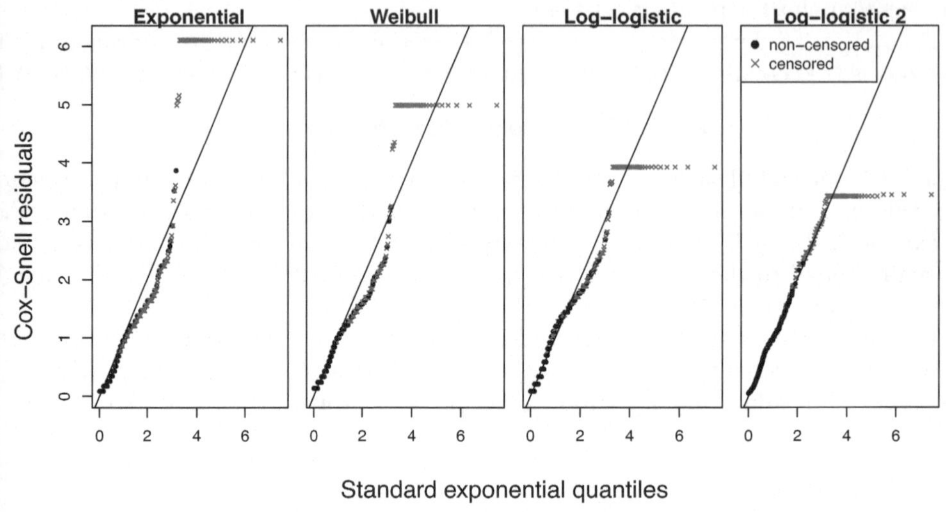

As we saw in the example above, misspecifications of the models are more clear when using the appropriate residuals than if the survival curves are compared directly to the Kaplan-Meier survival curve (see Example 7.11). Further, if regressors are included, then the survival curves cannot be compared directly since every point has their own curve (mean value model). Hence, an important part of the residual analysis is to take into account the difference in distributions, i.e. μ (defined by the regressors).

Just as we saw for generalized linear models, there are a number of different residuals that can be used for model checking. These include Martingle and deviance residuals, but the basic idea is the same for all, i.e. do a transformation such that the distribution (when the assumptions are correct) is the same for all residuals and compare the residuals to that distribution. We will not discuss those further here but a small discussion is given in Dobson and Barnett (2008).

7.9 Mixed effect

We saw in Chapter 6 that the family of generalized linear models can be extended to include mixed effects with some chosen distribution and further that the Laplace approximation is used to approximate the integral involved in evaluating the integral. In Chapter 6, we also saw that easy-to-use wrappers for so-called generalized mixed effect models (i.e. Gaussian random effects) exist, e.g. as implemented in `glmmTMB`. The package `tramME` (Tamási and Hothorn, 2021) includes survival models with mixed effects. `tramMe` is a quite general package including many of the distributions and models, we have discussed so far in this book (including e.g. ordinal regression). The package is an extension of the package `tram` (Hothorn et al., 2018; Hothorn, 2022), which also includes modelling of survival data and other type of models, for example ordinal regression.

For mathematical details and other models of other types of data than survival data, the readers are referred to Tamási and Hothorn (2021). We will however give a short presentation of the model framework with focus on the interpretation of the parameters in `tramME` and `survival`. The models in `tram` are formulated as (the formulation is slightly simplified as it does not cover all classes covered by `tram`)

$$P(Y \leq y|\boldsymbol{x}) = F(\lambda(y) - \boldsymbol{x}^T\boldsymbol{\beta}), \tag{7.62}$$

where Y is the response variable, $\boldsymbol{x}$ is the covariate, F is related to the distribution function (may include a link function) and h is a "base line" transformation (i.e. a transformation of data), which defines the parametric model. So far we have named the outcome t i.e. $y = t$ and we will follow that notation in the following, e.g. in `tram` the exponential survival model we would have $\lambda(t) = \alpha_0 + \log(y)$ and $F(\cdot) = 1 - \exp(-\exp(\cdot))$, and therefore

$$P(T \leq t) = 1 - e^{-e^{\alpha_0 + \log(t) - \boldsymbol{x}\boldsymbol{\beta}}} = 1 - e^{-t\frac{e^{\alpha_0}}{e^{\boldsymbol{x}\boldsymbol{\beta}}}}. \tag{7.63}$$

Hence for the exponential, the interpretation of the estimated slopes is the same for the two models, but the intercept is included in the baseline transformation in `tram` and $\alpha_0 = -\beta_0$. If we consider the Weibull regression model, the baseline transformation in `tram` is set to $\lambda(t) = \alpha_0 + \alpha_1 \log(t)$ and hence

$$P(T \leq t) = 1 - e^{-e^{\alpha_0 + \alpha_1 \log(t) - \boldsymbol{x}\boldsymbol{\beta}}} = 1 - e^{-t^{\alpha_1}\frac{e^{\alpha_0}}{e^{\boldsymbol{x}\boldsymbol{\beta}}}}, \tag{7.64}$$

whereas the Weibull regression model covered in the last sections is given by

$$P(T \leq t) = 1 - e^{-t^{1/\sigma} e^{\frac{\boldsymbol{x}\boldsymbol{\beta}}{\sigma}}}. \tag{7.65}$$

Hence, this is a slightly different interpretation of the parameters. While the direction is preserved between the transformations, the effect sizes have different meanings (there is a one-to-one transformation between them though). Since $P(T \leq t) = 1 - P(T > t) = 1 - S(t)$ and $-\log(S(t)) = H(t)$, we can also write this as the accumulated hazard function

$$H(t) = t^{\alpha_1}\frac{e^{\alpha_0}}{e^{\boldsymbol{x}\boldsymbol{\beta}}}. \tag{7.66}$$

The relation is exemplified in the example below.

Example 7.16 (Comparing `tram` and `survreg`) *In this section we will use a larger dataset including multiple rooms to exemplify how the mixed effect models work. As a start, we will compare the output from fixed effect models using* `survreg` *and* `tramME`*, so we estimate the models for each of the formulations (for the purpose of illustration we only use 4 rooms in this example);*

```
logLik(fit0Surv.wei)

## 'log Lik.' -5913 (df=5)

coef(fit0Surv.wei)

## (Intercept)    room4206    room5201    room5202
##        2.01       -0.62        1.24        0.23

(fit1Tram.wei<-SurvregME(Surv(time, status) ~ 1 + room,
                       data = dat.survReduced, dist = "weibull"))

##
## Weibull Linear Regression Model
##
##  Formula:
## Surv(time, status) ~ 1 + room
##
##  Fitted to dataset dat.survReduced
##
##  Fixed effects parameters:
##  ==========================
##
## room4206 room5201 room5202
##    -0.53     1.06     0.20
##
##  Log-likelihood: -5913 (npar = 5)
```

From the output above, we see that **tramME** *does not per default print parameters related to the base line transformation (i.e. α_0 and α_1). Further we see that the likelihood in the 2 cases is exactly the same and hence these are identical models. At the same time, the parameters are different due to the interpretation that was discussed earlier. All parameters from the* **tramME** *model can be extracted by*

```
fit1Tram.wei$opt$par
```

Here the first two parameters are related to the baseline transformation (i.e. α_0 and α_1); hence a direct comparison to **survreg** *can be made by;*

```
alpha <- fit1Tram.wei$opt$par[1:2]
beta <- fit1Tram.wei$opt$par[-(1:2)]
c(-alpha[1], beta) / alpha[2]

## (Intercept)    room4206    room5201    room5202
##        2.01       -0.62        1.24        0.23
```

and hence

$$\boldsymbol{\beta}_{surv} = \frac{1}{\alpha_1}[-\alpha_0, \boldsymbol{\beta}_{tram}^T]^T. \tag{7.67}$$

Adding random effects is done in a similar fashion as for generalized mixed effect models, i.e.

$$P(Y \leq y|\boldsymbol{x}, \boldsymbol{z}, \boldsymbol{u}) = F(\lambda(y) - \boldsymbol{x}^T\boldsymbol{\beta} - \boldsymbol{z}^T\boldsymbol{u}); \quad \boldsymbol{u} \sim N(\boldsymbol{0}, \boldsymbol{\Sigma}), \tag{7.68}$$

and the marginal likelihood is calculated in a similar way as we saw in Chapter 6, integrating the joint likelihood over the random effect $\boldsymbol{u}$ using the Laplace approximation.

Example 7.17 (Using tramME) *Mixed effect survival models can be estimated using* `tramME` *by;*

```
fit1.llogis<-SurvregME(Surv(time, status) ~ 1 + (1 | room),
                       data = dat.surv.4rooms, dist = "loglogistic")
```

That is a syntax similar to `lme4`*, meaning that we see the rooms as random effects and use the logistic distribution. That is*

$$P(T > t|room) = 1 - F(\lambda(t) - u_{room}); \quad u_{room} \sim N(0, \sigma^2_{room}), \tag{7.69}$$

where the intercept and the scale parameters are included in the function λ, as we also saw for the Weibull distribution above, and in the specific case of the log-logistic distribution this implies that ($u_{room} \sim N(0, \sigma^2_{room})$)

$$S(t|room) = 1 - \frac{e^{log(t)\alpha_1 + \alpha_0 log(t) - u_{room}}}{1 + e^{log(t)\alpha_1 + \alpha_0 log(t) - u_{room}}} \tag{7.70}$$

$$= \frac{1}{1 + e^{log(t)\alpha_1 + \alpha_0 log(t) - u_{room}}}. \tag{7.71}$$

The posterior estimates of the random effects can be extracted from the model by

```
ranef(fit1.llogis)

## $room
##      (Intercept)
## 4202        -0.12
## 4205        -0.25
## 4206        -0.81
## 5201         1.18
##
## attr(,"class")
## [1] "ranef.tramME" "list"
```

The figure below shows the survival curves for each of the rooms, using the estimated random effects, i.e. $\hat{u}_{room}$.

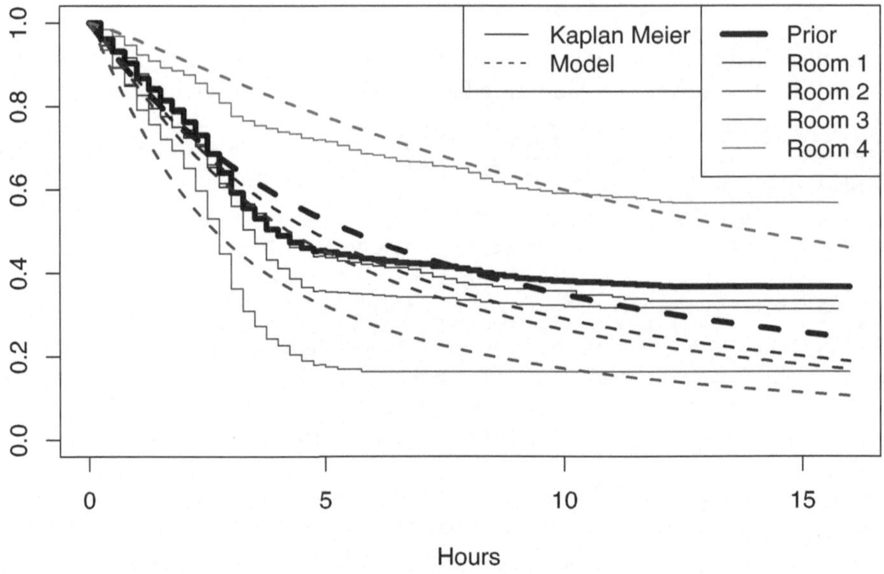

We see a huge variation between the rooms. The models can be compared by e.g. AIC

```
c(AIC(fit0m.llogis),AIC(fit1.llogis))

## [1] 9077 8734
```

and we see that including mixed effects (`fit1.llogis`*) has indeed improved the likelihood.*

As we saw in the example above, random effects can be included in survival models as well. The same discussion as we have covered earlier of course also applies here, namely that one should consider if we are interested in a population of rooms or in the particular rooms. In the latter case, we should use fixed effects models (including room as a factor).

7.10 Summary

This chapter presents a brief summary of survival models and their application in the occupant behaviour and occupancy modelling literature, and the more general theory for survival data and models. In summary,

- A survival model of a random process is the complement of the cumulative density function. It can be used as an empirical density function or can be approximated as a continuous probability distribution (e.g. a Weibull distribution). In the context of occupant modelling, a survival model predicts the life-time of a building component's state with which occupants frequently interact (e.g. blinds, windows, lighting) or the life-time of an occupancy state (presence or absence periods).

- Survival models are suitable to characterise occupant behaviour types, which cannot be predicted by common environmental indicators (e.g. indoor temperature, illuminance, relative humidity). As an example, the process to develop a survival blinds opening behaviour model is presented.

- Survival models can be used to characterise the occupancy and vacancy durations. An example survival occupancy model to predict the duration of vacancy periods is shown. The use cases for survival occupancy models are discussed by providing two behaviour model examples from the literature. These models require estimates for the duration of absence periods to make predictions of the plug load intensity and the light switch-off probability at departure.

8

Linear time–series models

8.1 Introduction

In this chapter we introduce the concepts of stochastic processes and linear time–series models. Thereby, we do not attempt to make a detailed account of stochastic processes, but in general we refer to the stochastic process as the data-generating system and the model as an object, we build to make a good description of the data at hand (i.e. approximate the data-generating process). The description is rather condensed and the justification for that is that the chapter is intended for giving a brief overview only. For a more detailed treatment of linear time–series models, we refer to e.g. Box and Jenkins (1976), Brockwell and Davis (1991), Shumway (1988) or Madsen (2008). The present chapter builds on the presentation in Madsen (2008). The models considered in this chapter are suitable for modelling linear and time–invariant processes. Further, the methods described here assume that data is sampled equidistantly (i.e. same time difference between measurements).

Linear time–series modelling is concerned with continuous outcomes (output), while occupant behaviour is often related to a discrete action, e.g. a window opening. Time-series models describes how the output or outcome (e.g. CO_2) depends on some control variable (e.g. window position) that can be controlled and/or input variables that are observed (e.g. ambient temperature). Hence in this chapter we are mostly concerned with modelling the continuous outcomes that are affected by discrete occupant actions (input). The example we will use throughout the chapter is modelling of CO_2 using an occupant action (window opening) as an input. This implies modelling an output (here CO_2) by using an occupant action, the resulting window position, as an input. Models that can take both measurements into account are treated in later chapters (Chapters 10 and 11), multivariate time–series models can also be formulated, see e.g. Madsen (2008), but will not be covered here.

We have seen examples of the importance of modelling the correlation correctly in previous chapters, and in particular this was the main theme of Chapter 5, where part of the correlation between clothing insulation level was modelled as an AR(1)process within subjects and day (see Example 5.14). The AR(1) (AutoRegresive of order 1) model, which in a sense is the most simple time-series model, will be generalised in this chapter. We will focus on identification and tools needed for that. Further, we cover methods for parameter estimation and prediction.

As outlined in the example below, time–series methods should be used when the assumption of time–independent residuals is clearly violated.

Example 8.1 (Motivating example CO_2 level) *As a motivating example we will use CO_2 levels in a bedroom. The data is a sample, from 1 dwelling covering about 35 days (i.e. a subset of the full dataset), of the data used for modelling window openings in Chapters 4 and 6, but here we focus on the CO_2 levels rather than window openings. The data is aggregated from 10 minutes to half-hourly resolution by averaging. This*

DOI: 10.1201/9781003340812-8

also implies that the window position which is originally binary can now take values 0, 1/3, 2/3, and 1. See Andersen et al. (2013) for a more comprehensive description of the data–set.

The figure below shows plots of the entire time–series used for modelling. In addition to log-CO_2 levels, the figure also shows the position of the window (1 means closed and 0 means open). Just by looking at the figure, we can see large variation and also what seems to be daily patterns.

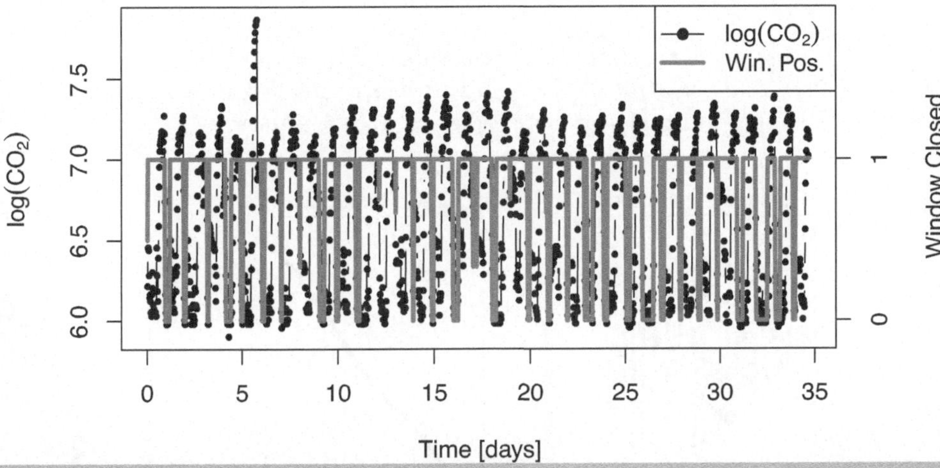

To further investigate these patterns, we could look at a smaller time window and plot CO_2 levels as a function of time of day; this is seen in the figure below.

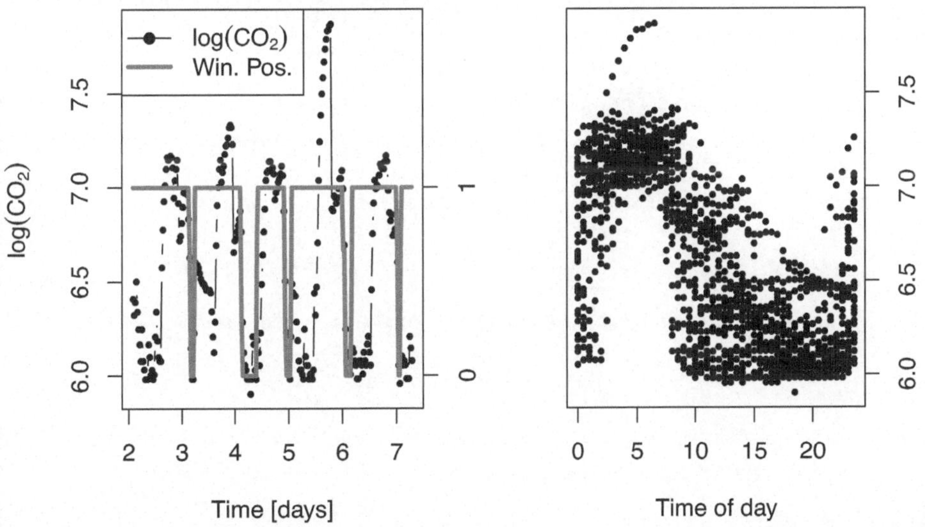

The figure clearly indicates that there is a daily pattern and that window opening will cause a decreased CO_2 level. This may lead us to a model of the form (`ToD` is short

for "Time of day")

$$\log(CO_2)_t = \beta_0(WinClo_t) + \beta_1(ToD_t) + \epsilon_t; \quad \epsilon_t \sim N(0, \sigma^2), \tag{8.1}$$

where ϵ_t are the iid residuals, and $\beta_0(WinClo_t)$ can take 2 levels depending on the window being opened or closed , $\beta_1(ToD_t)$ could be modelled as a harmonic function of the time of day. Such a model could be estimated in R by;

```
fitCO2.1 <- lm(log(CO2C) ~ WindowClosed + sin(2 * pi * Hour / 24) +
                 cos(2 * pi * Hour / 24) + sin(4 * pi * Hour / 24) +
                 cos(4 * pi * Hour / 24), data = win1.agr)
```

We will not go further into the discussion of the linear model here, but the reader is referred to Chapter 3 for a discussion. In the figures below, some residual plots are given and in addition a plot of observed ϵ_t vs. ϵ_{t-1} is shown.

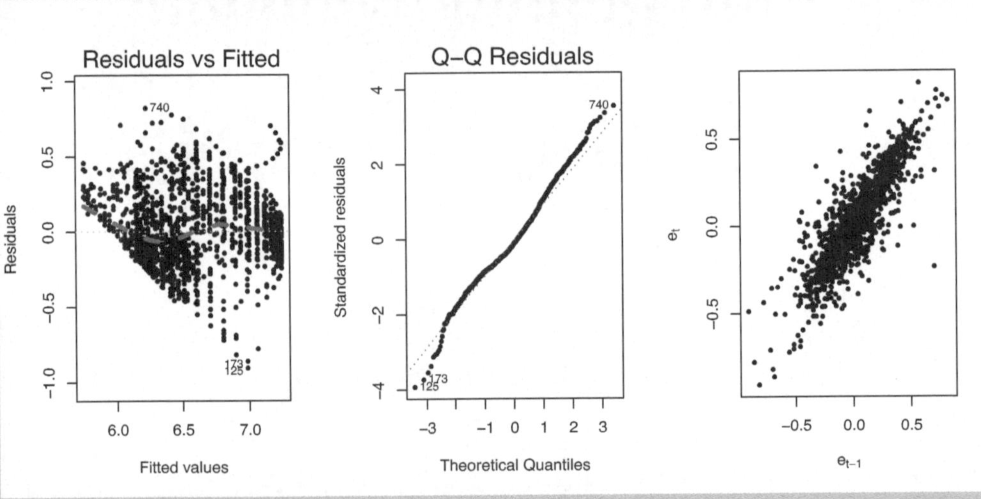

We see significant problems in the diagnostic plot: there are clear patterns in the residuals vs. fitted, and also we see a high correlation between ϵ_t and ϵ_{t-1}. The last figure clearly indicates that the independence assumption is violated.

The figure indicates that a model for the residuals could be something like

$$\epsilon_t = \epsilon_{t-1} + u_t; \quad u_t \sim N(0, \sigma_u^2). \tag{8.2}$$

This is called a random walk and starting from $\epsilon_0 = 0$ the variance of ϵ_t is

$$V[\epsilon_t] = \sigma_u^2 \cdot t, \tag{8.3}$$

that is, the variance increases with time, which would violate our intuition of CO_2 levels (that they are bounded). Another model is

$$\epsilon_t = \phi\epsilon_{t-1} + u_t, \tag{8.4}$$

with $|\phi| < 1$. Such a model is called an AR(1) model. Also, data may suggest that there is a period of one day in the residual corresponding to

$$\epsilon_t = \Phi\epsilon_{t-24h} + u_t. \tag{8.5}$$

Which model to choose should rely on analysis of the data at hand, and a simple analysis is to measure the observed correlation between lagged residuals (this is the first lag in the so-called autocorrelation function, see Section 8.2);

```
res <- residuals(fitCO2.1)
n <- length(res)
cor(res[-1], res[-n])

## [1] 0.86
```

Which suggest the model

$$\epsilon_t = 0.86\epsilon_{t-1} + u_t, \tag{8.6}$$

where $u_t \sim N(0, \sigma^2)$. Such a model includes the (auto-) correlation between observations, which is an estimated value and further the variance (σ^2) should be estimated. The model is called an AutoRegressive model of order 1 (AR(1)-model), and we will discuss generalizations of that model throughout this chapter.

The example above illustrates that in some cases the assumption about independent observations is violated, and hence we need structures related directly to the time-series at hand. We will treat a family of such models in Section 8.3, but first we will introduce the key tools for structural identification of time–series models. These are the so-called autocorrelation and partial autocorrelation functions.

The models we consider in this chapter are of the form

$$Y_t = \phi_1 Y_{t-1} + \phi_2 Y_{t-2} + ... + \phi_p Y_{t-p} + \epsilon_t + \theta_1 \epsilon_{t-1} + \theta_q \epsilon_{t-q}, \tag{8.7}$$

with $\epsilon_t \sim N(0, \sigma^2)$ and iid. This model is called an AutoRegressive Moving Average model of order (p, q) (ARMA(p,q)). Further, some regressors could also be included, for example using (8.7), on the residuals from a linear regression model (e.g. the model formulated in Example 8.1).

In principle, the models we consider in this chapter can be written as

$$\boldsymbol{Y} = \boldsymbol{X\beta} + \boldsymbol{\epsilon}; \quad \boldsymbol{\epsilon} \sim N(\boldsymbol{0}, \boldsymbol{\Sigma}), \tag{8.8}$$

where the structure of $\boldsymbol{\Sigma}$ is determined by the parameters in an ARMA(p,q) model as described above ($\boldsymbol{\Sigma} = \boldsymbol{\Sigma}(\phi_1, ..., \phi_p, \theta_1, ..., \theta_q, \sigma^2)$), and $\boldsymbol{X}$ is a collection of input variables, implying that the models are a subset of the models considered in Chapter 5, see also Example 8.4. In time–series analysis, the dimension of the times–series and hence $\boldsymbol{\Sigma}$ is however usually large, and therefore the likelihood is usually constructed recursively, by

$$L(\boldsymbol{\theta}; \boldsymbol{y}) = \prod_{t=1}^{n} L(\boldsymbol{\theta}; y_t | y_{t-1}, .., y_1), \tag{8.9}$$

for large t the recursion is straight forward as we know the past values. But if we want to use all data, the first terms have to be treated separately. We illustrate this point with a small example.

Example 8.2 (Likelihood estimation in an AR(1) process) *As we saw above, the AR(1) process is defined by*

$$Y_t = \phi_1 Y_{t-1} + \epsilon_t; \quad \epsilon_t \sim N(0, \sigma^2), \tag{8.10}$$

with ϵ_t iid. The likelihood of $\boldsymbol{\theta} = [\phi, \sigma^2]^T$ is

$$\begin{aligned} L(\boldsymbol{\theta}; \boldsymbol{y}) &= \prod_{t=1}^{n} L(\boldsymbol{\theta}; y_t|y_{t-1}) \\ &= L(\boldsymbol{\theta}; y_1|y_0) \prod_{t=2}^{n} L(\boldsymbol{\theta}; y_t|y_{t-1}), \end{aligned} \tag{8.11}$$

by construction we have

$$Y_t|Y_{t-1} = y_{t-1} \sim N(\phi y_{t-1}, \sigma^2), \tag{8.12}$$

which is well defined for all $t > 1$, while we need some other representation of $L(y_1|y_0)$. Conditioning on the first observation, we can write the likelihood as

$$\begin{aligned} L(\boldsymbol{\theta}; y_n, ..., y_2|y_1) &= \prod_{t=2}^{n} L(\boldsymbol{\theta}; y_t|y_{t-1}) \\ &= \prod_{t=2}^{n} \frac{1}{\sqrt{2\pi\sigma^2}} e^{-\frac{1}{2}\frac{(y_t - \phi y_{t-1})^2}{\sigma^2}}, \end{aligned} \tag{8.13}$$

or using the log-likelihood

$$l(\boldsymbol{\theta}) = -\frac{n-1}{2}\log(\sigma^2) - \frac{1}{2\sigma^2}\sum_{t=2}^{n}(y_t - \phi y_{t-1})^2, \tag{8.14}$$

which can be implemented in R by;

```
nll.fun <- function(theta, y){
    phi <- theta[1]; sigma.sq <- theta[2]
    ll <- 0
    for(t in 2 : length(y)){
        ll <- ll + dnorm(y[t], mean = phi * y[t-1],
                         sd = sqrt(sigma.sq), log = TRUE)}
    -ll
}
```

Using the residuals from the linear model in Example 8.1, we can estimate the parameters using the negative log-likelihood function defined above, using a standard optimiser;

```
opt.ar1 <- nlminb(c(0, 1), nll.fun, y=res,lower=c(-Inf,0))
```

The parameters of our model are;

```
phi <- opt.ar1$par[1]; sigma <- sqrt(opt.ar1$par[2])
c(phi = phi, sigma = sigma)

##   phi sigma
##  0.86  0.12
```

We see that the value of ϕ is close to the correlation we estimated in Example 8.1. The variance σ^2 of the white noise process is related to the variance of the process, but $V[Y]$ is not equal σ^2. It is rather (as we will see later)

$$V[Y_t] = \frac{\sigma^2}{1-\phi^2}, \tag{8.15}$$

and these can also be compared by

```
c(var(res), sigma^2 / (1 - phi^2))

## [1] 0.053 0.053
```

which as we see are the same numbers.

In the following, we will discuss ARMA(p,q)–models and some of their properties and how to identify the appropriate structure for the problem at hand. We will start by definitions of stochastic processes and the autocorrelation functions, which we will use for identification in ARMA(p,q)–processes.

8.2 Stochastic processes and autocorrelation

A stochastic process is defined by

$$\{Y(t); t \in \mathbb{T}\}, \tag{8.16}$$

where $\mathbb{T}$ is some space, in time–series usually $\mathbb{R}_+$ or $\mathbb{N}$. The stochastic process is characterised by its finite dimensional distribution

$$f(Y(t_1), ..., Y(t_n)). \tag{8.17}$$

In the discrete time setting that we consider here, the stochastic process can be defined by

$$\{Y_t; t \in \mathbb{N}\}, \tag{8.18}$$

where $\mathbb{N}$ is the set of natural numbers, and the finite dimensional distribution can be written as

$$f(Y_1, ..., Y_n). \tag{8.19}$$

Further for the Gaussian processes we consider in the chapter, the finite dimensional distribution can be given as

$$\boldsymbol{Y} \sim N(\boldsymbol{\mu}, \boldsymbol{\Sigma}), \tag{8.20}$$

where $\boldsymbol{Y}$ is a finite dimensional collection of points, and $\boldsymbol{\mu}$ and $\boldsymbol{\Sigma}$ are determined by the process.

A fundamental process for structural identification is white noise, which is closely related to the concept of independence and iid random variables (see Section 2.1).

Definition 8.1 (White noise) *A sequence $\{\epsilon_t\}$ is said to be a white noise process if all finite dimensional distributions are independent, with mean equal zero, i.e. if*

$$f(\epsilon_{t_1},...,\epsilon_{t_n}) = f_1(\epsilon_{t_1})\cdots f_n(\epsilon_{t_n}), \tag{8.21}$$

and $E[\epsilon_t] = 0$, if further $f_i(\cdot) = f_j(\cdot)$, it is said to be a strictly stationary white noise process.

Throughout this chapter, we will use the term white noise process for a strictly stationary Gaussian white noise process and we will only consider Gaussian models. Within this context, weak stationarity (as defined below) implies stationarity.

Example 8.3 *In Chapter 3, we defined residuals (ϵ_i) as being iid random variables with $\epsilon_i \sim N(0,\sigma^2)$, which is an example of a strictly stationary white noise process.*

A stochastic process is said to be stationary if

$$f(Y(t_1),...,Y(t_n)) = f(Y(t_1+h),...,Y(t_n+h)), \tag{8.22}$$

for all t_i and h. This implies that the finite dimensional distribution of the process is time-invariant. A stochastic process is said to be weak stationary if the first and second moment of the process are both time-invariant, that is

$$\begin{aligned} E[Y(t)] =& E[Y(t+h)] \\ Cov[Y(t_1),Y(t_1+h)] =& Cov[Y(t_2),Y(t_2+h)]. \end{aligned} \tag{8.23}$$

In application, we do not see the actual process, but rather a single realisation of the stochastic process. The realisation

$$\{y_{t_1}, y_{t_2}, ..., y_{t_n}\}, \tag{8.24}$$

is referred to as a time-series. Furthermore, in this chapter we assume that $t_{i+1} - t_i = \Delta$, i.e. equidistant sampling, and in that case we can write the time-series by

$$\{y_1,...,y_n\}. \tag{8.25}$$

In the following, we will assume that the processes are stationary. The discussion above implies that it suffices to consider the first and second order moment. The first moment (the mean) is usually zero or we can subtract the mean and hence obtain a zero first order moment. Consequently, the main analysis concentrates on the second moment, which is called the autocovariance function. This function is defined by

$$\gamma_k = Cov[Y_t, Y_{t+k}]. \tag{8.26}$$

For structural identification, we usually consider the autocorrelation function, which is defined by

$$\rho_k = \frac{Cov[Y_t, Y_{t+k}]}{\sqrt{V[Y_t]V[Y_{t+k}]}} = \frac{\gamma_k}{\gamma_0}. \tag{8.27}$$

Let us illustrate these aspects with a small example.

Example 8.4 (Autocorrelation for the AR(1)–process) *Consider the process*

$$Y_t = \phi Y_{t-1} + \epsilon_t; \quad \epsilon_t \sim N(0, \sigma^2), \tag{8.28}$$

with ϵ_t iid (i.e. white noise) and $|\phi| < 1$. The autocovariance function in lag 1 is given by

$$\gamma_1 = Cov[Y_t, Y_{t-1}] = Cov[\phi Y_{t-1} + \epsilon_t, Y_{t-1}] = \phi V[Y_{t-1}] = \phi \gamma_0, \tag{8.29}$$

and the autocovariance function in lag k is

$$\gamma_k = Cov[Y_t, Y_{t-k}] = Cov[\phi Y_{t-1} + \epsilon_t, Y_{t-k}] = \phi \gamma_{k-1}, \tag{8.30}$$

and hence $\gamma_k = \phi^k \gamma_0$. γ_0 can be calculated by

$$\gamma_0 = Cov[Y_t, Y_t] = Cov[\phi Y_{t-1} + \epsilon_t, \phi Y_{t-1} + \epsilon_t] = \phi^2 \gamma_0 + \sigma^2. \tag{8.31}$$

Solving (8.31) for γ_0 gives

$$\gamma_0 = \frac{\sigma^2}{1 - \phi^2}, \tag{8.32}$$

and the autocorrelation is therefore

$$\rho(k) = \phi^k. \tag{8.33}$$

If $\phi \geq 1$, the process is not stationary, and γ_0 will not be constant.

Referring to the example above and (8.8), the distribution of all observations from an AR(1) process can be written as

$$\boldsymbol{Y} \sim N(\boldsymbol{0}, \boldsymbol{\Sigma}), \tag{8.34}$$

with $\boldsymbol{\Sigma}_{ij} = \frac{\sigma^2 \phi^{|i-j|}}{1-\phi^2}$.

8.2.1 Partial autocorrelation function

The partial autovariance function at lag k is defined by

$$\begin{aligned} \gamma_{kk} =& E[(Y_t - E[Y_t | Y_{t+1}, .., Y_{t+k-1}])(Y_{t+k} - E[Y_{t+k} | Y_{t+1}, .., Y_{t+k-1}])] \\ =& Cov[Y_t, Y_{t+k} | Y_{t+1}, .., Y_{t+k-1}], \end{aligned} \tag{8.35}$$

i.e. the covariance between Y_t and Y_{t+k} is given by all observations in between. The partial autocorrelation function in lag k is

$$\phi_{kk} = \frac{\gamma_{kk}}{\sqrt{V[Y_t | Y_{t+1}, .., Y_{t+k-1}] V[Y_{t+k} | Y_{t+1}, .., Y_{t+k-1}]}}. \tag{8.36}$$

It follows directly from the definition that $\rho_1 = \phi_{11}$.

Again we illustrate this aspect with a small example.

Example 8.5 (Partial autocorrelation function for AR(1)) *For illustration purposes, we consider only 3 observations, as this makes it possible to do all calculations by hand. First, note that it follows from Example 8.4, that if $\{Y_t\}$ is an AR(1) process,* *then* $[Y_t, Y_{t+1}, Y_{t+2}]^T \sim N(\mathbf{0}, \sigma_y^2 \mathbf{R})$ *with* $\sigma_y^2 = \frac{\sigma^2}{1-\phi^2}$, *and* $\mathbf{R}_{ij} = \phi^{|i-j|}$. *Now using the projection theorem (Theorem 5.3), we can write*

$$\begin{aligned} E[Y_t|Y_{t+1}] &= \frac{\gamma_1}{\gamma_0} Y_{t+1} = \phi Y_{t+1} = E[Y_{t+2}|Y_{t+1}] \\ V[Y_t|Y_{t+1}] &= \gamma_0 - \frac{\gamma_1^2}{\gamma_0} = V[Y_{t+2}|Y_{t+1}], \end{aligned} \tag{8.37}$$

and hence

$$\begin{aligned} Y_t - E[Y_t|Y_{t+1}] &= Y_t - \phi Y_{t+1} \\ Y_{t+2} - E[Y_{t+2}|Y_{t+1}] &= Y_{t+2} - \phi Y_{t+1} = \epsilon_{t+2}. \end{aligned} \tag{8.38}$$

Therefore, since ϵ_s *and* Y_t *are independent for* $s > t$, $\phi_{22} = 0$. *These connections do apply for all lags* $k > 1$, *and the lag 1 partial autocorrelation is* $\phi_{11} = \rho_1$.

In general, the partial autocorrelation function can be calculated as the solution to the Yule-Walker equations

$$\begin{bmatrix} 1 & \rho_1 & \cdots & \rho_{k-1} \\ \rho_1 & 1 & \cdots & \rho_{k-2} \\ \vdots & \vdots & \ddots & \vdots \\ \rho_{k-1} & \rho_{k-2} & \cdots & 1 \end{bmatrix} \begin{bmatrix} \phi_{k1} \\ \phi_{k2} \\ \vdots \\ \phi_{kk} \end{bmatrix} = \begin{bmatrix} \rho_1 \\ \rho_2 \\ \vdots \\ \rho_k \end{bmatrix}. \tag{8.39}$$

As we will see later in this chapter, the result illustrated above generalises to more general time–series models, and the (empirical) autocorrelation and partial autocorrelation functions are key tools for identification in time-series models. In order to use the autocorrelation for identification, we need an estimator, including sampling distribution, of the autocorrelation functions.

8.2.2 Estimators for autocorrelation functions

The usual estimator for the covariance between random variables X and Y is given by

$$S_{xy} = \frac{1}{n-1} \sum_{i=1}^{n} (X_i - \bar{X})(Y_i - \bar{Y}), \tag{8.40}$$

and the usual estimator for the variance of a random variable Y is

$$S_y^2 = \frac{1}{n-1} \sum_{i=1}^{n} (Y_i - \bar{Y})^2. \tag{8.41}$$

The autocovariance function is estimated in a similar way, by

$$\hat{\gamma}_k = \frac{1}{n} \sum_{t=k+1}^{n} (Y_{t-k} - \bar{Y})(Y_t - \bar{Y}), \tag{8.42}$$

where division by $n-1$ is replaced by division by n regardless of the lag. This implies that the estimator is biased (though asymptotically consistent), but most importantly, the estimator is non-negative definite just as the true autocovariance function, see e.g. Madsen (2008). The estimator for the autocorrelation function is given by

$$\hat{\rho}_k = \frac{\hat{\gamma}_k}{\hat{\gamma}_0}, \tag{8.43}$$

and hence calculated in the same way as the theoretical autocorrelation function. The estimates of the partial autocorrelation function can be calculated from the Yule-Walker equation using the estimated autocorrelation function.

When using the estimated autocorrelation function for identification, it is important to know when it is reasonable to assume that $\rho_k = 0$. To that end, we need the variance of the estimator $\hat{\rho}_k$, and hence we need the properties of $\hat{\gamma}_k$, under the assumption that $\{Y_t\}$ is a white noise process. We start by providing the result:

Theorem 8.1 (Distribution of $\hat{\rho}_k$ and ϕ_{kk} for white noise) *Assume that $\{\epsilon_t\}$ is a white noise process with $\epsilon_t \sim N(0, \sigma^2)$. Then, the asymptotic distribution of the estimators for the autocorrelation and partial autocorrelation is*

$$\hat{\rho}_k \sim N(0, 1/n); \quad \hat{\phi}_{kk} \sim N(0, 1/n); \quad k > 0. \tag{8.44}$$

We will not go through all detail of the arguments here, but instead go through arguments for the mean and variance of the autocorrelation function. As a first approximation, we ignore the fact that $\bar{\epsilon}$ is a random variable and also assume $E[\epsilon_t] = 0$. Hence we have (assuming that n is large)

$$\hat{\gamma}_k \approx \frac{1}{n} \sum_{t=k+1}^{n} \epsilon_{t-k}\epsilon_t, \tag{8.45}$$

and since $\{\epsilon_t\}$ is a white noise process, we have

$$E[\hat{\gamma}_0] \approx \frac{1}{n} \sum_{t=1}^{n} E[\epsilon_t \epsilon_t] = \frac{1}{n} \sum_{t=1}^{n} \sigma^2 = \sigma^2, \tag{8.46}$$

and as $\{\epsilon_t\}$ is a white noise, (i.e. ϵ_t and ϵ_{t+k} are uncorrelated when $k \neq 0$), the expected values of $\hat{\gamma}_k$ can be written as

$$E[\hat{\gamma}_k] \approx \frac{1}{n} \sum_{t=k+1}^{n} E[\epsilon_{t-k}] E[\epsilon_t] = 0; \quad k \geq 1. \tag{8.47}$$

Using error propagation (i.e. first–order Taylor expansion around the expected values), we can approximate the variance of the autocorrelation estimator by

$$V[\hat{\rho}_k] \approx \left(\frac{\partial \rho_k}{\partial \gamma_k}\right)^2 V[\hat{\gamma}_k] + \left(\frac{\partial \rho_k}{\partial \gamma_0}\right)^2 V[\hat{\gamma}_0] + 2 \frac{\partial \rho_k}{\partial \gamma_0} \frac{\partial \rho_k}{\partial \gamma_k} Cov[\hat{\gamma}_0, \hat{\gamma}_k], \tag{8.48}$$

where the partial derivatives should be evaluated at the expected values and hence

$$V[\hat{\rho}_k] \approx \frac{1}{E[\hat{\gamma}_0]^2} V[\hat{\gamma}_k] + \frac{E[\hat{\gamma}_k]^2}{E[\hat{\gamma}_0]^4} V[\hat{\gamma}_0] + 2 \frac{E[\hat{\gamma}_k]}{E[\hat{\gamma}_0]^3} Cov[\hat{\gamma}_0, \hat{\gamma}_k]. \tag{8.49}$$

Still using the assumption that $\{\epsilon_t\}$ is a white noise sequence, we have that $E[\hat{\gamma}_k] = 0$, and hence the only missing part is the variance of $\hat{\gamma}_k$, which is ($k > 0$)

$$\begin{aligned} V[\hat{\gamma}_k] &= \frac{1}{n^2} V\left[\sum_{t=k+1}^{n} \epsilon_{t-k}\epsilon_t\right] \\ &= \frac{1}{n^2} \sum_{t=k+1}^{n} V\left[\epsilon_{t-k}\epsilon_t\right] + \frac{2}{n^2} \sum_{t \neq s} Cov[\epsilon_{t-k}\epsilon_t, \epsilon_{s-k}\epsilon_s]. \end{aligned} \tag{8.50}$$

Because of independence, the second term is zero and $V[\epsilon_{t-k}\epsilon_t] = V[\epsilon_{t-k}]V[\epsilon_t] = \sigma^4$ and hence

$$V[\hat{\gamma}_k] = \frac{1}{n^2}(n-k)\sigma^4 \approx \frac{\sigma^4}{n}, \tag{8.51}$$

and therefore

$$V[\hat{\rho}_k] \approx \frac{1}{\sigma^4}\frac{\sigma^4}{n} = \frac{1}{n}. \tag{8.52}$$

Using the above, an approximate 95% confidence interval for either of the autocorrelation functions is $\pm z_{0.975}/\sqrt{n}$. As we will see in the next sections, the empirical autocorrelation function can, combined with knowledge of theoretical the autocorrelation function, be used to identify structures in the observed data.

Example 8.6 (ACF of residuals) *We continue with the CO_2 example that we have analysed so far in this chapter. The autocorrelation and partial autocorrelation function can be estimated and plotted in R by* `acf` *and* `pacf`*. The figure below shows the autocorrelation and partial autocorrelation function for the residuals from the linear model in Example 8.1. The vertical lines show the autocorrelation and partial autocorrelation in different lags, and the horizontal lines show the 95% confidence interval for the autocorrelations. In our case, $n = 1701$ and hence the extend of the 95% confidence interval (under the assumption of white noise residuals) is approximately $2/\sqrt{1701} = 0.048$.*

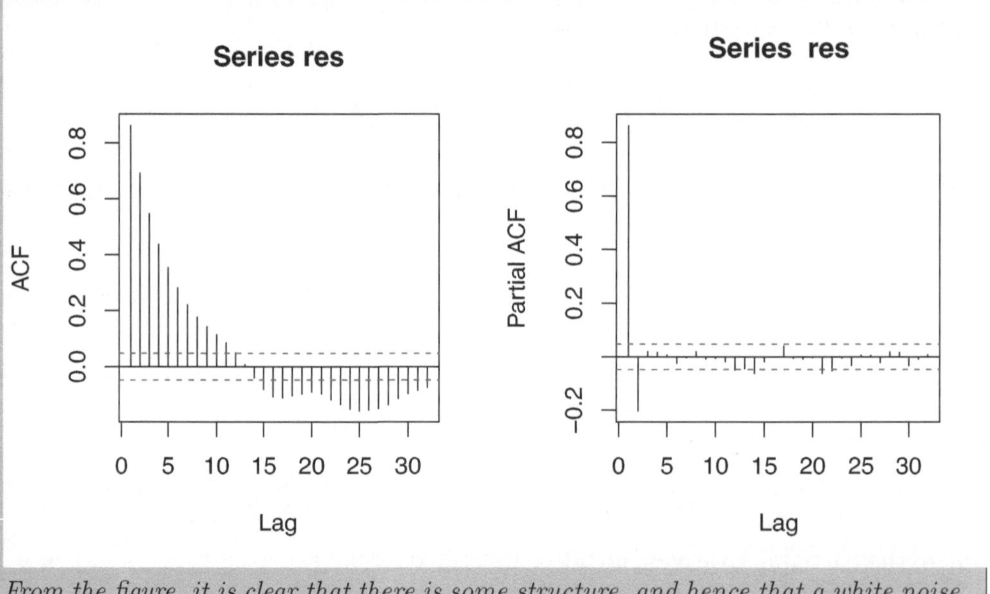

From the figure, it is clear that there is some structure, and hence that a white noise process is not sufficient. The first part of the autocorrelation function looks like an exponential decay (left figure), but there is also some further structure present. The partial autocorrelation is close to 0 (i.e. not significant or just outside the confidence interval) after lag 2 (right figure). Hence based on Examples 8.4 and 8.5, an AR(1) model is not enough to describe the structures present in the time-series (since $\phi_{22} \neq 0$). However, we can still compare the autocorrelation functions of the original series with the autocorrelation functions of the 1-step prediction error from the model estimated in Example 8.2. The 1-step prediction error for an AR(1) process is given by

$$e_{t|t-1} = y_t - \hat{\phi} y_{t-1}. \tag{8.53}$$

Hence we can calculate the 1-step prediction error by;

```
n <- length(res)
e <- res[-1] - phi * res[-n]
```

The autocorrelation functions for the 1-step prediction errors is seen in the figure below and even though the lag 1 autocorrelation is significant for both the autocorrelation and the partial autocorrelation, it is also clear that most of the structure is removed when including the AR(1) model.

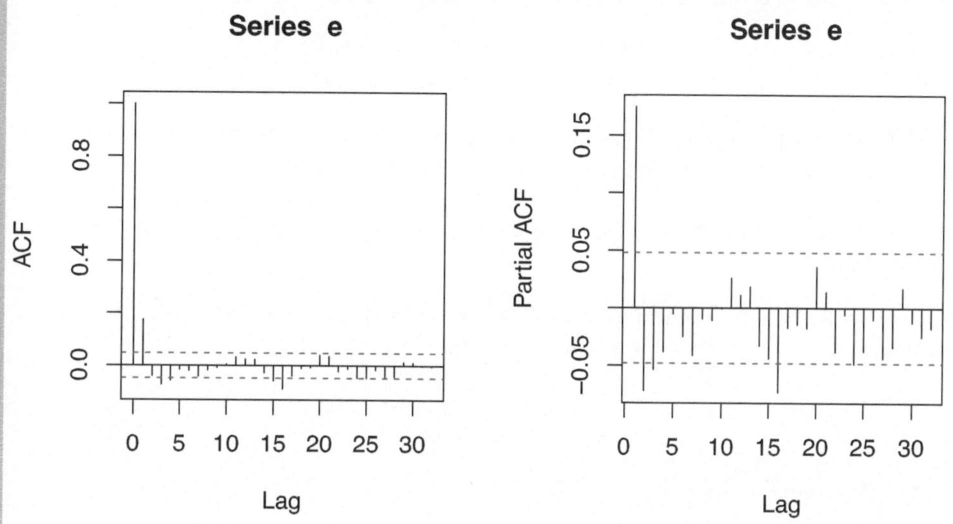

The example above illustrates how the autocorrelation functions can be used for identifying a simple structure and how it can also be used for checking if the residuals can be assumed independent. As we saw in the example, the AR(1) structure might not be enough and in the next section we will treat a more general class of time series models.

8.3 ARIMA models

As we have seen in the previous examples, we often need more structure than the AR(1) process. The most widely applied generalisation is the ARMA(p,q), which is defined by

Definition 8.2 (ARMA(p,q)–process) *The process $\{Y_t\}$ defined by*

$$Y_t + \phi_1 Y_{t-1} + \cdots + \phi_p Y_{t-p} = \epsilon_t + \theta_1 \epsilon_{t-1} + \cdots + \theta_q \epsilon_{t-q}, \tag{8.54}$$

where $\{\varepsilon_t\}$ is white noise, is called an ARMA(p,q)–process.

The definition above is not unique, and statistical software packages (e.g. R) use the formulation

$$Y_t = \phi_1 Y_{t-1} + \cdots + \phi_p Y_{t-p} + \epsilon_t + \theta_1 \epsilon_{t-1} + \cdots + \theta_q \epsilon_{t-q}, \tag{8.55}$$

and hence there is a shift in sign of the AR parameters when comparing (8.54) and the output from R.

The ARMA(p, q) process is often written in terms of the back-shift operator B, defined by

$$B^k Y_t = Y_{t-k}, \tag{8.56}$$

and using this, the ARMA(p, q)–process can be written in the more compact form

$$\phi(B)Y_t = \theta(B)\epsilon_t, \tag{8.57}$$

where $\phi(B)$ and $\theta(B)$ are polynomials in the back-shift operator

$$\begin{aligned} \phi(B) &= (1 + \phi_1 B + \ldots + \phi_p B^p) \\ \theta(B) &= (1 + \theta_1 B + \ldots + \theta_q B^q). \end{aligned} \tag{8.58}$$

The ARMA(p, q)-process is stationary if all roots of $\phi(z^{-1})$ are inside the unit circle in the complex domain, and it is said to be invertible if all roots of $\theta(z^{-1})$ are inside the unit circle.

Special cases of the ARMA(p, q) processes are the AR(p) ($\theta(B) = 1$), and the MA(q) process ($\phi(B) = 1$),

Example 8.7 (Stationarity of AR(2) process) *For the AR(2) process solving the polynomial $\phi(z^{-1}) = 0$ is the same as solving*

$$1 + \phi_1 z^{-1} + \phi_2 z^{-2} = 0, \tag{8.59}$$

or

$$z^2 + \phi_1 z + \phi_2^2 = 0, \tag{8.60}$$

with the solution

$$z_0 = -\frac{\phi_1}{2} \pm \frac{1}{2}\sqrt{\phi_1^2 - 4\phi_2}. \tag{8.61}$$

If z_0 is real (i.e. $\phi_1^2 > 4\phi_2$), we must have

$$\begin{aligned} -\phi_1 + \sqrt{\phi_1^2 - 4\phi_2} &< 2 \\ -\phi_1 - \sqrt{\phi_1^2 - 4\phi_2} &> -2, \end{aligned} \tag{8.62}$$

or

$$\begin{aligned} \phi_1^2 - 4\phi_2 < (2 - \phi_1)^2 &= 4 + \phi_1^2 - 4\phi_1 \\ \phi_1^2 - 4\phi_2 < (2 + \phi_1)^2 &= 4 + \phi_1^2 + 4\phi_1, \end{aligned} \tag{8.63}$$

which is fulfilled for $\phi_2 > -1 + \phi_1$ and $\phi_2 > -1 - \phi_1$. If the roots are complex, then the condition $|z_0| < 1$ implies

$$\frac{\phi_1^2}{4} - \frac{1}{4}(\phi_1^2 - 4\phi_2) = \phi_2 < 1. \tag{8.64}$$

Hence the AR(2) process is stationary for all values defined by the triangle $\phi_2 < 1$, $\phi_2 > -1 + \phi_1$, and $\phi_2 > -1 - \phi_1$; further, the roots are complex for $\phi_2 > \frac{\phi_1^2}{4}$.

An example of a non-stationary AR(2) process is $\phi_1 = \phi_2 = -1/2$ ($\phi_1 + \sqrt{\phi_1^2 - 4\phi_2} = 2$). Examples of stationary AR(2) processes are $\phi_1 = -0.7$, $\phi_2 = -0.1$ (real roots), and $\phi_1 = -1.3$, $\phi_2 = 0.6$ (complex roots). The autocorrelation for the latter two is shown in Figure 8.1, which illustrates the difference between real and complex roots.

The above example and discussion tell us when an ARMA process is stationary, but for identification we need results on the autocorrelation and partial autocorrelation. Such calculation can be done by inserting the process in the general formulation of the autocorrelation functions. For an ARMA(p,q)–process, this implies (considering autocovariance)

$$\begin{aligned}\gamma_k =& Cov[Y_{t+k}, Y_t] \\ =& Cov[-\phi_1 Y_{t+k-1} - \cdots \phi_p Y_{t+k-p} + \epsilon_{t+k} + \theta_1 \epsilon_{t+k-1} + \cdots + \theta_q \epsilon_{t+k-q}, Y_t] \\ =& -\phi_1 \gamma_{k-1} - \cdots - \gamma_{k-p} + \theta_1 \gamma_{\epsilon Y}(-k) + \cdots + \theta_q \gamma_{\epsilon Y}(-k+q),\end{aligned} \tag{8.65}$$

where $\gamma_{\epsilon Y}(k) = Cov[\epsilon_t, Y_{t+k}]$, it follows from the assumption on ϵ_t that $\gamma_{\epsilon Y}(k) = 0$ for $k < 0$. The properties of the autocorrelation and partial autocorrelation functions will be discussed below, but first we consider an example to illustrate the concrete calculations.

Example 8.8 (Auto correlation functions) *Consider the AR(2) process, the autocorrelation can be calculated by*

$$\begin{aligned}\gamma_0 =& Cov[Y_t, Y_t] \\ =& Cov[-\phi_1 Y_{t-1} - \phi_2 Y_{t-2} + \epsilon_t, Y_t] \\ =& -\phi_1 \gamma_1 - \phi_2 \gamma_2 + \sigma^2 \\ \gamma_1 =& Cov[-\phi_1 Y_{t-1} - \phi_2 Y_{t-2} + \epsilon_t, Y_{t-1}] \\ =& -\phi_1 \gamma_0 - \phi_2 \gamma_1 \\ \gamma_2 =& -\phi_1 \gamma_1 - \phi_2 \gamma_0.\end{aligned} \tag{8.66}$$

Hence the first lags in the autocorrelation function can be calculated from the linear systems of equations

$$\begin{bmatrix} 1 & \phi_1 & \phi_2 \\ \phi_1 & 1+\phi_2 & 0 \\ \phi_2 & \phi_1 & 1 \end{bmatrix} \begin{bmatrix} \gamma_0 \\ \gamma_1 \\ \gamma_2 \end{bmatrix} = \begin{bmatrix} \sigma^2 \\ 0 \\ 0 \end{bmatrix}, \tag{8.67}$$

and higher lags can be calculated recursively using $\gamma_k = -\phi_1 \gamma_{k-1} - \phi_2 \gamma_{k-2}$.

To illustrate differences, we also calculate the autocorrelation function for an ARMA(1,1) process. To that end, we need the cross-covariance function ($\gamma_{\epsilon Y}(k)$) between ϵ_{t-k} and Y_t, which can be found by

$$\begin{aligned}\gamma_{\epsilon Y}(0) =& Cov[\epsilon_t, Y_t] \\ =& Cov[\epsilon_t, -\phi_1 Y_{t-1} + \epsilon_t + \theta_1 \epsilon_{t-1}] \\ =& \sigma^2 \\ \gamma_{\epsilon Y}(1) =& Cov[\epsilon_{t-1}, -\phi_1 Y_{t-1} + \epsilon_t + \theta_1 \epsilon_{t-1}] \\ =& -\phi_1 \gamma_{\epsilon Y}(0) + \theta_1 \sigma^2 \\ =& \sigma^2(-\phi_1 + \theta_1),\end{aligned} \tag{8.68}$$

and in general $\gamma_{\epsilon Y}(k) = \sigma^2(-\phi_1)^{k-1}(-\phi_1 + \theta_1); \quad k > 0$.

Table 8.1: Some characteristics of the ACF and the PACF for ARMA–processes. The exponential decays and sine functions might be overlaid by initial behaviour (related to difference between p and q).

	AR(p)	MA(q)	ARMA(p, q)
ACF	Linear combination of exponential decays and sine functions.	0 after lag q.	Linear combination of exponential decays and sine functions.
PACF	0 after lag p.	Linear combination of exponential decays and sine functions.	Linear combination of exponential decays and sine functions.

The first two lags of the autocorrelation function of an ARMA(1,1) process can now be found as (we omit detailed derivations)

$$\begin{aligned}\gamma_0 =&\phi_1^2\gamma_0 + \sigma^2(1-\phi_1^2+(\theta_1-\phi_1)^2)\\ \gamma_1 =&\phi_1\gamma_0+\theta_1\sigma^2,\end{aligned} \tag{8.69}$$

and finally the autocorrelation is

$$\begin{aligned}\gamma_0 =&\sigma^2\left(1+\frac{-\phi_1+\theta_1}{1-\phi_1^2}\right)\\ \gamma_1 =&\phi_1\gamma_0+\theta_1\sigma^2\\ \gamma_2 =&\phi_1\gamma_1\\ \gamma_k =&\phi_1\gamma_{k-1}; \quad k\geq 2.\end{aligned} \tag{8.70}$$

Figure 8.1 shows some examples of the autocorrelation function and the partial autocorrelation function for two AR(2) processes (one of which has complex roots of $\phi(z^{-1})=0$*) and two ARMA(1,1) processes. We see that the partial autocorrelation function of the AR(2) processes is 0 after lags 2, while the partial autocorrelation function for the ARMA(1,1) processes decays towards zero.*

We saw in the example above that the autocorrelation function can be calculated recursively for the AR(2) and the ARMA(1,1) processes. This statement is true for general ARMA(p, q)–processes. We will however not go through the details of those calculations here. The interested reader can consult Madsen (2008). The partial autocorrelation can also be calculated by recursive methods, using the projection theorem as indicated in Example 8.7, but again we refer the reader to Madsen (2008) for details.

For model identification, the learning from the example above is that the autocorrelation and partial autocorrelation functions can be used for identification of the order of the appropriate time–series model. The specifics of the models we saw in the example generalise and the result is stated in Table 8.1. Hence, if the observed autocorrelation function decays slowly towards zero, while the observed partial autocorrelation function drops to zero after some lag p, then a good guess for the process is an AR(p) process.

8.3.1 Identification

When identifying structures (i.e. p and q), the main tool is Table 8.1 and the observed autocorrelation function and partial autocorrelation function (the inverse autocorrelation is also sometimes used, see e.g. Madsen (2008)). Hence, we rely on estimates of the autocor-

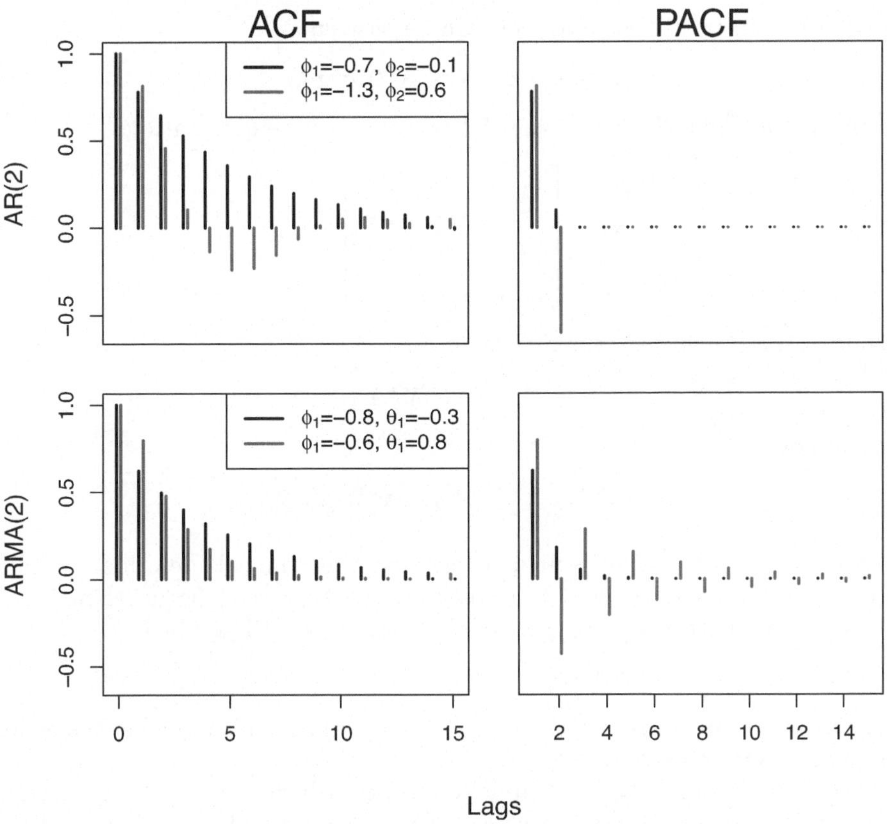

Figure 8.1: ACF and PACF for two AR(2) processes, and two ARMA(1,1) processes with different choices of parameters.

relation and partial autocorrelation functions. This also implies that one should consider the distribution of the estimators of the two functions. We will not go into details here, but just note that the estimator for autocovariance is itself autocorrelated. The practical implications of that is that we would expect to see some random noise in the autocorrelation functions and one should start by considering the first few lags and try to capture that, and let the residual analyses guide us towards more complicated models.

Example 8.9 (Identification for CO_2) *Based on the first autocorrelation functions presented in Example 8.6, we would expect that an appropriate model is an AR(2) model.*

With the tools for identification in place, we will discuss estimation in the next section.

8.4 Estimation in ARMA models

The simplest model in terms of estimation is the AR(p) model. In that case, we have the model

$$Y_t = -\phi_1 Y_{t-1} - \phi_2 Y_{t-2} - \ldots - \phi_p Y_{t-p} + \epsilon_t \quad \epsilon_t \sim N(0, \sigma^2). \tag{8.71}$$

Conditioning on the first p observations this can be written as

$$\boldsymbol{Y} = \boldsymbol{X}\boldsymbol{\phi} + \boldsymbol{\epsilon}; \quad \boldsymbol{\epsilon} \sim N(\boldsymbol{0}, \sigma^2 \boldsymbol{I}), \tag{8.72}$$

where $\boldsymbol{Y} = [y_{p+1}, ... y_n]$ and $\boldsymbol{X}$ contain the shifted values of the observations

$$\boldsymbol{X} = \begin{bmatrix} y_1 & \cdots & y_p \\ y_2 & \cdots & y_{p+1} \\ \vdots & & \vdots \\ y_{n-p} & \cdots & y_{n-1} \end{bmatrix}, \tag{8.73}$$

and the $\boldsymbol{\phi} = [\phi_1, ..., \phi_p]^T$ can be estimated by

$$\hat{\boldsymbol{\phi}} = (\boldsymbol{X}^T\boldsymbol{X})^{-1}\boldsymbol{X}^T\boldsymbol{Y}, \tag{8.74}$$

the ML estimate of σ^2 is

$$\hat{\sigma}^2 = \frac{1}{n-p}(\boldsymbol{Y} - \boldsymbol{X}\hat{\boldsymbol{\phi}})^T(\boldsymbol{Y} - \boldsymbol{X}\hat{\boldsymbol{\phi}}). \tag{8.75}$$

Even though the estimators are similar to general linear models described in Chapter 3, these are not linear projections of the observations. However, the usual methods for estimating the standard error and variance of the parameters still apply, i.e

$$V[\hat{\boldsymbol{\phi}}] \approx \sigma^2(\boldsymbol{X}^T\boldsymbol{X})^{-1}, \tag{8.76}$$

and when the assumptions of the model (i.e. $\epsilon_i \sim N(0, \sigma^2)$ and iid) are true, the estimator is consistent and asymptotically efficient.

The method is compared with the R-function `arima`, which can be used for estimation of the ARIMA processes we discuss in this chapter, in the example below (we will discuss the "I" part later in Section 8.5).

Example 8.10 (Compare estimation for AR(2)) *The simple method for estimation of AR–parameters as outlined above can be implemented as (we exclude the intercept as we know the residuals have average equal zero);*

```
n <- length(res)
Y <- res[-(1:2)]
X <- cbind(res[-c(1, n)], res[-c(n - 1, n)])
(phi <- coef(lm(Y ~ -1 + X)))

##   X1   X2
##  1.0 -0.2
```

The estimates can be compared with the usual R-function for estimating time-series models;

```
fit.ar2 <- arima(res,order=c(2, 0, 0),include.mean = FALSE)
coef(fit.ar2) - phi

##         ar1      ar2
## -0.00043  0.00017
```

We see that the estimates are almost identical. Also in estimating the ARIMA–process, `order` *refers to p, d and q (d is the order of the difference operator, which we will discuss later), and we exclude the mean value estimation as we know that the average of the time-series is zero.*

As shown in the example above, the linear regression estimates are very close, but not identical to the estimates using the build in **arima** method. They are not identical since **arima** uses all the available data to estimate the non-observed observations (i.e. for $t \leq 0$) in order to use all data. For long time–series, the estimates will be close.

For more general models, including MA-terms, the estimation problem is no longer linear, but it is also in that case possible to construct conditional models that condition on the first p observation. The simplest solution is to consider the conditional likelihood

$$L(\boldsymbol{\theta}; Y_{p+1}, ..., Y_n | Y_1, .., Y_p) = \prod_{t=p+1}^{n} f(Y_t | Y_{t-1}, ..., Y_{t-p}; \boldsymbol{\theta}), \tag{8.77}$$

where $\boldsymbol{\theta} = [\phi_1, ..., \phi_p, \theta_1, ..., \theta_p]$, we can write the residuals as

$$\epsilon_t(\boldsymbol{\theta}) = Y_t + \phi_1 Y_{t-1} + ... + \phi_p Y_{t-p} - \theta_1 \epsilon_{t-1}(\boldsymbol{\theta}) - ... - \theta_q \epsilon_{t-q}(\boldsymbol{\theta}), \tag{8.78}$$

we see that the residuals depend on $\boldsymbol{\theta}$, previous values of the observations and the previous values of the residuals. A simple solution is to set $\epsilon_t = 0$ for $t \leq p$. Doing so makes it possible to find the maximum likelihood estimates; this method is called the "conditional sum of squares". We illustrate this with a small example.

Example 8.11 (Conditional sum of squares) *In order to show the simplicity of the procedure, we can write the conditional sum of squares as a loss function;*

```
css.arma11 <- function(theta,y){
    n <- length(y)
    e <- numeric(n)
    for(i in 2:n){
        e[i] <- (y[i] - theta[1] * y[i - 1] - theta[2] * e[i - 1])
    }
    sum(e^2)
}
```

which can be estimated in the usual way using some optimiser;

```
n <- length(res)
opt.arma11 <- nlminb(c(0, 0), css.arma11, y = res)
```

and we can compare with the usual estimation using **arima**

```
arma11.fit <- arima(res,order=c(1, 0, 1), include.mean = FALSE)
coef(arma11.fit) - opt.arma11$par

##        ar1       ma1
## -0.00013 -0.00019
```

and again, since the time-series is quite long we get very similar results.

More general methods for estimation include estimations of the initial values of Y_t and ϵ_t (i.e. for $t < 1$), but we will not pursue this further here. Even though the general framework is likelihood estimation, the mean value parameters ($[\phi_1, ..., \phi_p, \theta_1, .., \theta_q]^T$) are usually estimated separately from σ^2. This implies that rather than maximising the log-likelihood, we minimise the residual sum of squares

$$RSS = \sum \epsilon_t^2(\boldsymbol{\theta}), \tag{8.79}$$

and set the variance equal

$$\hat{\sigma}^2 = \frac{RSS(\hat{\boldsymbol{\theta}})}{df}. \tag{8.80}$$

Here, df depends on the method used for estimation, for example for the conditional sum of squares we only use $n-p$ observations from which the number of estimated parameters should be subtracted. In more general methods that use all data, one should use n minus the number of estimated parameters (here $p+q$).

Example 8.12 (Variance estimates) *Using the estimates above, we can find an estimate of σ^2 by*

```
opt.arma11$objective / (n - 1 - 2)

## [1] 0.013
```

Where we divide by $n-3$ since we condition on the first observation and estimate two mean value parameters. The variance can of course also be extracted from the `arima` *fit by;*

```
arma11.fit$sigma2

## [1] 0.013
```

We conclude this subsection with a small discussion related to the CO_2 example we have used throughout this chapter. The focus will be on applying the autocorrelation functions for identification and determining if the 1-step residuals are white noise.

Example 8.13 (Autocorrelation of residuals) *Considering the ACF and PACF of the residuals from the linear model in Example 8.6 and Table 8.1, it seems plausible that a good candidate model would be an AR(2) model, which was estimated in Example 8.10. If this is indeed the correct model structure, then the 1-step ahead residuals should be independent, i.e. the autocorrelation and partial autocorrelation should be 0 in all lags.*

This is, as seen in the figure below, indeed the case (except for a single point in the autocorrelation functions that exceeds the 95% confidence interval), at least when looking at lags up to 32 (corresponding to 16 hours).

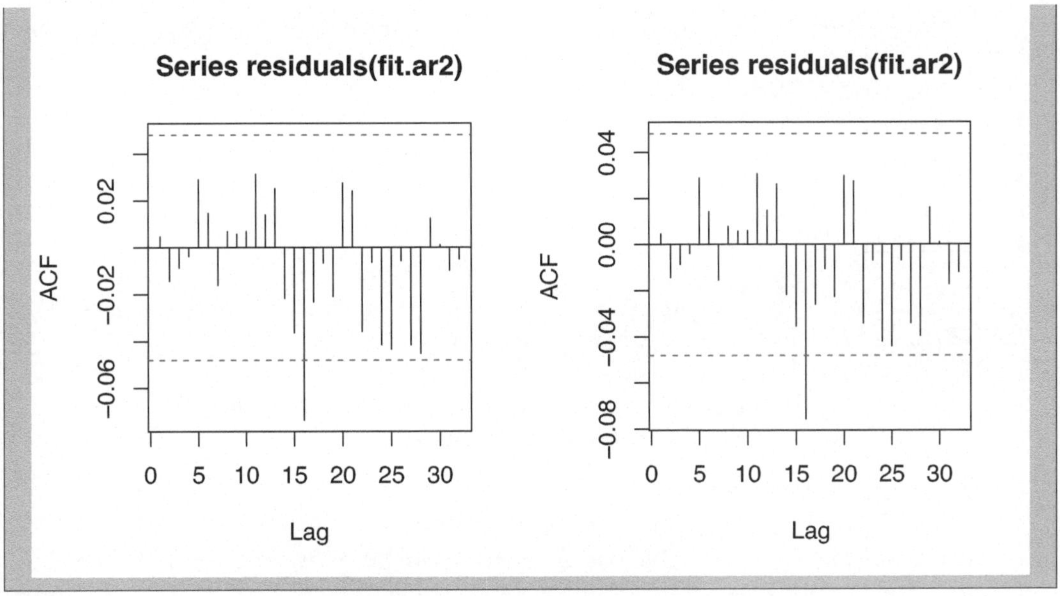

8.4.1 Parameter uncertainty and testing

Regardless of the method used for estimation, the parameter uncertainty can be estimated by the observed Hessian matrix of the log-likelihood, and confidence intervals and Wald test can be constructed in the usual way (see Chapter 2). that is, the standard errors of the parameters are given by

$$se(\boldsymbol{\theta}_i) = \sqrt{\boldsymbol{H}_{ii}}, \tag{8.81}$$

with

$$\boldsymbol{H}_{ij} = \left.\frac{\partial^2 \log L}{\partial \theta_i \theta_j}\right|_{\boldsymbol{\theta}=\hat{\boldsymbol{\theta}}}. \tag{8.82}$$

Models may be compared by likelihood based information criteria, e.g. AIC or by the likelihood ratio test for nested models.

Example 8.14 *The fitted* `arima` *object contains the estimated values along with the standard deviations, for example;*

```
fit.ar2

##
## Call:
## arima(x = res, order = c(2, 0, 0), include.mean = FALSE)
##
## Coefficients:
##          ar1      ar2
##        1.037   -0.202
## s.e.   0.024    0.024
##
## sigma^2 estimated as 0.0131:  log likelihood = 1248,  aic = -2491
```

Hence we see that both parameters are highly significant (the smaller estimate is around 9 times the standard error). The AR(2) and the ARMA(1,1) are not nested and hence we cannot do a likelihood ratio test, but we can compare these models by AIC;

```
AIC(fit.ar2, arma11.fit)

##            df   AIC
## fit.ar2     3 -2489
## arma11.fit  3 -2488
```

We see that there is a small difference in favour of the AR(2). The hypothesis that the AR(2) can be reduced to an AR(1) could be tested by;

```
fit.ar1 <- arima(res, order=c(1, 0, 0), include.mean = FALSE)
(Q <- 2 * (logLik(fit.ar2) - logLik(fit.ar1)))

## 'log Lik.' 70 (df=3)
```

Which should (i.e. $2 \cdot 70 = 140$*) be compared to an* χ^2*-distribution with 1 degree of freedom, and the resulting p-value is therefore extremely small.*

8.4.2 Model validation

The statistical analysis above is used to compare models and to test the significance of individual parameters. Such analysis does not give any indication of the model being appropriate or not. The most important tool for such analysis is the autocorrelation function and the partial autocorrelation function of the 1-step ahead prediction errors. If the model is appropriate, the 1-step ahead residuals should be white noise and hence both the autocorrelation and the partial autocorrelation function should be zero in all lags. More structure should be added to the model; if this is not the case, the extra structure can be identified using Table 8.1. As a small example, assume that we have fitted an AR(1) model to some data, but the autocorrelation functions suggest that the residuals are also an AR(1) model. In that case, we should expand the model to an AR(2) model. Hence the autocorrelation functions is our primary tool for model development. However, more methods related to model validation are covered in Section 8.6.

8.5 Generalisation of the ARMA–process

So far, we have considered ARMA processes for the residuals from a general linear model, that is we have fixed the parameters for the linear model and estimated the dynamics of the residuals. However, as we also discussed in Chapter 5, estimation of structures in the residuals and the regression parameters should be done simultaneous, such models are called ARMAX models. Further in the example we have considered here, it is reasonable to assume that the residual have a diurnal (corresponding to lag 48) variation. Models that include these type of effects are discussed in this section.

One of the basic assumptions we have made so far is that the time–series are stationary. For some applications, this is not a reasonable assumption, for example asset prices, and electricity consumption. In such case, a reasonable model should include a detrending, that can be imposed by differentiating the time series, the difference operator is defined as

$$\nabla = (1 - B). \tag{8.83}$$

The difference operator can be included multiple times by introducing ∇^k. Such models are called ARIMA models (I of integrated) and are defined by

$$\phi(B)\nabla^d Y_t = \theta(B)\epsilon_t. \tag{8.84}$$

Another generalisation is the seasonal ARIMA model, defined by

$$\phi(B)\Phi(B^s)\nabla^d\nabla_s^D Y_t = \theta(B)\Theta(B^s)\epsilon_t, \tag{8.85}$$

where

$$\begin{aligned}\Phi(B^s) &= 1 + \Phi_1 B^s + \ldots + \Phi_P B^{Ps}\\ \Theta(B^s) &= 1 + \Theta_1 B^s + \ldots + \Theta_Q B^{Qs}\\ \nabla_s &= (1 - B^s).\end{aligned} \tag{8.86}$$

The model is written as ARIMA$((p, d, q)\times((P, D, Q)_s))$, and the process is stationary if $D = d = 0$ and the root of the polynomials $\phi(z^{-1})$ and $\Phi(z^{-1})$ is inside the unit circle.

Example 8.15 (ARIMA $(1,0,1)\times(1,0,1)$) *The ARIMA$((1,0,1)\times(1,0,1)_s)$ is given by*

$$(1 - \phi_1 B)(1 - \Phi_1 B^s)Y_t = (1 + \theta_1 B)(1 + \Theta B^s)\epsilon_t, \tag{8.87}$$

which can be written as

$$Y_t + \phi_1 Y_{t-1} + \Phi_1 Y_{t-s} + \phi_1\Phi_1 Y_{t-s-1} = \epsilon_t + \theta_1\epsilon_{t-1} + \Theta\epsilon_{t-s} + \theta\Theta\epsilon_{t-s-1}, \tag{8.88}$$

and hence the model can be written as an ARMA$(s+1, s+1)$, with AR (MA) parameters ϕ_1 (θ_1), Φ_1 (Θ_1), and $\phi_1\Phi_1$ $(\theta_1\Theta_1)$ in lag 1, s and $s+1$, respectively, and all other parameters equal zero.

Identification in seasonal models is similar to the identification in ARIMA models, i.e., one should just focus on lag s, $2s$ and so on. We can illustrate the required steps using the example taken earlier in this chapter.

Example 8.16 (Seasonal model for CO_2) *The CO_2 data that we have considered so far is given in half hourly resolution and hence a reasonable period is 48 lags. This implies that for identification we would need at least 48 lags in the autocorrelation functions. We use residuals from the unidentified AR(2) model to check for seasonality. The figure below shows the ACF and PACF up to lag $3 \cdot 48$, and lag, 48, $2 \cdot 48$, and $3 \cdot 48$ are clearly marked.*

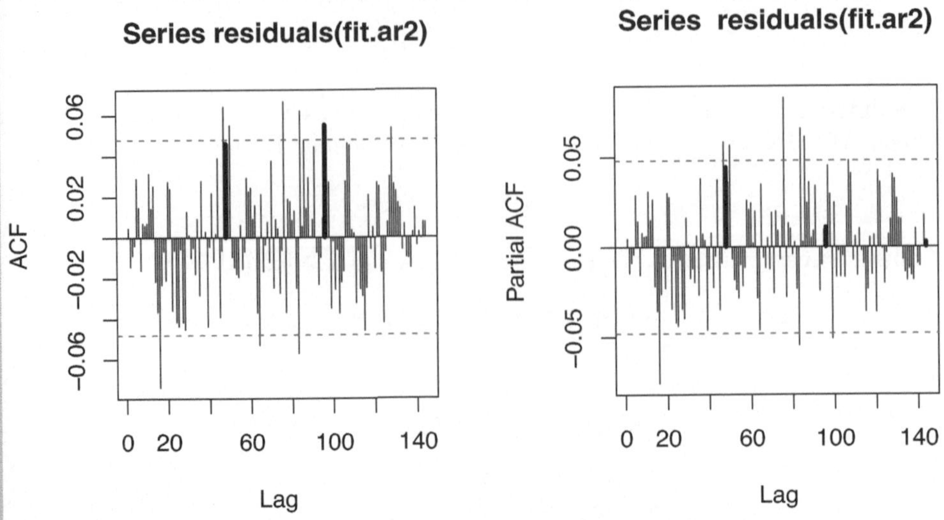

The value of the autocorrelation function in lag 48, 2·48, *and* 3·48 *are all small, even though some are outside the confidence interval an autocorrelation of 0.05 is extremely weak and would not be visible from visual inspection. For illustrative purposes, we will however include it in the model. Also, it is not clear if the seasonal part is an AR(1) or an MA(1). If one wishes to compare different models, this can be done using likelihood based methods (e.g. AIC). Here we illustrate these points using a seasonal MA term;*

```
ar2Sma1.fit <- arima(res, order = c(2, 0, 0),
                     seasonal = list(order = c(0 ,0, 1), period = 48),
                     include.mean=FALSE)
```

The ACF and PACF for the resulting 1-step residuals are shown in the figure below. It is clear that the ACF does not change much. This can to some extend be attributed to the fact that the residuals are actually quite close to uncorrelated.

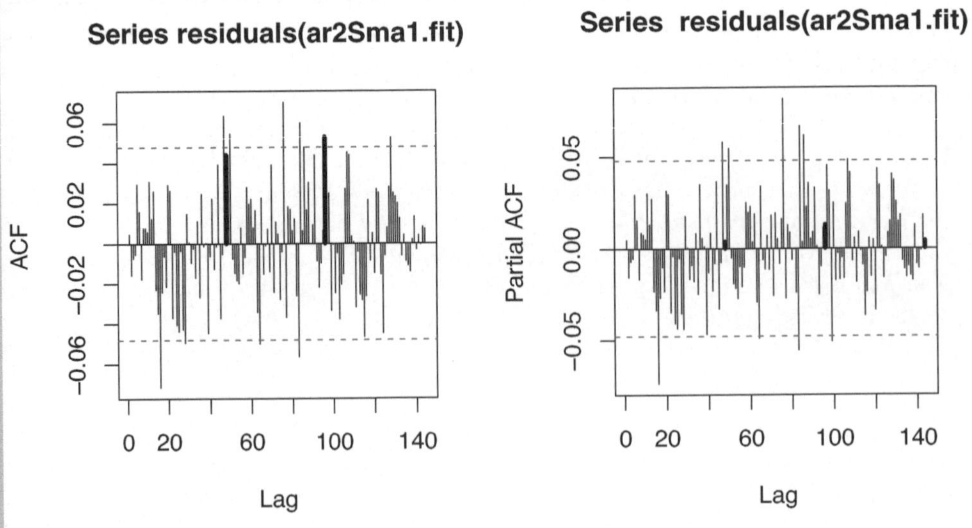

The last family of models we look at in this chapter is including regressors in the model.

8.5.1 The ARIMAX–process

The most general model we consider here is defined as

$$\phi(B)\Phi(B^s)\nabla^d\nabla_s^D(Y_t - \boldsymbol{x}_t\boldsymbol{\beta}) = \theta(B)\Theta(B^s)\epsilon_t. \tag{8.89}$$

The model usually referred to as the ARIMAX is slightly more general, as it allows different order for the regression and the AR part, see Madsen (2008). We will restrict ourselves to the definition above as that is the one included in the standard method in R.

In this chapter, we have not considered input variables in the examples we have considered so far. Instead we have used the residuals from a general linear model, but actually the parameters for the dynamical model and the regression parameters should be estimated together. This type of estimation is handled by the ARIMAX setup.

The model can be formulated as

$$\boldsymbol{e} = \boldsymbol{Y} - \boldsymbol{X}\boldsymbol{\beta}, \tag{8.90}$$

and estimating parameters for a seasonal ARIMAX using the time–series $\boldsymbol{e}$. The important part here is that the dynamic parameters related to the ARMA process and the static parameters related to the input parameters are estimated simultaneously, we illustrate this by extending the example we have considered earlier.

Example 8.17 (ARIMAX model for CO_2) *So far, we have used the residuals from the model in Example 8.1. However, as pointed out above, the parameters for the static and the dynamic model should be estimated together. As we have seen, it seems that a seasonal part should be included, but also that the seasonal part is not very important for the model. Finally, it does not seem important if the seasonal part is an AR(1) or an MA(1). Hence, we estimate ARIMAX(2,0,0) models without any seasonal terms as well as models with seasonal AR(1) and seasonal MA(1) terms. All cases will be using the same input variables (the "X" part) as the initial GLM model (i.e. variables related to diurnal variation and window opening, see also Example 8.1). These can be estimated in R by;*

```
X <- model.matrix(fitCO2.1)
ar2Sma1x.reg <- arima(log(win1.agr$CO2), order = c(2, 0, 0),
                      seasonal = list(order = c(0, 0, 1), period = 48),
                      xreg = X,include.mean = FALSE)
ar2Sar1x.reg <- arima(log(win1.agr$CO2), order = c(2, 0, 0),
                      seasonal=list(order = c(1, 0, 0), period = 48),
                      xreg = X, include.mean = FALSE)
ar2x.reg <- arima(log(win1.agr$CO2),order = c(2, 0, 0),
                  xreg = X, include.mean = FALSE)
```

Not all models are nested and hence we compare them using AIC, rather than likelihood ratio test;

```
c(AIC(ar2Sma1x.reg), AIC(ar2Sar1x.reg), AIC(ar2x.reg))

## [1] -2488 -2489 -2486
```

We see an improvement from including the seasonal model, and the AR(1) seasonal part have the lower AIC. At the same time, it is also clear that the difference between the models is small, and since the model with non–seasonal component is nested to the other models, we can compare them using the usual likelihood ratio test;

```
Q <- 2 * (logLik(ar2Sar1x.reg)-logLik(ar2x.reg))
1 - pchisq(Q, df = 1)

## 'log Lik.' 0.036 (df=10)
```

hence the improvement is borderline significant on the usual 5% level, and it could be argued that the simple ARX(2) model is actually appropriate for this example.

Having estimated a model, one should test that the assumptions of the model are adequate, which is the subject of the next section.

8.6 Diagnostic

Diagnostics in ARIMA models are based on the 1-step ahead prediction errors as those are the ones that are white noise (iid sequence). The notes on diagnostic plots covered in Chapter 3 (Section 3.4.3) still apply, but in addition we should investigate if the residuals can be assumed independent.

In the model formulations, we have assumed that ϵ_t is iid (white noise), and this assumption has to be checked. If the residuals are not white, noise then the model is not able to describe the available observations, and the model can be further improved.

All the methods below assume that $\{\epsilon_t\}$ is a realisation of an iid process, and the list gives a selection of tests that can be used for testing (in all cases, n is the length of the time series)

- Using the autocorrelation function, we can visually check that

$$\hat{\rho}_\epsilon(k) \in N(0, 1/n), \tag{8.91}$$

  we have already used this for model development.

- Under the assumption we would expect that the residuals change sign on average, every second time step, this implies that (with C being the number of change of sign) $C \sim Binom\left(\frac{1}{2}, n-1\right)$, which can be approximated by a normal distribution with mean $\frac{n-1}{2}$ and variance $\frac{n-1}{4}$, and hence a hypothesis test can be performed by comparing

$$Z_c = \frac{C_{obs} - \frac{n-1}{2}}{\sqrt{(n-1)/4}}, \tag{8.92}$$

  to the standard normal distribution.

- The distribution of the empirical autocorrelation implies that

$$Q = n \sum_{k=1}^{m} \hat{\rho}_\epsilon(k)^2, \tag{8.93}$$

will approximately follow a χ^2 distribution (under the iid assumption); if no parameters were estimated, the degrees of freedom would be m. But, as we have estimated some parameter, the degrees of freedom should be $m-p$, where p is the number of estimated parameters. m is often chosen as 15-25, but it is also clear that if some period is expected it makes sense to choose m larger than the expected period.

- Cumulated periodogram. The cumulated periodogram is calculated from the periodogram

$$\hat{I}(\nu_i) = \frac{2}{n}\left[\left(\sum_{t=1}^{n}\epsilon_t\cos(2\phi\nu_i t)\right)^2 + \left(\sum_{t=1}^{n}\epsilon_t\sin(2\phi\nu_i t)\right)^2\right], \tag{8.94}$$

which can be calculated for the frequencies $\nu_i = i/n$, $i = 0, ..., \lfloor n/2 \rfloor$. If ϵ_t is a white noise sequence then ($E[\epsilon_t\epsilon_s] = 0$ for $t \neq s$), and therefore

$$\begin{aligned} E[\hat{I}(\nu_i)] =& \frac{2}{n}\left[\sum_{t=1}^{n}E[\epsilon_t^2]\cos^2(2\phi\nu_i t) + \sum_{t=1}^{n}E[\epsilon_t^2]\sin^2(2\phi\nu_i t)\right] \\ =& 2\sigma^2, \end{aligned} \tag{8.95}$$

hence constant independent of ν_i, the scaled cumulated periodogram is defined as

$$\hat{C}(\nu_j) = \frac{\sum_{i=1}^{j}\hat{I}(\nu_i)}{\sum_{i=1}^{\lfloor n/2 \rfloor}\hat{I}(\nu_i)}. \tag{8.96}$$

Using the above, we can write the scaled cumulative periodogram for a white noise process as

$$\hat{C}_{white}(\nu_j) = \frac{j}{\lfloor n/2 \rfloor}, \tag{8.97}$$

that is, a straight line with slope equal 2. The theoretical scaled periodogram can be compared to the observed scaled periodogram by visual inspection in a plot. It is also possible to include confidence limits in a similar fashion as the autocorrelation functions. Such confidence limits are based on a Kolmogorov-Smirnov test (see e.g. Madsen (2008)). The cumulated scaled periodogram can be calculated and plotted in R using;

```
cpgram(x)
```

Which is similar to the definitions used above except it uses the truncated periodogram (controlled by `taper`), that is related to robust estimation of the periodogram (see again e.g. Madsen (2008)).

Example 8.18 (Residual analysis of CO_2 model) *To illustrate the diagnostic procedures we go through the test described above and compare it with the initial model in Example 8.1.*

We start by the change of sign test.

```
e0 <- residuals(fitCO2.1)
e <- residuals(ar2Sar1x.reg)
e.sign0 <- sign(e0)
e.sign <- sign(e)
n <- length(e)

sig.shift0 <- sum(abs(e.sign0[-1] != e.sign0[-n]))
sig.shift <- sum(abs(e.sign[-1] != e.sign[-n]))
zc <- (c(sig.shift0, sig.shift)  - (n - 1) / 2) /
    sqrt((n -1 ) / 4)
zc

## [1] -28.6  -6.7
```

Hence the number of changes in sign is clearly too small (acceptance region is approx. $[-2, 2]$*), but it is also clear that there is a large improvement (-6.6 closer to zero than -29).*

For the test using the autocorrelation function, we have to choose the number of lags. Since we expect a period of 48, it is natural to look at lag up to 48, and the test statistics is calculated by;

```
e.acf <- acf(e, lag.max = 48, plot = FALSE)$acf
Q <- c(n * sum(e0.acf^2), n * sum(e.acf^2))
```

which should be compared to a χ^2*-distribution with appropriate degrees of freedom. The p-value is calculated by*

```
p <- c(length(coef(fitCO2.1)),  length(coef(ar2x.reg)))
m <- length(e.acf)
(1 - pchisq(Q, df = m - p))

## [1] 0.000 0.024
```

Hence, we will reject the null–hypothesis on the usual 5% level that the residuals are white noise for the ARIMAX model. However, the test statistics is again much higher for the initial model (and hence very strong evidence against the null hypothesis).

The cumulative periodogram can be calculated using

```
cpgram(e)
```

where `e` *is a vector of residuals. The figure below shows the autocorrelation function upto lag* $3 \cdot 48$ *and the cumulated periodogram, for the initial regression model and the developed model. It is clear that the residuals from the general linear model are not white noise, while nothing suggests that the residuals from the ARIMAX model are not white noise.*

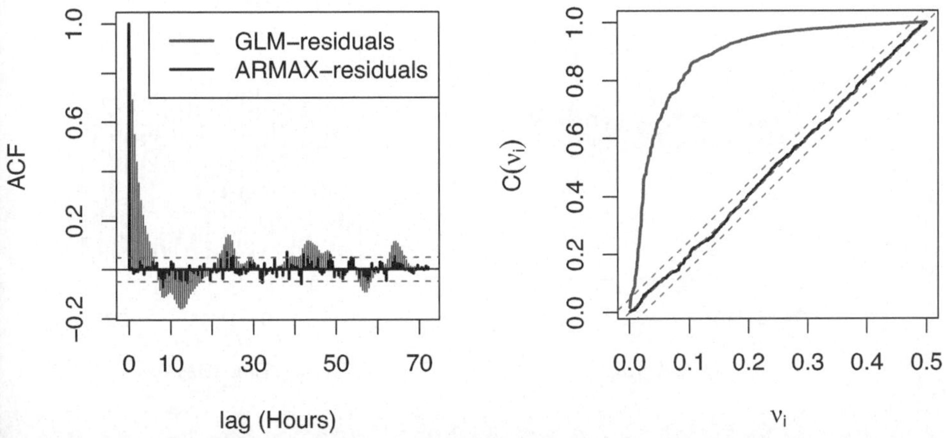

Finally, we check for normality of the residuals using the usual qq-plot, and further compare residuals and fitted values. It is clear that the residuals deviate from the normal distribution with heavier tails (in particular there is one extreme observation) than expected, and also there are some patterns left in the relation of residuals vs. fitted (though less than for the original regression model).

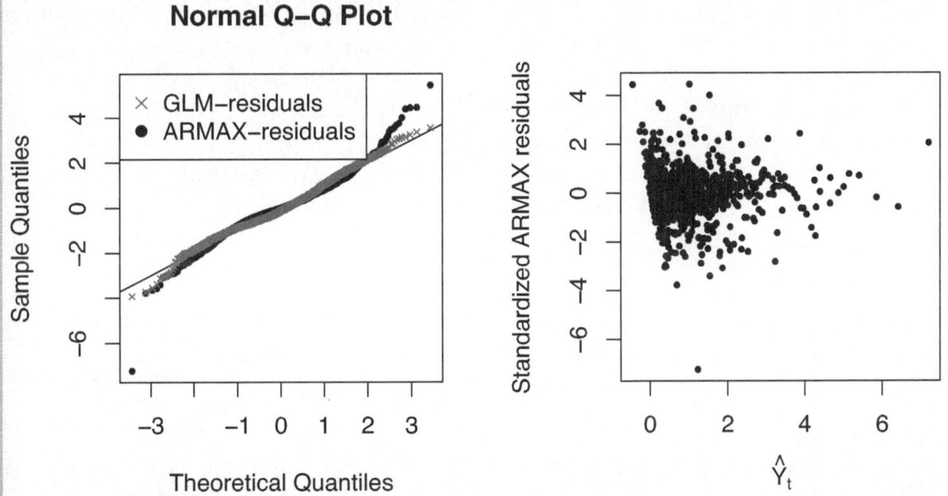

The most extreme residual (around 6 days, see the figure below) is associated with a period of very high CO_2 concentrations followed by a sudden drop in concentration.

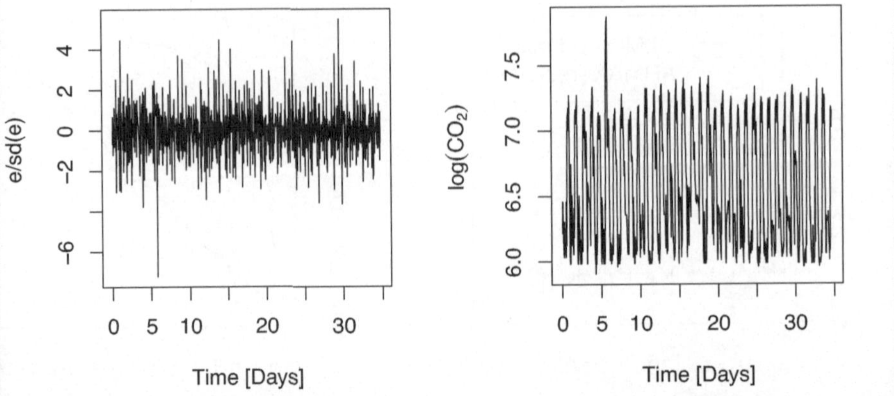

In order to understand the dynamics around extreme residuals, time–series plots for log CO_2 concentration around the most extreme positive (around day 28) and most extreme negative residuals (around day 6) are shown in the figures below. In both cases, the large residual is associated with a sudden change in CO_2 level without a change in window status. In the first case, the level drops without a window opening. In this case, the CO_2 level is also unusually high before the drop. In the second case, there is a sudden increase in the CO_2 level while the windows have been closed for several hours. These types of behaviour in the CO_2 levels cannot be captured by the model, since there is nothing to explain these. At the same time, they might be associated with occupant behaviour, for example in the first case some airing of the room might occur (e.g. a door opened and occupants leaving the room) while the second one might be explained by occupants entering the room. The models described in this chapter cannot accommodate such features directly since we would need a hidden state to describe the underlying mechanisms (e.g. occupants). We will get back to models that can handle this type of data in Chapters 10 and 11.

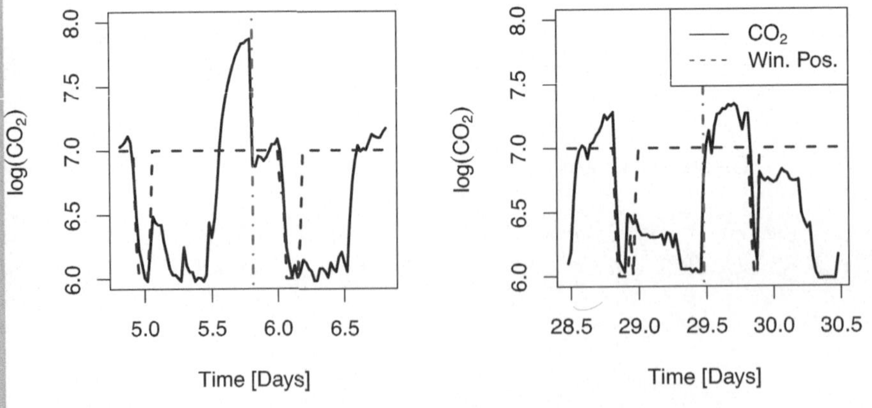

8.7 Prediction

It is often of interest to make predictions ahead of time using the estimated ARIMA models. We will use the notation

$$\begin{aligned}\hat{Y}_{t+k|t} &= E[Y_{t+k}|Y_t, Y_{t-1}, ...]\\ \hat{\epsilon}_{t+k|t} &= E[\epsilon_{t+k}|Y_t, Y_{t-1}, ...].\end{aligned} \tag{8.98}$$

It follows directly that

$$\begin{aligned}\hat{Y}_{t+k|t} &= Y_{t+k}; \quad \text{for } k \leq 0\\ \hat{\epsilon}_{t+k|t} &= \epsilon_{t+k}; \quad \text{for } k \leq 0\\ \hat{\epsilon}_{t+k|t} &= 0; \quad \text{for } k \geq 1.\end{aligned} \tag{8.99}$$

Further, for a model with an input, we will use

$$\hat{\mu}_t = \boldsymbol{x}_t\hat{\boldsymbol{\beta}}. \tag{8.100}$$

Hence k-step prediction in an ARMAX model can be written as (using the R-notation for ARMAX models)

$$\begin{aligned}\hat{Y}_{t+k|t} =& \mu_t + \phi_1(\hat{Y}_{t+k-1|t} - \mu_{t-1}) + \phi_p(\hat{Y}_{t+k-p|t} - \mu_{t-p}) +\\ & \epsilon_{t+k|t} + \theta_1\epsilon_{t+k-1|t} + ... + \theta_q\epsilon_{t+k-q|t}.\end{aligned} \tag{8.101}$$

This formulation is general in the sense that both seasonal and differenced series can be written as ARMAX models (see Example 8.15). When ignoring the parameter uncertainty (which is the usual way of handling the prediction uncertainty), the variance is calculated by

$$\begin{aligned}V[Y_{t+k|t}] =& V[\phi_1 Y_{t+k-1|t} + \phi_p Y_{t+k-p|t} + \epsilon_t + \theta_1\epsilon_{t+k-1|t} + ... + \theta_q\epsilon_{t+k-q|t}]\\ =& \sum_{i=1}^{p}\phi_i^2 V[Y_{t+k-i|t}] + \sigma^2 + \sum_{i=1}^{q}\theta_i^2 V[\epsilon_{t+k-i|t}] +\\ & 2\sum_{i=1}^{p-1}\sum_{j=i+1}^{p-1}\phi_i\phi_j Cov[Y_{t+k-i|t}, Y_{t+k-j|t}] +\\ & 2\sum_{i=1}^{p}\sum_{j=i}^{q}\phi_i\theta_j Cov[Y_{t+k-i|t}, \epsilon_{t+k-j|t}],\end{aligned} \tag{8.102}$$

where we have used that

$$\begin{aligned}Cov[\epsilon_t, \epsilon_s] =& 0; \quad \text{for } s \neq t\\ Cov[Y_t, \epsilon_s] =& 0; \quad \text{for } s > t,\end{aligned} \tag{8.103}$$

all values above can be calculated recursively, starting from $k = 1$, noting that

$$\begin{aligned}V[Y_{t+k-i|t}] =& 0; \quad \text{for } i \geq k,\\ V[\epsilon_{t+k-i|t}] =& 0; \quad \text{for } i \geq k,\\ V[\epsilon_{t+k-i|t}] =& \sigma^2; \quad \text{for } i < k,\\ Cov[Y_{t+k-i|t}, Y_{t+k-j|t}] =& 0; \quad \text{for } i \geq k \text{ or } j \geq k,\\ Cov[Y_{t+k-i|t}, \epsilon_{t+k-j|t}] =& 0; \quad \text{for } i \geq k \text{ or } j \geq k.\end{aligned} \tag{8.104}$$

Example 8.19 (Prediction in ARMAX(1,1)) *As a small example, we consider the model*

$$Y_t - \mu_t = \phi(Y_{t-1} - \mu_{t-1}) + \epsilon_t + \theta\epsilon_{t-1}, \tag{8.105}$$

The 1- and 2-step predictions are given by

$$\begin{aligned} \hat{Y}_{t+1|t} &= \mu_{t+1} + \phi(Y_t - \mu_t) + \theta\epsilon_t \\ \hat{Y}_{t+2|t} &= \mu_{t+2} + \phi(\hat{Y}_{t+1|t} - \mu_{t+1}), \end{aligned} \tag{8.106}$$

and in general

$$\hat{Y}_{t+k|t} = \mu_{t+k} + \phi(\hat{Y}_{t+k-1|t} - \mu_{t+k-1}); \quad \text{for } k > 1. \tag{8.107}$$

The variance is found by

$$V[Y_{t+k|t}] = \phi^2 V[Y_{t+k-1|t}] + \sigma^2 + \theta^2 V[\epsilon_{t+k-1|t}] + 2\phi\theta Cov[Y_{t+k-1|t}, \epsilon_{t+k-1|t}], \tag{8.108}$$

the last two terms ($k > 2$) are calculated by

$$\begin{aligned} V[\epsilon_{t+k-1|t}] &= \sigma^2 \\ Cov[Y_{t+k-1|t}, \epsilon_{t+k-1|t}] &= Cov[\phi Y_{t+k-2|t} + \epsilon_{t+k-1} + \theta\epsilon_{t+k-2}, \quad \epsilon_{t+k-1|t}] \\ &= \sigma^2, \end{aligned} \tag{8.109}$$

and hence

$$V[Y_{t+k|t}] = \phi^2 V[Y_{t+k-1|t}] + \sigma^2(1 + \theta^2 + 2\phi\theta); \quad k > 1, \tag{8.110}$$

with initial value

$$V[Y_{t+1|t}] = \sigma^2, \tag{8.111}$$

which also immediately gives us the stationary variance as

$$V[Y_t] = \frac{\sigma^2(1 + \theta^2 + 2\phi\theta)}{1 - \phi^2}. \tag{8.112}$$

Figure 8.2 shows an example of prediction in an ARMA(1,1) process ($\phi = 0.9$, $\theta = 0.5$, and $\sigma^2 = 0.1^2$). We see the variance converging to the stationary variance (which is 1.1 in this case), and therefore the prediction interval width converge, and finally for illustration one realization is also shown.

We conclude this chapter by treating again our CO_2 example that we have used throughout this chapter.

Example 8.20 (Prediction of CO_2 levels) *In the case of our CO_2 example, we have a model including a seasonal component of 24 hours (48 lags), as well as a set of regression variables related to the time of day and the window position. In order to make predictions, the regressors need to be known, and hence the prediction can be calculated for a scenario of the window position. The variance can be calculated in the same way as we saw in the previous example, but is also available using the predict function*

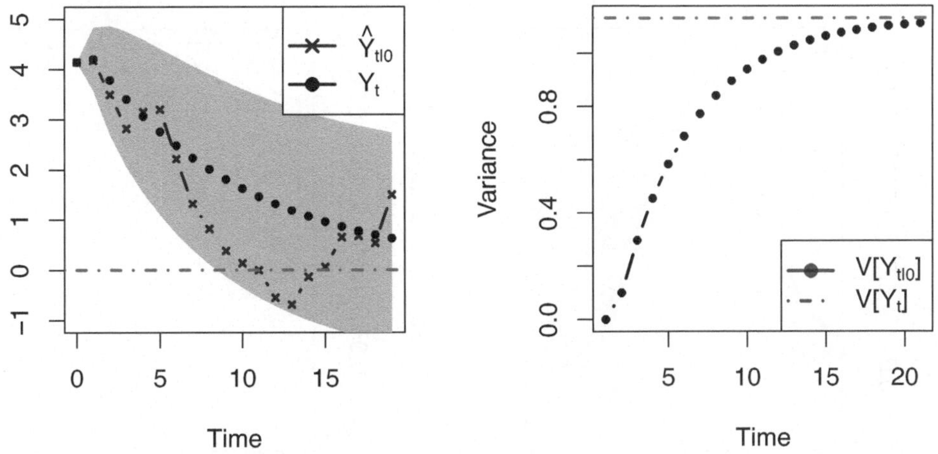

Figure 8.2: Example of prediction in an ARMA(1,1)–model; grey area indicates 95% prediction interval.

```
pred <- predict(ar2Sar1x.reg, newxreg = Xnew, n.ahead = 48,
                se.fit = TRUE)
```

Where `Xnew` *is a data.frame with* `n.ahead` *rows and the same number of columns as the regressors in the model, in this case including harmonics of time of day and window position. We do not calculate the stationary standard deviation here, but note that we expect it to be similar to the estimated variance of the linear model (the horizontal line). The standard deviation is extracted by*

```
summary(fitCO2.1)$sigma

## [1] 0.23
```

Figure 8.3 shows a 7–hour prediction (starting at 0) for the CO_2 *level based on a scenario for the window position. Also, we see in the right column that the variance is increasing very fast towards the stationary variance and that the standard deviation is only different from the stationary values in an important way for the first couple of hours.*

Also it is seen that the prediction from the ARMAX model is very close to the unconditional prediction (i.e. a prediction very far into the future) after a few time steps, implying that information on the current CO_2*-level does not have any impact on the prediction after a few hours.*

8.8 Summary and further perspectives

In this chapter, we have focused on the classical ARIMAX models, but of course the covered methods are a choice that we think is a good introduction to the methods and relevant for occupant–related modelling. As time–series analysis is a broad topic with many comprehensive text books, there are many topics that are left out of this introduction. We have, for

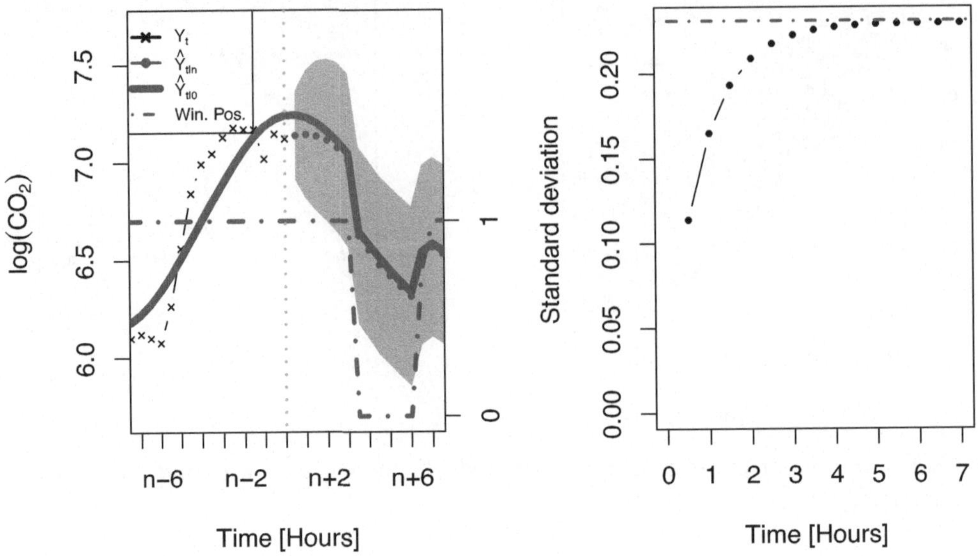

Figure 8.3: Out-of-sample predictions for CO_2 levels.

example completely left out spectral methods, multivariate time–series analysis, a number of generalisation of the ARIMAX model, for example non-linear time-series models such as GACH and Markov switching models, and time series models with varying coefficients. Some generalizations are covered in Chapter 10, where we also cover the state space formulation of the models we have presented in this chapter.

We do however hope that this introduction serves as a good starting point for identification and model building in time–series model, that can help the reader to start analysing their time series data themselves.

The R-package `forecast` (Hyndman and Khandakar, 2008; Hyndman et al., 2023) adds many additional features to the build in `arima` method, for example automatic model selection. Further, it is used for some of the figures in this chapter. The R-package `timeSeries` (Wuertz et al., 2022) includes further methods for visualisation and data handling for time–series.

9

Markov chain models

Markov chains are used in a wide range of applications in occupant behaviour, for instance in presence models, window opening and blind usage models, light switching as well as in occupant activities. An example of modelling occupancy in offices is Andersen et al. (2014). The defining assumption of a Markov chain is that each state is dependent only on the current state together with the probabilities of the state changing i.e. future states are independent of the past (when the present is known).

Time–series whose quantities take a finite number of states, can be modelled with Markov chains, we only consider processes with finite number of states here. Examples for quantities modelled in practice in the field of occupant behaviour are listed below.

- Occupancy (present, absent, number of people present)
- Window states over time (open, closed, tilted)
- Blind usage (open, closed, fraction of opening)
- Activity level (working, sleeping, resting, laundry, cooking, absent)

As in Chapter 8, we assume that the process is directly observed, but the outcomes are now from a discrete state space.

Example 9.1 *We again use a window opening example that was used in Chapter 8 as the motivating example throughout the chapter. The data will not be aggregated here, but rather given in 10–minute resolution (rather than the 30–minute resolution in Chapter 8). For further details on the data, see Example 8.1 and (Andersen et al., 2013). Further, we limit the data–set to 1 room in 1 dwelling for a period of approximately 35 days.*

The overall statistics of the window opening is given below;

```
(n <- length(win.dat2$WindowClosed))

## [1] 4996

table(win.dat2$WindowClosed) / n

##
##    0    1
## 0.17 0.83
```

Hence, there are around 5000 observations and the window is closed about 83% of the time. The figure below shows a time–series plot of the window position.

DOI: 10.1201/9781003340812-9

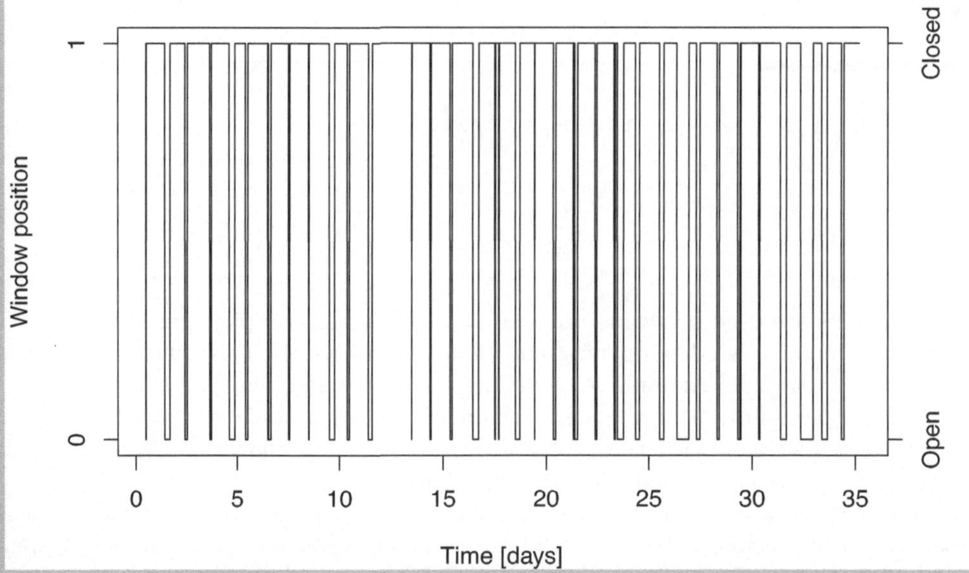

The figure above indicates that there is some autocorrelation and also that there may be some periodic behaviour in the data. These questions are addressed in the figure below, where we see that there is indeed a very strong autocorrelation, and also that there is a strong diurnal variation.

The right panel shows that a large part of the variation can be explained by the hour of the day.

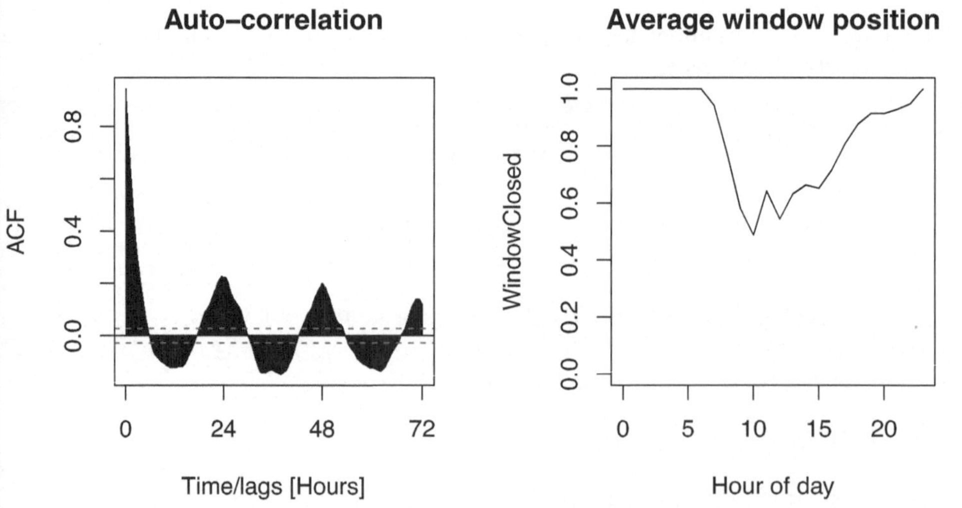

Considering the example above, we clearly need a model to capture autocorrelation in binary data, and that model should be able to include covariates to capture the clear diurnal variation in data. This can be included in the so-called Markov models, that we cover in this chapter. We start by the definition of Markov models.

9.1 Markov chains

The Markov chain technique requires the derivation of a of transition probability matrix from the source data, which details the probability of a transition from one level of the state to another, at each time step. The definition and calculations of these properties are given below:

Definition 9.1 (Markov chain) *Let $\mathcal{M}$ be a finite set and T be an index set. A collection of $\mathcal{M}$-valued random variables $\{Y_t\}$ with $t \in T$ is called Markov chain if*

$$P(Y_t|Y_{t-1}, Y_{t-2}, \ldots, Y_0) = P(Y_t|Y_{t-1}). \tag{9.1}$$

The set $\mathcal{M}$ is called the state space of the Markov chain.

For $i, j \in \mathcal{M}$ the conditional probability

$$P(Y_t = i|Y_{t-1} = j) = p_{ij}(t) \tag{9.2}$$

is called transition probability from state j to state i (at time t). The matrix $\mathbf{\Gamma}(t)$ (with elements $\mathbf{\Gamma}(t) = p_{ij}(t)$) is called the *transition probability matrix* (TPM). As a stochastic matrix, it has the following properties:

- $\mathbf{\Gamma}$ has non-negative entries, that is $p_{ij} > 0$,
- The rows sum up to one, $\sum_j p_{ij} = 1$ for all i.

Equation (9.1) is called the *Markov property*. It expresses the memory-less property of the Markov chain, i.e. that the probability distribution of the following state depends only on the current state. If the transition probabilities do not depend on time, that is if the transition probability matrix is constant over time, $\mathbf{\Gamma}(t) = \mathbf{\Gamma}$, the Markov chain is called *homogeneous*. Otherwise, it is called in-homogeneous. The s-step transition probability is defined by

$$p_{ij}^{(s)} = P(Y_{t+s} = i|Y_t = j), \tag{9.3}$$

with a corresponding s-step transition probability matrix $\mathbf{\Gamma}^{(s)}$. In the homogeneous case, the Chapman-Kolmogorov equation holds

$$\mathbf{\Gamma}^{(s+t)} = \mathbf{\Gamma}^{(s)}\mathbf{\Gamma}^{(t)}. \tag{9.4}$$

A direct conclusion from this is that the s-step transition probability matrix is given by the sth power of the (one-step) transition probability matrix

$$\mathbf{\Gamma}^{(s)} = \mathbf{\Gamma}^s. \tag{9.5}$$

A realisation of a Markov chain is a time-series of discrete values such that $y_t \in \{1, ..., m\}$, and prediction is made in form of a predictive distribution $\boldsymbol{u}_t \in [0, 1]^m$ such that $\sum_{i=1}^m \boldsymbol{u}_t = 1$. If the distribution at time t is known as $\boldsymbol{u}_t$, then the distribution at time $t+k$ is calculated by

$$\boldsymbol{u}_{t+k} = \boldsymbol{u}_t\mathbf{\Gamma}^k. \tag{9.6}$$

Using usual conventions for Markov models, $\boldsymbol{u}_t$ is a row–vector rather than a column vector.

Example 9.2 *Consider the transition probability matrix*

$$\mathbf{\Gamma} = \begin{bmatrix} 0.9 & 0.1 \\ 0.2 & 0.8 \end{bmatrix}, \tag{9.7}$$

and let $y_1 = 1$*, implying that* $\boldsymbol{u}_1 = [P(Y_1 = 1) \quad P(Y_1 = 2)] = [1 \quad 0]$*; in this case, we have*

$$\boldsymbol{u}_2 = [1 \quad 0] \begin{bmatrix} 0.9 & 0.1 \\ 0.2 & 0.8 \end{bmatrix} = [0.9 \quad 0.1], \tag{9.8}$$

and

$$\boldsymbol{u}_3 = [0.9 \quad 0.1] \begin{bmatrix} 0.9 & 0.1 \\ 0.2 & 0.8 \end{bmatrix} = [0.83 \quad 0.17], \tag{9.9}$$

this can be applied recursively to get

$$\begin{aligned} \boldsymbol{u}_4 &= [0.78 \quad 0.22] \\ \boldsymbol{u}_5 &= [0.75 \quad 0.25], \end{aligned} \tag{9.10}$$

and the distribution $\boldsymbol{u}_k$ *will converge to the so-called stationary distribution, which in this case is* $\boldsymbol{u}_\infty = [2/3 \quad 1/3]$.

With the definitions given above, we can take a first look at likelihood estimation. In this chapter, we will rely on R-functions which we define as we go along presenting the theory and examples, rather than relying on R-packages as we have done in previous chapters.

9.2 Maximum likelihood estimation of the transition probability matrix

In this section, we calculate the maximum likelihood estimators and estimates for the transition probabilities of a Markov chain. We will see that this formal derivation actually leads to what we would expect by intuition. The probability of an observed Markov chain $\mathbf{y} = [y_1, y_2, \ldots, y_{t-1}, y_T]^T$ is given by

$$P(\mathbf{y}) = P(Y_1 = y_1) \prod_{t=1}^{T} P(Y_t = y_t | Y_{t-1} = y_{t-1}). \tag{9.11}$$

Therefore, the likelihood of this observed sequence with respect to $\mathbf{\Gamma}$ is given by

$$L(\mathbf{\Gamma}; \mathbf{y}) = P(Y_1 = y_1) \prod_{t=1}^{T} p_{y_t, y_{t-1}}. \tag{9.12}$$

Alternatively, the likelihood can be expressed by

$$L(\mathbf{\Gamma}; \mathbf{y}) = P(Y_1 = y_1) \prod_{i=1}^{m} \prod_{i=1}^{m} p_{ij}^{n_{ij}}, \tag{9.13}$$

where m is the number of states and n_{ij} is the number of times i is followed by j in the observed sequence. As usual, we use the log-likelihood for derivations of the maximum likelihood estimators/estimates (MLE)

$$l(\mathbf{\Gamma};\mathbf{y}) = \log L(\mathbf{\Gamma};\mathbf{y}) = \log P(Y_1 = y_1) + \sum_{i=1}^{m}\sum_{j=1}^{m} n_{ij}\log p_{ij}. \tag{9.14}$$

Before we can maximise this function, we should take the following constraints related to the property of the transition probabilities into account:

$$\sum_{j=1} p_{ij} = 1. \tag{9.15}$$

We can account for it by replacing p_{ii} by $1 - \sum_{j\neq i}^{m} p_{ij}$ in the log-likelihood (9.14), and hence the parameters of the model is $\boldsymbol{\theta} = (p_{12}, p_{13}, .., p_{21}, p_{23}, ..., p_{m,m-1})$, and the log-likelihood function is

$$\begin{aligned} l(\boldsymbol{\theta};\mathbf{y}) =& \log P(Y_1 = y_1) + \sum_{i=1}^{m} n_{ii}\log\left(1 - \sum_{j\neq i} p_{ij}\right) + \\ & \sum_{i=1}^{m}\sum_{j\neq i}^{m} n_{ij}\log p_{ij}. \end{aligned} \tag{9.16}$$

For now, we will ignore the contribution for the first observation ($\log P(Y_1 = y_1)$), and hence calculate the likelihood conditional on the first observation.

Differentiating the log-likelihood function with respect to p_{ij} (for $j \neq 1$) leads to

$$\frac{l(\boldsymbol{\theta})}{\partial p_{ij}} = -\frac{n_{ii}}{1 - \sum_{j\neq i} p_{ij}} + \frac{n_{ij}}{p_{ij}} = -\frac{n_{ii}}{p_{ii}} + \frac{n_{ij}}{p_{ij}}, \tag{9.17}$$

and solving for $\frac{\partial log(L)}{\partial p_{ij}} = 0$, we get $\hat{p}_{ij} = \frac{n_{ij}}{n_{ii}}\hat{p}_{ii}$, inserting in (9.15) we get

$$1 = \sum_{j=1}^{m} \hat{p}_{ij} = \hat{p}_{ii} + \sum_{j\neq i} \frac{\hat{p}_{ii}}{n_{ii}} n_{ij} = \hat{p}_{ii}\left(1 + \frac{\sum_{j\neq i}^{m} n_{ij}}{n_{ii}}\right), \tag{9.18}$$

now if we set $n_i = \sum_{k=1}^{m} n_{ik}$, i.e. the total number of observations in state i we get

$$\hat{p}_{ii} = \frac{n_{ii}}{n_i}, \tag{9.19}$$

and for $j \neq i$

$$\hat{p}_{ij} = \frac{n_{ij}}{n_i}. \tag{9.20}$$

Hence, the MLE for the transition probability from state i to j is, as we would intuitively expect, the relative (to all transitions from state i) frequency of transitions from i to j.

The usual statistical methods, for example likelihood ratio test and Wald statistics, related to likelihood inference still apply and we can e.g. calculate the observed Fisher

Information matrix as ($j \neq i$, and $j_k \neq i$)

$$\begin{aligned}
I_{p_{ij}p_{ij}}(\hat{\boldsymbol{\theta}}) =& \left.\frac{\partial^2 l}{\partial p_{ij}^2}\right|_{\boldsymbol{\theta}=\hat{\boldsymbol{\theta}}} \\
=& \left.\frac{n_{ii}}{(1-\sum_{j\neq i} p_{ij})^2} + \frac{n_{ij}}{p_{ij}^2}\right|_{p_{ij}=\hat{p}_{ij}} \\
=& \left.\frac{n_{ii}}{p_{ii}^2} + \frac{n_{ij}}{p_{ij}^2}\right|_{p_{ii}=\hat{p}_{ii}} \\
I_{p_{ij_1}p_{ij_2}}(\hat{\boldsymbol{\theta}}) =& \left.\frac{\partial^2 l}{\partial p_{ij_1}\partial p_{ij_2}}\right|_{\boldsymbol{\theta}=\hat{\boldsymbol{\theta}}} \\
=& \left.\frac{n_{i,1}}{(1-\sum_{j\neq i} p_{i,j})^2}\right|_{p_{ij}=\hat{p}_{ij}} \\
=& \left.\frac{n_{ii}}{p_{ii}^2}\right|_{p_{ii}=\hat{p}_{ii}} \\
I_{p_{i_1j_1}p_{i_2j_2}}(\hat{\boldsymbol{\theta}}) =& 0, \quad i_1 \neq i_2,
\end{aligned} \tag{9.21}$$

and hence the observed Fisher information matrix at the MLE is given by

$$\begin{aligned}
I_{p_{ij}p_{ij}}(\hat{\boldsymbol{\theta}}) =& n_i^2 \left(\frac{1}{n_{ii}} + \frac{1}{n_{ij}}\right) \\
I_{p_{ij_1}p_{ij_2}}(\boldsymbol{\theta}) =& \frac{n_i^2}{n_{ii}}.
\end{aligned} \tag{9.22}$$

We see that information is increasing with the number of observations in each state. We illustrate this with a small example.

Example 9.3 (Two state Markov) *Consider the transition count matrix for a two state Markov process*

$$\boldsymbol{F} = \begin{bmatrix} n_{11} & n_{12} \\ n_{21} & n_{22} \end{bmatrix}, \tag{9.23}$$

in this case we have $n_1 = n_{11} + n_{12}$ *and* $n_2 = n_{21} + n_{22}$*, and the MLE's are*

$$\hat{p}_{12} = \frac{n_{12}}{n_1}; \quad \hat{p}_{21} = \frac{n_{21}}{n_2}; \tag{9.24}$$

and the observed Fisher information is

$$\begin{aligned}
\boldsymbol{I}(\hat{p}_{12}, \hat{p}_{21}) =& \begin{bmatrix} n_1^2\left(\frac{1}{n_{11}} + \frac{1}{n_{12}}\right) & 0 \\ 0 & n_2^2\left(\frac{1}{n_{22}} + \frac{1}{n_{21}}\right) \end{bmatrix} \\
=& \begin{bmatrix} n_1\left(\frac{1}{\hat{p}_{11}} + \frac{1}{\hat{p}_{12}}\right) & 0 \\ 0 & n_2\left(\frac{1}{\hat{p}_{22}} + \frac{1}{\hat{p}_{21}}\right) \end{bmatrix},
\end{aligned} \tag{9.25}$$

and hence the standard errors of $\hat{p}_{ij}$ can be written as $(i \neq j)$

$$se(\hat{p}_{ij}) = \sqrt{\frac{\hat{p}_{ii}(1-\hat{p}_{ii})}{n_i}}, \tag{9.26}$$

which is the well-known standard error for Binomial models.

As we can see, the estimates and the observed information dependent on the observed number of transitions, and hence that is a key number for the statistical analysis. The observed number of transitions can be calculated in R by the R-function;

```
trans.count <- function(y, m){
    ## y take values 1,...,m
    F <- matrix(0, ncol = m, nrow = m)
    n <- length(y)
    for(i in 1:m){
        for(j in 1:m){
            F[i, j] <- sum(y[-n] == i & y[-1] == j)
        }
    }
    F
}
```

which simply loops over all possible transitions counting the number of observed transitions.

Example 9.4 *If we again turn to our example, the estimated transition probability matrix can be calculated by (*`win.dat2$WinClo1` *and is a sequence taking values 1 and 2 rather than 0 and 1);*

```
win.dat2$WinClo1 <- win.dat2$WindowClosed+1
(F <- trans.count(win.dat2$WinClo1, 2))

##      [,1] [,2]
## [1,]  800   39
## [2,]   39 4117
```

and the transition probability matrix can be calculated by;

```
(Gamma1 <- F / rowSums(F))

##        [,1]  [,2]
## [1,] 0.9535 0.046
## [2,] 0.0094 0.991
```

Hence, we see that if the window is open, we observe a 95% probability that it will stay open in the next time-step, and the corresponding probability if the window is closed is about 99%.

The likelihood can be calculated directly from the definitions above, but as we have seen before, the likelihood is only really relevant relative to some other hypothesis. In this case, a relevant hypothesis would be that the probability of the next observation is independent from the present and past observations. This would mean that $P(Y_t = i|Y_{t-1} = j) = P(Y_t = i)$*, and an estimate of that probability is*

$$\hat{P}(Y_t = i) = \frac{n_i}{n}, \tag{9.27}$$

where n *is the total number of observations and* n_i *is the number of observations of state* i*. Expressed as a transition probability matrix this is*

$$\hat{\mathbf{\Gamma}} = \frac{1}{n}\begin{bmatrix} n_1 & n_2 \\ n_1 & n_2 \end{bmatrix}, \tag{9.28}$$

also we would expect that the vector $\boldsymbol{\delta} = \frac{1}{n}[n_1 \quad n_2]$ *is the long–term distribution (which we will refer to as the stationary distribution). The models are compared below;*

```
ll1 <- sum(F * log(Gamma1))
## null model
y <- win.dat2$WinClo1
delta <- 1 / length(y) * c(sum(y == 1), sum(y == 2))
Gamma0 <- rbind(delta, delta)
ll0 <- sum(F * log(Gamma0))
## Compare likelihoods
ll1 - ll0

## [1] 1882
```

This means that the log-likelihood increased with 1882 on 1 degree of freedom (which is significant on any reasonable level).

As the number of observations is relatively high, it is reasonable to assume that the likelihood function is well approximated by the quadratic approximation (i.e. it is regular), and confidence intervals can be constructed from the observed Fisher information matrix;

```
I1 <- F[1,1] / Gamma1[1,1] ^ 2 + F[1,2] / Gamma1[1,2] ^ 2
I2 <- F[2,2] / Gamma1[2,2] ^ 2 + F[2,1] / Gamma1[2,1] ^ 2
(se <- c(sqrt(1 / I1), sqrt(1 / I2)))

## [1] 0.0073 0.0015
```

Wald confidence intervals can be constructed based on these standard errors;

```
rbind(gamma12 = c(1,1) * Gamma1[1,2] + c(-1, 1) * 2 * se[1],
      gamma21 = c(1,1) * Gamma1[2,1] + c(-1, 1) * 2 * se[2])

##           [,1]  [,2]
## gamma12 0.0319 0.061
## gamma21 0.0064 0.012
```

Hence we see that the probability of transitioning out of state 1 (window open) is between 3 and 6% (using the usual confidence level).

9.2.1 Reformulation of the parameters

We saw in Chapter 4 that we would often prefer unbounded parameters for estimation, in particular if some kind of regression on the parameters is included. A simple way of introducing that in the case of Markov models is to use the formulation

$$\begin{aligned}\mathbf{\Gamma}_{ii} =& \frac{1}{1+\sum_{j\neq i} e^{\tau_{ij}}} \\ \mathbf{\Gamma}_{ij} =& \frac{e^{\tau_{ij}}}{1+\sum_{k\neq i} e^{\tau_{ik}}}; \quad i \neq j.\end{aligned} \tag{9.29}$$

In this set up, we have $\tau_{ij} \in \mathbb{R}$ and by construction $\sum_j \mathbf{\Gamma}_{ij} = 1$. The log-likelihood can be written as

$$l(\boldsymbol{\theta}) = -\sum_{i=1}^{m} n_i \log\left(1+\sum_{k\neq i} e^{\tau_{ik}}\right) + \sum_{i=1}^{m}\sum_{j\neq i} n_{ij}\tau_{ij}, \tag{9.30}$$

the derivative of the log-likelihood is

$$\begin{aligned}\frac{\partial l}{\partial \tau_{ij}} =& -\frac{n_i e^{\tau_{ij}}}{1+\sum_{ik\neq i} e^{ik}} + n_{ij} \\ =& - n_i e^{\tau_{ij}} p_{ii} + n_{ij},\end{aligned} \tag{9.31}$$

due to the invariance principle (see Section 2.6), the MLE must fulfil the equation

$$n_i e^{\tau_{ij}} = \frac{n_{ij}}{\hat{p}_{ii}}, \tag{9.32}$$

or

$$\hat{\tau}_{ij} = \log\left(\frac{n_{ij}}{\hat{p}_{ii} n_i}\right) = \log\left(\frac{n_{ij}}{n_{ii}}\right). \tag{9.33}$$

The diagonal elements of the Hessian matrix are calculated by

$$\begin{aligned}\frac{\partial^2 l}{\partial \tau_{ij}^2} =& -\frac{n_i e^{\tau_{ij}}}{1+\sum_{ik\neq i} e^{ik}} + \frac{n_i e^{2\tau_{ij}}}{\left(1+\sum_{ik\neq i} e^{ik}\right)^2} \\ =& - n_i p_{ij} + n_i p_{ij}^2 \\ =& - n_i p_{ij}(1-p_{ij}),\end{aligned} \tag{9.34}$$

and the off–diagonal elements are calculated by

$$\begin{aligned}\frac{\partial^2 l}{\partial \tau_{ij_1}\tau_{ij_2}} =& \frac{n_i e^{\tau_{ij_1}+\tau_{ij_2}}}{1+\sum_{ik\neq i} e^{ik}} \\ =& n_i p_{ij_1} p_{ij_2}.\end{aligned} \tag{9.35}$$

The advantages of the formulation above are that the parameters can take values in $\mathbb{R}$ and are hence well suited for non-linear optimisation, and are particular useful for regression type models. Before we go on to more general model formulations, we discuss some properties of Markov models.

9.3 Properties of Markov models

We will give a brief account of some of the properties that are used to characterise Markov chains. In all of the models we consider here, the construction ensures that the process can move around between all states and that a stationary distribution exists. Some of the main properties of Markov chains are:

- Persistent states: A state is said to be persistent if the process will revisit the state at some time with probability one.
- Transient states: If a state is not persistent, it is said to be transient.
- Irreducibly: Implies that all states communicate with each other
- Aperiodicity: A state is said to be aperiodic, if starting at time 0 it can, for some T, visit the state at times $T+1, T+2, \ldots$.
- Stationary distribution: If a Markov chain has a stationary distribution it is given as the solution to

$$\boldsymbol{\delta} = \boldsymbol{\delta}\boldsymbol{\Gamma}, \tag{9.36}$$

with the constraint $\sum_i \delta_i = 1$.

For a more in–depth treatment of properties of Markov chains, the reader is referred to more dedicated textbooks, for example Grimmit and Stirzaker (2001).

In general, we will only consider aperiodic, irreducible Markov chains, that only have persistent states and also have a stationary distribution. To get a feeling of the meaning of the different properties, we consider a few examples.

Example 9.5 *Consider the transition probability matrix*

$$\boldsymbol{\Gamma} = \begin{bmatrix} 0 & 1 \\ 1 & 0 \end{bmatrix}, \tag{9.37}$$

both states are periodic. Since starting at time $t = 0$ at state e.g. 1 the process will revisit the state at times $2, 4, 6, ..,$ but not at times $3, 5, ...$

Consider the Markov chain

$$\boldsymbol{\Gamma} = \begin{bmatrix} 1-p_{12} & p_{12} & 0 \\ 0 & 1-p_{23} & p_{23} \\ 0 & p_{32} & 1-p_{32} \end{bmatrix}. \tag{9.38}$$

In this case, state 1 is transient and the process is not irreducible since state 1 cannot be reached from state 2 or 3.

As a final example consider the Markov process characterised by

$$\boldsymbol{\Gamma} = \begin{bmatrix} 1-p_{12} & p_{12} \\ p_{21} & 1-p_{21} \end{bmatrix}, \tag{9.39}$$

with $p_{12} > 0$, and $p_{21} > 0$. Here, both states are persistent and aperiodic. Further, the *process is irreducible and it can be verified that the stationary distribution is*

$$\boldsymbol{\delta} = \frac{1}{p_{12} + p_{21}} \begin{bmatrix} p_{21} & p_{12} \end{bmatrix}. \tag{9.40}$$

An important property of a Markov chain is the stationary distribution, which can be used both to explore the long–term properties of the chain, but also for estimation using the unconditional likelihood.

Theorem 9.1 *If a Markov chain model has a stationary distribution, then it can be calculated by*

$$\boldsymbol{\delta} = \mathbf{1}^T(\boldsymbol{I} - \boldsymbol{\Gamma} + \boldsymbol{E})^{-1}, \tag{9.41}$$

where $\boldsymbol{E}$ is a matrix of ones.

We will not go through the proof here, but just note that the first part follows directly from the definition of the stationary distribution and the matrix $\boldsymbol{E}$ is introduced in order to ensure that the stationary distribution sums to 1, see Zucchini and MacDonald (2009).

We will use the stationary distribution from time to time throughout this chapter and hence we define an R-function to calculate that;

```
stat.dist <- function(Gamma){
    m <- dim(Gamma)[1]
    E <- matrix(1, ncol = m, nrow = m)
    rep(1, m) %*% solve(diag(m) - Gamma + E)
}
```

Example 9.6 *For the model estimated in Example 9.4, the stationary distribution is;*

```
(delta <- stat.dist(Gamma1))

##       [,1] [,2]
## [1,] 0.17 0.83
```

This is implying that the window is open 17% of the time (4.1 hours per day).

As we saw in Chapter 8, the autocorrelation function is an important tool for identifying the process and also for checking if the appropriate model structure has been identified. For a Markov process with a stationary distribution, the autocovariance function is given by, see Zucchini and MacDonald (2009),

$$\gamma_k = \boldsymbol{\delta} \boldsymbol{V} \boldsymbol{\Gamma}^k \boldsymbol{v} - (\boldsymbol{\delta} \boldsymbol{v})^2, \tag{9.42}$$

where $\boldsymbol{v} = [1, 2, ..., m]$ and $\boldsymbol{V} = \text{diag } \boldsymbol{v}$, and the autocorrelation function can be calculated by $\rho_k = \gamma_k / \gamma_0$

The autocorrelation function can be calculated by the R-function below. The function uses the R-package `expm` (Maechler et al., 2023) to calculate powers of matrices;

```
acf.markov <- function(k,Gamma){
    m <- dim(Gamma)[1]
```

```
    v <- 1:m
    V <- diag(v)
    delta <- stat.dist(Gamma)
    gamma0 <- delta %*% V %*%  v - (delta %*% v)^2
    (delta %*% V %*% (Gamma %^% k) %*% v -
     (delta %*% v)^2) / gamma0
}
```

Example 9.7 (ACF of Markov models) *The autocorrelation function related to the model in Example 9.4 is shown in the figure below together with the empirical autocorrelation function. It is clear from the figure that the empirical autocorrelation function is captured well in the first few lags. The diurnal variation in data can of course not be captured by the model we have estimated here.*

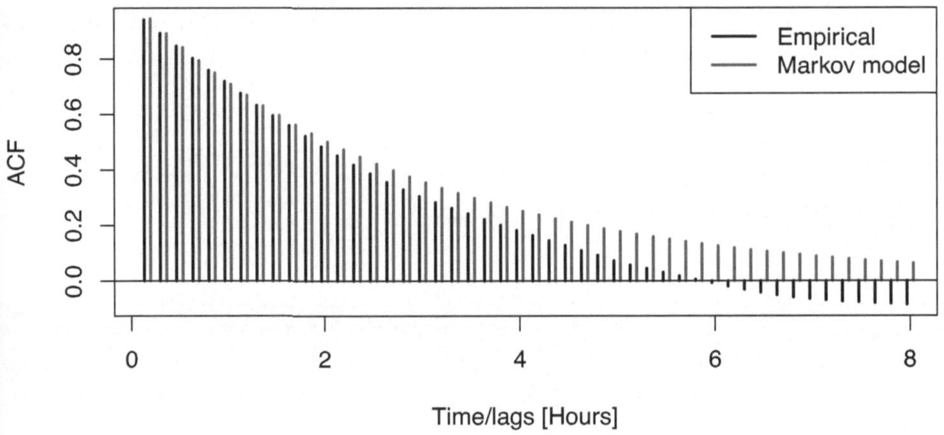

9.3.1 Simulation of Markov chains

Often it is of interest, e.g. for parametric bootstrapping (see Section 9.4), to simulate realisations of an estimated Markov model. This can be easily done using the transition probability matrix recursively, i.e.

$$Y_t = j; \quad \text{with probability } p_{ij} \text{ if } Y_{t-1} = i, \tag{9.43}$$

using either fixed initial condition $Y_1 = j_1$, or choosing the initial values from the stationary distribution. The R-function below simulates a Markov chain based on a given transition probability matrix.

```
sim.markov <- function(Gamma, n, y1 = NULL){
    m <- dim(Gamma)[1]; y <- numeric(n)
    if(is.null(y1)){
        delta <- stat.dist(Gamma)
        y1 <- sample(1:m, size = 1, prob = delta)
    }
```

```
    y[1] <- y1
    for(i in 2:n){
        y[i] <- sample(1:m, size = 1, prob = Gamma[y[i - 1], ])
    }
    y
}
```

The function takes as input the transition probability matrix `Gamma`, the length of the Markov chain to be simulated `n`, and the initial state `y1` (if not included the initial state will be drawn from the stationary distribution).

9.4 Maximum likelihood revisited

We already discussed likelihood estimation in Section 9.2 as the likelihood given (conditional on) the first observation. Using the stationary distribution, we can also set up the unconditional likelihood of all observations by using the stationary distribution for the first observation

$$P(Y_1 = y_1) = \boldsymbol{\delta}(y_1). \tag{9.44}$$

This will lead to a non-linear optimisation problem, that needs to be solved. The parameters of the model are still p_{ij}, that needs to be constructed such that

$$p_{ij} > 0; \quad \sum_j p_{ij} = 1. \tag{9.45}$$

We have previously used transformations (link-functions) in order to have an unbounded optimisation. The constraints here suggest a logit-type transformation and an example of such a transformation is (see Section 9.2.1)

$$p_{ij} = \begin{cases} \frac{e^{\tau_{ij}}}{1+\sum_{j\neq i} e^{\tau_{ij}}} & j \neq i \\ \frac{1}{1+\sum_{j\neq i} e^{\tau_{ij}}} & j = i, \end{cases} \tag{9.46}$$

following (Zucchini and MacDonald, 2009) we will refer to the parameters τ_{ij} $(i \neq j)$ as working parameters and p_{ij} as natural parameters. The inverse transformation is given by

$$\tau_{ij} = \log\left(\frac{p_{ij}}{1 - \sum_{i\neq j} p_{ij}}\right); \quad i \neq j. \tag{9.47}$$

Other parameterisation do exist, for example based on continuous time description of the Markov model using matrix exponentials. However, the description given above has a clear reference to the use of link–functions in Chapter 4.

In order for the formulation above to be useful, we need the transformation implemented as R-functions. The transformation (9.46) is implemented by

```
Markov.Inv.Link <- function(pars,m){
    exp.pars <- exp(pars)
    Gamma <- diag(m)
```

```
    Gamma[Gamma != 1] <- exp.pars
    t(Gamma) / (colSums(Gamma))
}
```

We will use this function throughout this chapter. Note also here that there is no check that dimensions match (related to the length of pars).

When estimating the parameters of the Markov chain, we have so far used the conditional likelihood and hence ignored the information from the first observation. This may however be included in same way as we saw in Chapter 8. This implies that using the stationary distribution and giving up the opportunity of a closed–form solution, the likelihood is in this case given by

$$L(\boldsymbol{\theta}; \mathbf{y}) = P(Y_1 = y_1) \prod_{i=1}^{m} \prod_{i=1}^{m} p_{ij}^{n_{ij}}, \tag{9.48}$$

where $P(Y_1 = y_1) = \boldsymbol{\delta}(y_1)$, see (9.41), and $p_{ij}^{n_{ij}}$ are functions of the parameters, which are natural or working parameters (see (9.46) and (9.47) above). The R-function below gives an example for one implementation of the log-likelihood using working parameters

```
nll1.markov <- function(pars,y,m){
    F <- trans.count(y, m)
    Gamma <- Markov.Inv.Link(pars, m)
    delta <- stat.dist(Gamma)
    ll <- log(delta[y[1]]) + sum(F * log(Gamma))
    - ll
}
```

Example 9.8 *Using the example of this chapter, we can estimate parameters using the stationary distribution by*

```
opt1.markov <- nlminb(c(0, 0), nll1.markov,
                      y = win.dat2$WinClo1, m = 2)
opt1.markov$par

## [1] -3.0 -4.7
```

or given as natural parameters;

```
Markov.Inv.Link(opt1.markov$par,2)

##          [,1]  [,2]
## [1,] 0.9533 0.047
## [2,] 0.0093 0.991
```

The parameters are very close to the estimate using the conditional distribution (Example 9.4), but for shorter time–series, it may be worth using the unconditional likelihood.

As a final remark for the likelihood estimation, we note that if there are missing observations, then we can simply ignore those and take an extra step in the forward calculations of the transition probabilities by applying the multi-step transition probability matrix $\boldsymbol{\Gamma}^k$ (i.e. k-step transition probability matrix).

Example 9.9 (Missing observations) *Assume that we are given the (short) sequence of observations*

$$\boldsymbol{y} = [y_1 \quad y_2 \quad y_4 \quad y_7 \quad y_8]^T, \tag{9.49}$$

that is observations 3,5, and 6 are missing. In this case, we can calculate the likelihood (conditional on y_1) of the transition probability matrix $\boldsymbol{\Gamma}$ as

$$L(\boldsymbol{\Gamma}, \boldsymbol{y}) = (\boldsymbol{\Gamma})_{y_1 y_2} (\boldsymbol{\Gamma}^2)_{y_2 y_4} (\boldsymbol{\Gamma}^3)_{y_4 y_7} (\boldsymbol{\Gamma})_{y_7 y_8}. \tag{9.50}$$

Hence we simply take extra steps over the missing values.

9.4.1 Confidence intervals and uncertainty evaluation

Confidence intervals for the working parameters (τ_{ij}) can be constructed in the usual way, i.e., using the observed Fisher Information matrix (Wald confidence intervals). However, often the working parameters will not be of our primary interest, but rather either the transition probabilities or maybe even the stationary distribution.

Both are complicated functions of the working parameters, and while it is, in principle, possible to formulate the profile likelihood for these parameters, this is not an easy task in particular for high dimensional ($m >> 1$) Markov chains. The challenge is that each of the transition probabilities depends on more than one parameter (when $m > 2$). Some alternatives to the profile likelihood are;

1. Calculate the Hessian of the natural parameters using error propagation.
2. Simulate working parameters using the Information matrix and back-transform realisations.
3. Use parametric bootstrap.

The first two options are computationally cheap and option 3 is computationally expensive. The first option relies on the Wald statistics being appropriate, for both working an natural parameters. However, as transition probabilities are bounded between 0 and 1, this might not be the case and the usefulness is limited (Zucchini and MacDonald, 2009). In addition, it requires calculations of derivatives of the natural parameters with respect to the working parameters.

The second option uses the distribution implied by the Wald statistics of the working parameters and the inverse transformation. That is, simulate outcomes from the distribution

$$\boldsymbol{\theta} \sim N(\hat{\boldsymbol{\theta}}, \boldsymbol{I}(\hat{\boldsymbol{\theta}})), \tag{9.51}$$

and calculate realisations of the natural parameters. The advantage here is that we do not need to find any derivative and bounds on the natural parameters are respected. The method does however rely on the likelihood function for the working parameters being reasonably regular, which may not be the case if transition probabilities are close to 0 (and/or low number of observations).

For the third option, one simulates realisations of the Markov chain with the same number of observations as the original sequence for multiple times, using the estimated transition probability matrix. For each realisation, parameters are estimated, resulting in a set of realised parameters that constitute a distribution of the parameters. Confidence intervals can be calculated as quantiles of these distributions. As mentioned above, this is computationally expensive in particular if there is no closed–form solution to the estimation problem.

Example 9.10 (Confidence intervals) *For the example we are using in this chapter, the number of observations is high and the number of states is low and hence the choice of method is actually not crucial. However, for the sake of illustration we present both alternative 2 and 3 here and also we may be interested in the stationary distribution, which is a quite complicated function of the parameters.*

For the method based on simulation using the estimated variance-covariance of the working parameters, we need the parameter estimates and the observed Fisher information. These can be calculated using (9.33) *and* (9.34)*;*

```
tau <- c(log(F[1, 2] / F[1, 1]), log(F[2, 1] / F[2, 2]))
I <- diag(c(prod(Gamma1[1, ]) * sum(F[1, ]),
            prod(Gamma1[2, ]) * sum(F[2, ])))
V.tau <- solve(I)
```

The next step is to simulate outcomes of τ_{12} *and* τ_{21}*;*

```
k <- 1e4
Tau <- rmvnorm(k, mean = tau, solve(I))
```

The final step is to extract the features we are interested in. Here, it is p_{12}, p_{21} *and* δ *(we omit an initial definition of the matrices);*

```
for(i in 1:k){
    Gam.tmp <- Markov.Inv.Link(Tau[i, ], 2)
    Theta[i, ] <- c(Gam.tmp[1, 2], Gam.tmp[2, 1])
    Delta[i, ] <- stat.dist(Gam.tmp)
}
```

The confidence interval can now be extracted by;

```
apply(Theta, 2, quantile, prob = c(0.025, 0.975))

##         p12    p21
## 2.5%  0.034 0.0069
## 97.5% 0.063 0.0128

apply(Delta, 2, quantile, prob = c(0.025, 0.975))

##       delta1 delta2
## 2.5%    0.12   0.76
## 97.5%   0.24   0.88
```

The parametric bootstrap can be implemented by (again we omit initial definition of matrices);

```
k <- 100
for(i in 1:k){
    yi <- sim.markov(Gamma1, T)
    F.tmp <- trans.count(yi, 2)
    Tau2[i, ] <- c(log(F.tmp[1, 2] / F.tmp[1, 1]),
                   log(F.tmp[2, 1] / F.tmp[2, 2]))
    Gam.tmp <- Markov.Inv.Link(Tau2[i, ], 2)
    Theta2[i, ] <- c(Gam.tmp[1, 2], Gam.tmp[2, 1])
    Delta2[i, ] <- stat.dist(Gam.tmp)
}
```

Again the confidence interval can be extracted by;

```
apply(Theta2, 2, quantile, prob = c(0.025, 0.975))

##          p12    p21
## 2.5%  0.038 0.0072
## 97.5% 0.062 0.0115

apply(Delta2, 2, quantile, prob = c(0.025, 0.975))

##       delta1 delta2
## 2.5%     0.12   0.77
## 97.5%    0.23   0.88
```

We can note that the confidence intervals are quite similar in the presented case.

As is clear from the example above and other examples of this chapter, it is not very important which method is used for uncertainty evaluation when the time–series is long. However, for short series and parameters close to the boundary, the method might be important. Direct application of the Wald statistic relies on a good quadratic approximation of profile log-likelihood (see Section 2.5). Simulations from the Wald statistic of the working parameters rely on the profile log-likelihood of the working parameters being well approximated by a quadratic function. The parametric boot strap method does not rely on quadratic approximation of the likelihood, but as mentioned, it is in general expensive in terms of calculation time.

9.5 Higher order Markov

A higher order Markov chain is characterised by the 1-step transition depending on not only the current state, but also on the state of the chain some steps into the past. Formally a Markov chain of order l is defined as

$$P(Y_t|Y_{t-1}, Y_{t-2}, \ldots) = P(Y_t|Y_{t-1}, Y_{t-2}, \ldots, Y_{t-l}). \tag{9.52}$$

Hence, the process is not a Markov chain in the usual sense that we have considered so far. However, in a similar way as for the ARIMA processes (see Chapter 10 for the state space formulation of ARIMA-processes), we can define the state vector

$$\boldsymbol{Y}_t = [Y_t, \ldots, Y_{t-l}]. \tag{9.53}$$

In this formulation, $\boldsymbol{Y}_t$ is a first–order Markov chain of dimension l. One way of handling this is by redefining the state with values from 1 to m^l and then consider the redefined state values (with possible outcomes $1, ...m^l$). The dimension of the state space grows fast, but in such a redefined space not all transitions are possible, and hence the number of parameters does not grow as fast.

Example 9.11 (Second–order Markov chain) *We again consider the window opening example. We are interested in checking if the sequence is better described as a second-order Markov chain. The possible outcomes of the vector $\mathbf{Y}_t$ are $(0,0)$, $(1,0)$, $(0,1)$, and $(1,1)$. Renaming we can set*

$$Y_t^* = \begin{cases} 1 & \text{if } \mathbf{Y}_t = (0,0) \\ 2 & \text{if } \mathbf{Y}_t = (1,0) \\ 3 & \text{if } \mathbf{Y}_t = (0,1) \\ 4 & \text{if } \mathbf{Y}_t = (1,1). \end{cases} \tag{9.54}$$

From the definitions here, we see that not all transitions are possible, e.g. the transition from 1 to 3 is not possible, and actually there are only two possible transitions from each value of the state. These are

$$\begin{aligned} P(Y_t^* = 1 | Y_t^* = 1) =& 1 - p_{12} \\ P(Y_t^* = 2 | Y_t^* = 1) =& p_{12} \\ P(Y_t^* = 3 | Y_t^* = 2) =& p_{23} \\ P(Y_t^* = 4 | Y_t^* = 2) =& 1 - p_{23} \\ P(Y_t^* = 1 | Y_t^* = 3) =& 1 - p_{32} \\ P(Y_t^* = 2 | Y_t^* = 3) =& p_{32} \\ P(Y_t^* = 3 | Y_t^* = 4) =& p_{43} \\ P(Y_t^* = 4 | Y_t^* = 4) =& 1 - p_{43}, \end{aligned} \tag{9.55}$$

and hence we can set up the transition probability matrix as

$$\boldsymbol{\Gamma} = \begin{bmatrix} 1-p_{12} & p_{12} & 0 & 0 \\ 0 & 0 & p_{23} & 1-p_{23} \\ 1-p_{32} & p_{32} & 0 & 0 \\ 0 & 0 & p_{43} & 1-p_{43} \end{bmatrix}. \tag{9.56}$$

The transition from $Y_t^* = 1$ to $Y_{t+1}^* = 2$ implies that the status of the window changes from open to close. The transition from a second-order Markov chain with two states to a first order Markov chain with 4 states is illustrated in Figure 9.1.

The conditional likelihood can be calculated in a similar way as the first–order likelihood

$$\begin{aligned} L(\boldsymbol{\theta}) =& P(Y_1 = y_1, \ldots, Y_l = y_l) \prod_{t=l+1} P(Y_t = y_t | Y_{t-1} = y_{t-1}, \ldots, Y_{t-l} = y_{t-l}) \\ =& P(\boldsymbol{Y}_1 = \boldsymbol{y}_1) \prod_{t=l+1} P(\boldsymbol{Y}_t = \boldsymbol{y}_t | \boldsymbol{Y}_{t-1} = \boldsymbol{y}_{t-1}). \end{aligned} \tag{9.57}$$

Hence, the conditional likelihood ignores the first l observations. The unconditional likelihood can be calculated in the same way as we saw in (9.48).

The number of different states is related to the order of the Markov, that is the number of different combinations of length l of the numbers from 1 to m, that is m^l. A simple way to generate the possible states is given in the R-function below;

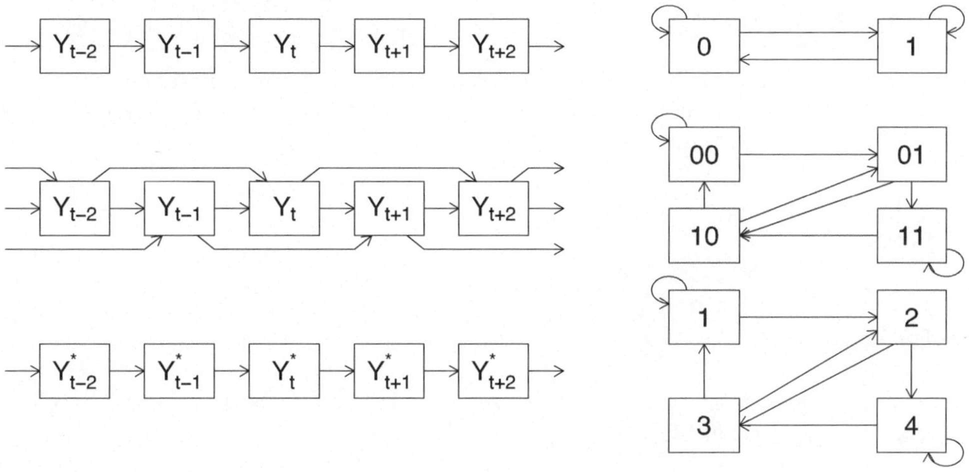

Figure 9.1: Schematic illustration of the second–order Markov chain presented in Example 9.11. Top row: First–order Markov with 2 states. Second row: Second–order Markov chain. Third row: Relabeling of states give a first–order Markov chain. Second column shows possible transitions in each case.

```
make.states <- function(m, l){
    state.mat <- matrix(1, ncol = l, nrow  =m ^ l)
    tpm <- matrix(0, ncol = m ^ l, nrow  =m ^ l)
    states <- 1 : m ^ l
    s <- 0
    for(i in 1:m^(l-1)){
        for(j in 1:m){
            state.mat[s+j, ] <- c(state.mat[i, -1], j)
        }
        s <- s + m
    }
    state.mat
}
```

The output from the function is a matrix representing the possible outcomes.

Example 9.12 *As an example we generate the possible outcomes of $\mathbf{Y}_t$ in our window opening example where $l = 2$;*

```
(state.mat2 <- make.states(2, 2))

##      [,1] [,2]
## [1,]    1    1
## [2,]    1    2
## [3,]    2    1
## [4,]    2    2
```

Hence, there are 4 different possible values of the state defined by the position of the window at time t and $t-1$.

In order to use the methods we have discussed so far, we need to redefine the outcomes to a time series with outcomes $1, ..., m^l$. This can be implemented in many ways. A simple (though quite inefficient) implementation is;

```
make.high.order.seq <- function(y, state.mat){
    y.new <- y * 0
    s <- 1:dim(state.mat)[1]
    l <- dim(state.mat)[2]
    for(i in 1:length(y)){
        for(j in s){
            if(sum(y[(i-l+1):i] == state.mat[j, ]) == l){
                y.new[i] <- j
            }
        }
    }
    y.new
}
```

The function takes a sequence y (that takes values in {1,...,m}) and returns the redefined sequence with states corresponding to the matrix `state.mat`, taking values in $\{1,...,m^l\}$ (note that the first l-1 entries is set as zero).

Models of different order can be compared directly using likelihood–based methods. However, we need to make sure that the likelihood is calculated based on the same data time-series, as the length of the part we condition on depends on the order. We can either deal with that by ignoring the first l observations (in both the simple and the higher order model) or using the unconditional likelihood. In the latter case, we need a structural matrix indicating the possible transitions. We will illustrate this with the next example.

Example 9.13 *One could expect that the window position is an Markov model of order 2. In this case, we can redefine the observations as;*

```
win.dat2$WinClo2 <- make.high.order.seq(win.dat2$WinClo1, state.mat2)
```

`WinClo2` *is now a sequence varying from 1 to 4, with possible transitions (0 implies that the transition is not possible and 1 and 2 are codings used for the estimation);*

```
(tpm.struc <- rbind(c(1, 2, 0, 0), c(0, 0, 2, 1),
                    c(1, 2, 0, 0), c(0, 0, 2, 1)))

##      [,1] [,2] [,3] [,4]
## [1,]    1    2    0    0
## [2,]    0    0    2    1
## [3,]    1    2    0    0
## [4,]    0    0    2    1
```

The corresponding transition count is given by;

```
(F2 <- trans.count(win.dat2$WinClo2,4))

##      [,1] [,2] [,3] [,4]
## [1,]  765   35    0    0
## [2,]    0    0    1   38
## [3,]   35    4    0    0
## [4,]    0    0   37 4079
```

We see that in most cases the window stays open/closed for at least 2 time steps (very low count from state 3 to 2 and from 2 to 3). The transition probability matrix can be implemented in R by;

```
Markov.Inv.Link.Q <- function(pars, tpm.struc){
    exp.pars <- exp(pars)
    m <- dim(tpm.struc)[1]
    Gamma <- matrix(0, ncol = m, nrow = m)
    s <- 1
    for(i in 1:m){
        r <- sum(tpm.struc[i, ] == 2)
        p <- exp.pars[s : (s + r - 1)] /
            (1 + sum(exp.pars[s:(s + r - 1)]))
        Gamma[i, tpm.struc[i, ] == 2] <- p
        Gamma[i, tpm.struc[i, ] == 1] <- 1 - sum(p)
        s <- s + r
    }
    Gamma
}
```

The function uses the definition of the working parameters as we have discussed above and ensure zeros in positions as defined by the matrix `tpm.struc`*. Finally the likelihood is implemented by;*

```
nll1.struc.markov <- function(pars, y, m, tpm.struc){
    F <- trans.count(y, m)
    Gamma <- Markov.Inv.Link.Q(pars, tpm.struc)
    delta <- as.numeric(stat.dist(Gamma))
    ll <- log(delta[y[1]]) + sum(F[tpm.struc != 0] *
                                 log(Gamma[tpm.struc != 0]))
    - ll
}
```

The second-order markov model can now be estimated by;

```
pars <- rep(0, 4)
opt2.markov <- nlminb(pars, nll1.struc.markov, y = win.dat2$WinClo2[-1],
                      m = 4, tpm.struc = tpm.struc)
```

Which can be compared to the first-order Markov chain by

```
-c(opt2.markov$objective, opt1.markov$objective)

## [1] -377 -379
```

Here we see an increase of about 2 in the log-likelihood with 2 extra degrees of freedom, which is not significant on the usual significance level. The model can of course be extended to a third-order Markov model, using the exact same methods. In this case, there is not a significant improvement in the likelihood compared to the second-order model. Even though the second-order model is not significant, we will use it to exemplify the analysis.

The estimated transition probability matrix is given by;

```
(Gamma2 <- Markov.Inv.Link.Q(opt2.markov$par,tpm.struc))

##      [,1]  [,2]   [,3] [,4]
## [1,] 0.96 0.044 0.0000 0.00
## [2,] 0.00 0.000 0.0262 0.97
## [3,] 0.90 0.103 0.0000 0.00
## [4,] 0.00 0.000 0.0092 0.99
```

From this transition probability matrix, we see that the probability that the window stays open given that it is open is around 96% (transition from state 1 to state 1). If the window is closed, it will stay closed with a probability of around 97% (transition from state 2 to state 4) and so on.

Based on an estimated transition probability matrix, we can answer questions about the dynamics of the observed sequence. Basically any analysis can be done based on simulation using the estimated transition probability matrix. In some cases, it is also quite simple to answer questions based directly on the estimated transition probability matrix. For example, we can answer questions like, given we are in state i, what is the expected time before we revisit state i? or given we just left state i what is the time before we revisit state i? For more general treatment of such questions we refer to more dedicated textbooks on Markov chains, e.g. (Grimmit and Stirzaker, 2001). Here we will just give one example.

Example 9.14 *As an example we may ask what the mean time for returning to the current state is (the recurrence time). This is given by*

$$T_{rec} = \frac{1}{\delta}, \tag{9.58}$$

where $\boldsymbol{\delta}$ is the stationary distribution. For the two state model we have considered so far these are;

```
1/delta

##      [,1] [,2]
## [1,]    6  1.2
```

For the second-order model, the times are;

```
delta2 <- stat.dist(Gamma2)
1/delta2

##       [,1] [,2] [,3] [,4]
## [1,]   6.3  128  128  1.2
```

This implies that the mean recurrence time for the position open is about 6 time steps (1 hour) for either of the models, and 1.2 time steps if the window is closed. For the second-order model, the transition states (state 2 and 3) have a mean recurrence time of 128 times steps (app. 21 hours). The transition step describes, in our case, transitions from open to closed or from closed to open and the time can be interpreted as the time between interactions with the window.

Another number that might be of interest and relates to the discussion of survival model in Chapter 7 is the probability of returning to state i given that we just left it. This probability is determined by the transition probability matrix. In the 2 state model, we have

$$P(Y_{t+k} = i, Y_{t+k-1} \neq i, \ldots, Y_{t+1} \neq i | Y_{t-1} = i, Y_t \neq i) = p_{jj}^{k-1} p_{ji}, \tag{9.59}$$

and, for example if the window changes from open to closed, we have

$$P(Y_{t+k} = 1, Y_{t+k-1} = 2, \ldots, Y_{t+1} = 2 | Y_{t-1} = 1, Y_t = 2) = p_{22}^{k-1} p_{21}. \tag{9.60}$$

For the second-order model, states $Y_t^* = 1$ and $Y_t^* = 3$ imply that the window is opened at time t; therefore the probability is calculated separately for $k = 1$ and $k > 1$

$$P(Y_{t+1} = 1 | Y_{t-1} = 1, Y_t = 2) = p_{21}; \quad k = 1. \tag{9.61}$$

This is the probability of opening the window immediately again after it was closed, and for $k > 1$ we have

$$P(Y_{t+k} = 1, Y_{t+k-1} = 2, \ldots, Y_{t+1} = 2 | Y_{t-1} = 1, Y_t = 2) = p_{24} p_{44}^{k-2} p_{43}, \tag{9.62}$$

Of course, the relation applies the other way around (by shifting index). Again relating to the results in Chapter 7, these transitions represent the transition from exponential survival times to more complicated survival times.

The example illustrates the connection between survival models and Markov chains. We will not go further into this here, but rather discuss models that include covariates.

9.6 Including covariates

In many situations, it is reasonable to assume that the transition probability matrix is a function of time in some way. This may be an explicit dependence on time or through some covariate ($\boldsymbol{x}_t$), that is measured or predicted. As a general model formulation, we consider a time in-homogeneous Markov chain with the state transition given by

$$\boldsymbol{u}_{t+1} = \boldsymbol{u}_t \boldsymbol{\Gamma}(t), \tag{9.63}$$

where the time dependence of $\boldsymbol{\Gamma}$ may be given in form of some input variables.

We start by considering a Markov chain with only two states. In this case, we can write the transition probability matrix as

$$\boldsymbol{\Gamma}(\boldsymbol{x}_t) = \begin{bmatrix} p_{11}(\boldsymbol{x}_t^1) & p_{12}(\boldsymbol{x}_t^1) \\ p_{21}(\boldsymbol{x}_t^2) & p_{22}(\boldsymbol{x}_t^2) \end{bmatrix}, \tag{9.64}$$

with $p_{11}(\boldsymbol{x}_t^1) + p_{12}(\boldsymbol{x}_t^1) = 1$ and $p_{21}(\boldsymbol{x}_t^2) + p_{22}(\boldsymbol{x}_t^2) = 1$. In the case of $m = 2$, the solution is quite straight forward. Using the usual link function for binomial data, we could choose

$$p_{ii}(\boldsymbol{x}_t^i) = \frac{e^{\boldsymbol{x}_t^i \boldsymbol{\beta}_{ii}}}{1 + e^{\boldsymbol{x}_t^i \boldsymbol{\beta}_{ii}}}, \tag{9.65}$$

and $p_{ij}(\boldsymbol{x}_t^i) = 1 - p_{ii}(\boldsymbol{x}_t^i)$, $(i \neq j)$. If $m > 2$, considerations on the number of parameters need to be taken. One suggestion is given by Liisberg et al. (2016), where the diagonal elements are modelled by (here $\boldsymbol{x}_t$ is free of the intercept)

$$p_{ii}(\boldsymbol{x}_t^i) = \frac{e^{\boldsymbol{x}_t^i \boldsymbol{\beta}_{ii}}}{e^{\boldsymbol{x}_t^i \boldsymbol{\beta}_{ii}} + \sum_{j \neq i} e^{\tau_{ij}}}, \tag{9.66}$$

and the off-diagonal elements are given by

$$p_{ij}(\boldsymbol{x}_t^i) = \frac{e^{\tau_{ij}}}{e^{\boldsymbol{x}_t^i \boldsymbol{\beta}_{ii}} + \sum_{j \neq i} e^{\tau_{ij}}}; \quad i \neq j. \tag{9.67}$$

The total number of parameters in this model is $m(m-1)$ (τ-parameters) plus mp ($\boldsymbol{\beta}$-parameters, here p is the dimension of $\boldsymbol{\beta}_{ii}$). For $m = 2$, there are no savings in terms of parameters compared to the most complex model, but for $m > 2$, there is a saving in number of parameters.

For higher order models, the number of parameters will still be a concern and we discuss one possible solution in Section 9.6.2.

9.6.1 Parameter estimation

When estimating parameters, the easiest implementation is a recursive estimation starting from the first observation and calculating the likelihood contribution (as the conditional likelihood) for each observation.

In order to calculate the likelihood, we need a function to calculate the transition probability matrix for each time step, see (9.66)–(9.67). This is done using the function below.

```
Markov.Inv.Link.Reg <- function(pars, m, X){
    p <- dim(X)[2]; n <- dim(X)[1]
    beta <- matrix(pars[1 : (p * m)], ncol = m)
    exp.eta <- exp(X %*% beta)
    exp.tau <- exp(pars[-(1: (p * m))])
    Tau.mat <- diag(m)
    Tau.mat[Tau.mat != 1] <- exp.tau
    Gamma <- array(dim=c(n, m, m))
    for(i in 1:n){
        diag(Tau.mat) <- exp.eta[i, ]
        Gamma[i, , ] <- t(Tau.mat) / colSums(Tau.mat)
    }
    Gamma
}
```

The vector pars contain all parameters (i.e. $\boldsymbol{\beta}_{ii}$ and τ_{ij}), and return an array, that contains transition probability matrices at all time points. The function assumes that $\boldsymbol{x}_t^i = \boldsymbol{x}_t^j$, but can be modified to more general cases.

Using the above, the likelihood can be calculated recursively using the observed Markov chain and the sequence of transition probability matrices (i.e. the conditional likelihood), as is seen in the R-function below.

```
nll.markov.reg <- function(pars, y, X, mm){
    n <- length(y)
    Gamma <- Markov.Inv.Link.Reg(pars, mm, X)
    lGamma <- log(Gamma)
    ll <- 0
    for(i in 2:n){
        ll <- ll + lGamma[i - 1, y[i - 1], y[i]]
    }
    -ll
}
```

where `mm` is the number of states and `ll` is the likelihood.

Example 9.15 *Based on the results in the exploitative analysis in Example 9.1, we could hypothesise that the transition from open to closed and the other way could depend on the time of day. One way of doing this is inspired by the Fourier expansion of time of day. This is the same approach as used in Chapter 8. Hence the model is*

$$p_{ii}(t) = \frac{e^{\eta_{ii}(t)}}{e^{\eta_{ii}(t)} + e^{\tau_{ij}}}, \tag{9.68}$$

for i equal 1 or 2 and $j \neq i$ (but also either 1 or 2), and $\eta_{ii}(t)$ given by

$$\eta_{ii}(t) = \sin(2\pi t)\beta_{ii}^1 + \cos(2\pi t)\beta_{ii}^2 + \sin(4\pi t)\beta_{ii}^3 + \cos(4\pi t)\beta_{ii}^4 \tag{9.69}$$

The superscripts refer to indices in the vector $\boldsymbol{\beta}_{ii}$, and t is measured in days. The model is estimated in R by;

```
X1 <- cbind(sin(2 * pi * win.dat2$Time), cos(2 * pi * win.dat2$Time),
            sin(4 * pi * win.dat2$Time), cos(4 * pi * win.dat2$Time))
m <- 2
pars <- c(rep(0, dim(X1)[2] * m), opt1.markov$par)
opt.markov.reg1 <- nlminb(pars, nll.markov.reg, y = win.dat2$WinClo1,
                          X = X1, mm = m)
```

Here, the parameters are initialised at the MLE of the time homogeneous model. It is reasonable to assume that the transition probability also depends on the CO_2 level in the room corresponding to

$$\eta_{ii}(t) = \sin(2\pi t)\beta_{ii}^1 + \cos(2\pi t)\beta_{ii}^2 + \sin(4\pi t)\beta_{ii}^3 + \cos(4\pi t)\beta_{ii}^4 + \log(CO_{2,t})\beta_{ii}^5, \tag{9.70}$$

that is the transition probability depends on the CO_2 level at the current time. The model can be implemented by (here the average log CO_2 level is subtracted);

```
X2 <- cbind(X1, log(win.dat2$CO2C) - 6.5)
pars <- c(opt.markov.reg1$par[1 : 4], 0,
          opt.markov.reg1$par[5 : 8], 0,
          opt.markov.reg1$par[9 : 10])
opt.markov.reg2 <- nlminb(pars, nll.markov.reg, y = win.dat2$WinClo1,
                          X = X2, mm = m)
```

Again, the results can be compared using the usual likelihood methods such as the likelihood ratio test;

```
Q <- -2 * (opt.markov.reg2$objective - opt.markov.reg1$objective)
1 - pchisq(Q, df = length(opt.markov.reg2$par) -
                   length(opt.markov.reg1$par))

## [1] 0.0000042
```

Hence, the difference in likelihood is significant on any reasonable level (e.g. p-value << 0.01). The effect of including regression can be assessed comparing the likelihood to the conditional likelihood calculated in Example 9.4.

```
c(ll1, -opt.markov.reg2$objective)

## [1] -379 -315
```

Hence a difference in likelihood of about 64 units on 10-2=8 degrees of freedom, the p-value could be calculated in a similar way as above and would again show a significant effect on any reasonable significance level. The uncertainty can be evaluated using the information matrix (calculated using `hessian`*). The table below shows the results in terms of parameter estimates, standard errors and test statistics.*

```
##            theta   se     z
## beta11.1    1.15 0.86  1.34
## beta11.2   -0.23 0.78 -0.30
## beta11.3    0.45 0.61  0.74
## beta11.4   -0.58 0.34 -1.71
## beta11.5   -3.43 0.73 -4.67
## beta22.1   -1.85 1.19 -1.56
## beta22.2    8.35 4.06  2.06
## beta22.3    0.35 0.84  0.42
## beta22.4    2.42 1.46  1.66
## beta22.5    1.75 0.81  2.17
## tau12      -2.22 0.65 -3.41
## tau21      -9.45 2.74 -3.45
```

From the table we see that the parameter estimates for transitions out of state 1 (i.e. closing the window) are for the most part not significant (`beta11.1-beta11.4`*). Also, the effect of CO_2 is significant for both states and actually, partial test would suggest that the model should be simplified to include fewer parameter (compare* `z` *with quantiles of the standard normal distribution). It seems that the diurnal part is over-parametrised and it is natural to remove the higher order term in the periodic function (i.e. the two last terms in* (9.69)*). This is done in the model below.*

```
X2Re <- X2[ , -(3 : 4)]
pars <- rep(0, 2 * dim(X2Re)[2] + 2)
opt.markov.reg2Re <- nlminb(pars, nll.markov.reg,
                            y = win.dat2$WinClo1, X = X2Re,
                            mm = m)
Q <- - 2 * (opt.markov.reg2$objective -
            opt.markov.reg2Re$objective)
1 - pchisq(Q, df = length(opt.markov.reg2$par) -
               length(opt.markov.reg2Re$par))

## [1] 0.033
```

The p-value is around 3% and therefore the usual significance level of 5% would suggest not to reduce the model. However, with a borderline significant p-value, one can argue that we should go with the simpler model. The parameter table for the simpler model is given below;

```
##          theta   se     z
## beta11.1  0.45 0.32  1.41
## beta11.2 -0.33 0.40 -0.83
## beta11.3 -3.18 0.72 -4.39
## beta22.1 -2.72 0.63 -4.34
## beta22.2  3.56 0.71  5.01
## beta22.3  1.68 0.80  2.11
## tau12    -1.94 0.36 -5.40
## tau21    -6.67 0.70 -9.56
```

With the reduced model, all parameters for the transition out of state 2 (i.e. opening the window) are significant, while only the parameter related to CO_2 concentration is significant for transition out of state 1 (i.e. closing the window). This actually suggests that the model should be different for the two states. We will not pursue this further here, but rather present the result. The figure below shows the transitions out of state 1 and 2 as a function of time of day for different values of CO_2 concentration.

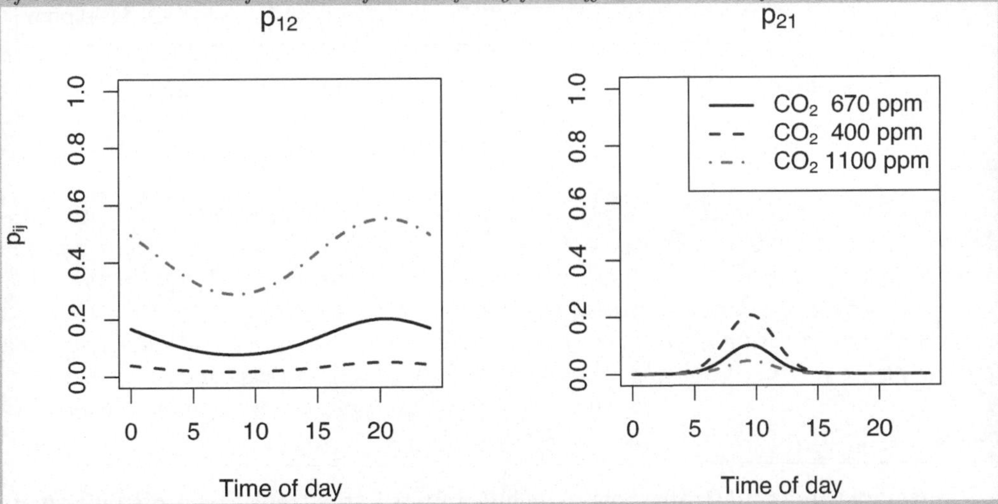

From the figure, we see that the transition out of state 2 depends heavily on the time

of day and also that increased CO_2 levels decrease the probability of transitioning out of state 2 (i.e. opening the window), which is of course counter intuitive. This is however a quite small sample from 1 room in 1 dwelling and hence no general conclusions should be drawn from that. A more in–depth analysis of the data using generalized linear model is given in (Andersen et al., 2013).

The interpretation of time-varying Markov chain models is a bit more difficult than time homogeneous models since there is not a stationary distribution that we can analyse. It is of course always possible to simulate from the estimated model, for example in order to analyse the probability distribution as a function of time.

Example 9.16 (Distribution as a function of hour of day) *By modifying the simulation function (`sim.markov` on page 264) slightly, the window position can be simulated over a long time and the probability of the window position as a function of time of day can then be plotted. The probability of the window being closed is plotted as a function of time of day for different levels of CO_2. Of course, the CO_2 level is affected by the window position, meaning that there are no data where the window is open and the CO_2–concentration is high (e.g. there are 2 observation where the window is open and the CO_2 concentration is above 1000ppm), basically implying that the figure extrapolate out of the range of data. That is ignored in the figure below (and in the developed model).*

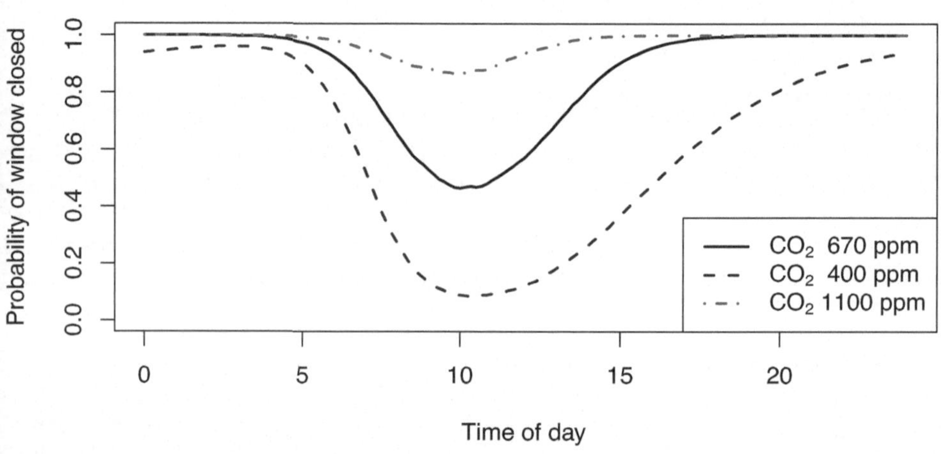

We see that the window is closed with a very high probability at night. When the CO_2 concentration is low, the probability that the window is open is higher and not zero at night, it should be mentioned that low CO_2 concentrations at night is not really observed in data and hence that part is doubtful.

Here we have used an input (CO_2 concentration) that is dependent on the variable that we are simulating. This is a good example that association does not equal causality. In this case, the causality is probably that the CO_2 concentration is low because the window is open (not the other way around).

The example above illustrates how to include inputs in the transition probability matrix, and also that care should be taken if inputs are affected by the output. In these cases, one should choose the model with care to ensure models that can be interpreted in a good way.

Models that use window position as an input was covered in Chapter 8 and extensions are covered in Chapter 10. Finally CO_2 is used for identifying states (that may be interpreted as window position) in Chapter 11.

9.6.2 Higher order Markov chains

As we have seen, higher order Markov chains can be written as first–order Markov models with an extended state space and as such the method for including covariates in the previous section can in principle be applied directly. However, the dimension of the parameter space will be a concern and considerations in that regard should be taken. We will not cover a very general setup here, but rather give an example of the type of considerations that should be taken by using our window opening example. The transition probability matrix is in the case we have discussed here

$$\mathbf{\Gamma} = \begin{bmatrix} 1-\gamma_{12} & \gamma_{12} & 0 & 0 \\ 0 & 0 & \gamma_{23} & 1-\gamma_{23} \\ 1-\gamma_{32} & \gamma_{32} & 0 & 0 \\ 0 & 0 & \gamma_{43} & 1-\gamma_{43} \end{bmatrix}. \tag{9.71}$$

Here, the transition from 1 to 2 implies that the window position changes from open to closed, the transition from 2 to 3 implies that the window position changes from closed to open and so on (but always taking the current position and the previous position into account, as opposed to considering only the current position). A direct application of the regression method above would imply an individual set of regression parameters for each value of the state (i.e. here 4 sets of regression parameters). However, with the interpretation just mentioned, a reasonable parametrisation is

$$\begin{aligned} \gamma_{12}(\boldsymbol{x}_t) &= \frac{e^{\tau_{12}}}{e^{\boldsymbol{x}_t \boldsymbol{\beta}_{11}} + e^{\tau_{12}}} \\ \gamma_{23}(\boldsymbol{x}_t) &= \frac{e^{\tau_{23}}}{e^{\boldsymbol{x}_t \boldsymbol{\beta}_{44}} + e^{\tau_{23}}} \\ \gamma_{32}(\boldsymbol{x}_t) &= \frac{e^{\tau_{32}}}{e^{\boldsymbol{x}_t \boldsymbol{\beta}_{11}} + e^{\tau_{32}}} \\ \gamma_{43}(\boldsymbol{x}_t) &= \frac{e^{\tau_{43}}}{e^{\boldsymbol{x}_t \boldsymbol{\beta}_{44}} + e^{\tau_{43}}}. \end{aligned} \tag{9.72}$$

Hence, only 2 extra parameter sets are needed.

As should be clear from the discussion in this chapter, there are many possibilities for modelling using Markov chains including time–varying transition probabilities and higher order models. One challenge in the modelling is that the number of parameters grows fast with the dimension and simplifications of the most general formulation should be considered.

The next section discusses an application of Markov chain models that combine the models discussed in this chapter with the models discussed in Chapter 8.

9.7 Markov switching model

We saw in Chapter 8 how linear time–series models can be used for modelling of CO_2 and we also saw that window opening was a primary input for the time series models. This combination makes it difficult to use such a model for forecasting since it includes a stochastic process (the window position). In the present chapter, we have estimated parameters in a Markov chain for the window position. These models may be combined in a so-called Markov (or regime) switching model. These models can be given in a quite general framework, but an example of a Markov switching AR model is

$$Y_t = \mu(C_t, \boldsymbol{x}_t) + \sum_{i=1}^{p} \theta_i(C_t, \boldsymbol{x}_t)(Y_{t-i} - \mu(C_{t-i}, \boldsymbol{x}_{t-i})) + \epsilon_t, \tag{9.73}$$

where $\epsilon_t \sim N(0, \sigma^2(C_t))$ and C_t is governed by a Markov chain process. Usually, the Markov chain is not observed, and in such cases it is called a Hidden Markov process (see Chapter 11). In the case we have covered in this chapter, the Markov chain is actually observed, and in the model we introduced in Chapter 8, the function μ was given as (ToD refer to time of day)

$$\mu(C_t, \boldsymbol{x}_t) = \beta_0(C_t) + f(ToD_t), \tag{9.74}$$

and further

$$\theta_i(C_t, \boldsymbol{x}_t) = \theta_i, \tag{9.75}$$

that is not a function of the state. Finally, $\sigma^2(C_t)$ was also constant. In order to evaluate the likelihood of such a model, we need so-called mixture distributions. A mixture distribution is in general defined by

$$p(y) = \sum_{i}^{m} \delta_i p_i(y), \tag{9.76}$$

where $p_i(y)$ is some probability density/mass function, with $\delta_i > 0$ and $\sum \delta_i = 1$. The connection to Markov models is made by

$$p(y_t|\boldsymbol{y}_{t-1}, \boldsymbol{C}_{t-1}, \boldsymbol{X}_t) = \sum_{i=1}^{m} p_{c_{t-1},i} p_i(y_t|\boldsymbol{y}_{t-1}, \boldsymbol{C}_{t-1}, \boldsymbol{X}_t). \tag{9.77}$$

Here, $\boldsymbol{y}_{t-1} = [y_{t-1}, \ldots, y_1]^T$ is the history of the process, $\boldsymbol{C}_{t-1} = [C_{t-1}, \ldots, C_1]^T$ is the history of the Markov chain, and $\boldsymbol{X}_t$ is the collection of all input variables.

As we have seen earlier, the mean and variance of the 1-step predictions are important for model validation and all moments (k) of Y_t given the past can be calculated by

$$E[Y_t^k|\boldsymbol{y}_{t-1}, \boldsymbol{C}_{t-1}, \boldsymbol{X}_t] = \sum_{i=1}^{m} p_{c_{t-1},i} E[Y_t^k|\boldsymbol{y}_{t-1}, \boldsymbol{C}_t, \boldsymbol{X}_t], \tag{9.78}$$

and the variance can be calculated from that. As a special case for a two–state Markov chain, the 1-step variance is given by

$$V[Y_t|\boldsymbol{y}_{t-1}, \boldsymbol{C}_{t-1}, \boldsymbol{X}_t] = \sum_{i=1}^{2} p_{c_{t-1},i}\sigma_{i,t}^2 + p_{c_{t-1},1}p_{c_{t-1},2}(\mu_{1,t} - \mu_{2,t})^2, \tag{9.79}$$

where $\sigma_{i,t}^2 = V[Y_t|\boldsymbol{y}_{t-1}, \boldsymbol{C}_{t-1}, \boldsymbol{X}_t, C_1 = i]$, and $\mu_{i,t} = E[Y_t|\boldsymbol{y}_{t-1}, \boldsymbol{C}_{t-1}, \boldsymbol{X}_t, C_t = i]$, so we see that the variance is greater than the weighted sum of the variance of the components.

The likelihood is given by

$$L(\boldsymbol{\theta}; \boldsymbol{y}_{t-1}, \boldsymbol{C}_{t-1}, \boldsymbol{X}_t) = \prod_{t=1}^{T} \sum_{i=1}^{m} p_{c_{t-1},i} p_i(y_t | \boldsymbol{y}_{t-1}, \boldsymbol{C}_{t-1}, \boldsymbol{X}_t), \tag{9.80}$$

and the log-likelihood is given by

$$l(\boldsymbol{\theta}; \boldsymbol{y}_{t-1}, \boldsymbol{C}_{t-1}, \boldsymbol{X}_t) = \sum_{t=1}^{T} \log \left(\sum_{i=1}^{m} p_{c_{t-1},i} p_i(y_t | \boldsymbol{y}_{t-1}, \boldsymbol{C}_{t-1}, \boldsymbol{X}_t) \right), \tag{9.81}$$

which may be estimated using the likelihood conditional on the first observation or full likelihood using a stationary distribution.

Example 9.17 *In Chapter 8, we discussed linear time-series models and found that an ARX(2) model was appropriate for modelling the CO_2-level. We used the window position as part of the input. Inspired by the results in this section, we could formulate a Markov switching model as*

$$Y_t = \beta_0(C_t) + f(ToD_t) + \sum_{i=1}^{2} \phi_i \left(Y_{t-i} - \beta_0(C_{t-i}) - f(ToD_{t-i})\right) + \epsilon_t, \tag{9.82}$$

where C_t (the window position) is described by the time in-homogeneous Markov chain that we found in Example 9.15. Assuming C_t is known, we can estimate the parameters using the `arima` *method by;*

```
Xar <- cbind(C = win.dat2$WindowClosed, X1)
colnames(Xar) <- c("C", "sin1", "cos1", "sin2", "cos2")
ar2x.reg <- arima(log(win.dat2$CO2C), order=c(2, 0, 0), xreg = Xar)
```

with the result

```
## [...]
## Coefficients:
##          ar1     ar2  intercept      C   sin1   cos1   sin2    cos2
##        1.429  -0.475      6.501  0.096  0.515  0.015  0.124  -0.082
## s.e.  0.013   0.013      0.018  0.006  0.023  0.003  0.018   0.018
##
## sigma^2 estimated as 0.00312:  log likelihood = 7319,  aic = -14622
## [...]
```

We see the same patterns as in Example 8.17, meaning higher CO_2-levels when the window is closed and a significant diurnal variation. The parameter values cannot be compared directly as the time resolution is not the same (10 minutes here and 30 minutes in Chapter 8).

In principle, the Markov switching model could be estimated using both CO_2 and window position as input resulting in a model, where parameter estimates are adapted to both parts simultaneously. For the purpose of illustration, we will however take a more simple approach and simply use an independently estimated transition probability matrix as input. The model can be implemented in R by

```
nll.markovSwich <- function(pars, y, C, Gamma, X, full = FALSE){
    ## parameters
    beta0 <- pars[1] + pars[2] * C
    beta <- pars[3 : (dim(X)[2] + 2)]
    phi <- pars[(dim(X)[2] + 3) : (dim(X)[2] + 4)]
    sigma <- exp(pars[dim(X)[2] + 5])
    ## initialization
    n <- length(y); mu <- rep(NA, n); sigma2 <- rep(NA, n)
    f <- X %*% beta
    ll <- 0
    ## main loop for likelihood
    for(i in 3:n){
        index <- (i-1):(i-2)
        tmp <- pars[1] + f[i] + sum(phi * (y[index] - beta0[index] - f[index]))
        ll <- ll + log(sum(Gamma[i, C[i - 1] + 1,1 : 2] *
                          dnorm(y[i], mean =tmp + c(0, pars[2]), sd = sigma)))
        if(full){
            mu[i] <- tmp + sum(Gamma[i, C[i - 1] + 1, 2] * pars[2])
            sigma2[i] <- sigma^2 + prod(Gamma[i, C[i - 1] + 1, 1 : 2]) *
                pars[2]^2
        }
    }
    if(full){return(list(mu = mu, sigma2 = sigma2))}
    -ll
}
```

Also to keep it simple, we will use the Markov model that only takes time of day as an input. The model is estimated below;

```
X <- Xar[ ,-1] ## remove win. pos. from X
## All transition probabilities
Gamma <- Markov.Inv.Link.Reg(opt.markov.reg1$par, 2, X)
pars <- c(coef(ar2x.reg)[3:4], coef(ar2x.reg)[5:8],
          coef(ar2x.reg)[1:2], -2)
opt2 <- nlminb(pars,nll.markovSwich, y = log(win.dat2$CO2C),
               C = win.dat2$WindowClosed, Gamma = Gamma, X = X)
```

The model can be compared in terms of likelihood by

```
logLik(ar2x.reg)

## 'log Lik.' 7319 (df=9)

-opt2$objective

## [1] 7485
```

We see a large increase in the log-likelihood, implying that the Markov switching model fits data much better. The parameters are;

```
opt2$par[1:2] ## intercept and effect of window pos.

## Intercept       beta
##      6.43       0.19

opt2$par[3:6] ## diurnal variation

##   sin1   cos1   sin2   cos2
##  0.509  0.014  0.126 -0.079

opt2$par[7:8] ## AR-parameters

##   ar1   ar2
##  1.40 -0.44

exp(opt2$par["lsig"] * 2) ## sigma^2

##   lsig
## 0.0026
```

The parameters are similar to the ARX(2) model, but with the variance slightly lower. One large difference between the models is that in the Markov switching model, the variance changes over time (for transition probabilities close to 1/2, the variance is largest, see (9.79)*). We will not go more into that subject here.*

The example above illustrates how models can be combined into models that include all the stochastic elements. Further, the models can of course be generalised in several ways.

9.8 Continuous time Markov chains and R

We have introduced Markov chains defining the R-functions that were needed as we introduced the theory and the models. There do exist some packages that estimate parameters in Markov models. One such example is the recent package `markovchain` (Spedicato, 2017). Also the R-package `msm` (Jackson, 2011) estimates parameters in continuous time Markov models.

A continuous time Markov chain is characterised by a transition intensity matrix rather than a transition probability matrix. The relation to the discrete time Markov chains that we have covered here is

$$\boldsymbol{\Gamma}(\Delta) = \text{Exp}(\boldsymbol{Q}\Delta), \tag{9.83}$$

where Δ is the time step of the discrete time Markov chain, Exp is the matrix exponential, and while the row-sum of $\boldsymbol{\Gamma}$ is 1, the row sums of $\boldsymbol{Q}$ are 0 with $Q_{ii} < 0$ and $Q_{ij} > 0$ $(i \neq j)$. Hence, the forward equation taking the state vector $\boldsymbol{u}_t$ to $\boldsymbol{u}_{t+\Delta}$ is given by

$$\boldsymbol{u}_{t+\Delta} = \boldsymbol{u}_t \text{Exp}(\boldsymbol{Q}\Delta). \tag{9.84}$$

We will not go into details with this, but rather just illustrate it with a small example.

Example 9.18 (Continuous time Markov chain) *We can estimate a continuous time Markov model, using* `msm`, *by;*

```
Q <- rbind(c(-0.5, 0.5), c(0.5,-0.5))
fit.msm <- msm(WinClo1 ~ time.min, data = win.dat2,
               qmatrix = Q )
```

The formula (`WinClo1 ~ time.min`*) defines the time step (here given in minutes) and* `Q` *is an initial guess of the intensity matrix (it also defines structures). An overview of the results can be given by;*

```
fit.msm

## [...]
## Transition intensities
##                   Baseline
## State 1 - State 1 -0.0047833 (-0.0065474,-0.0034945)
## State 1 - State 2  0.0047833 ( 0.0034945, 0.0065474)
## State 2 - State 1  0.0009656 ( 0.0007054, 0.0013219)
## State 2 - State 2 -0.0009656 (-0.0013219,-0.0007054)
##
## -2 * log-likelihood:  757
## [...]
```

That is including confidence intervals for the intensities. The 1-minute transition probability matrix can be calculated by;

```
(Gamma1min <- expm(fit.msm$Qmatrices$baseline))

##         State 1 State 2
## State 1 0.99523  0.0048
## State 2 0.00096  0.9990
```

Which can be compared to the results presented in Example 9.4 by ($\mathbf{\Gamma}_{10min} = \mathbf{\Gamma}_{1min}^{10}$)

```
Gamma1min %^% 10

##         State 1 State 2
## State 1  0.9535   0.046
## State 2  0.0094   0.991
```

Finally, we can check that the likelihoods are equal;

```
ll1 - logLik(fit.msm)

## 'log Lik.' 1.4e-09 (df=2)
```

Hence, these are the same models.

The `msm` package also includes the possibility of covariates, but we will not go further into continuous time Markov chain here.

9.9 Summary

In this chapter, we have presented and discussed discrete time Markov models. The presented models all assume that the stochastic process is directly observed. By including covariate patterns in the transition probability matrix, dynamics of the stochastic process can be learned, and higher order processes are sometimes also appropriate. As such, the models covered here have a lot in common with the time–series models covered in Chapter 8, though the models covered here have a discrete state space. Also, we have seen in Section 9.7 that Markov chains can be used as drivers for other processes. As we will see in Chapter 11, the Markov model might even be hidden and other processes can be used to infer the hidden process.

Part III

Time–Dependent Observations, Indirectly Observed States

10

State space models

Let us assume that the phenomena of interest (e.g. CO_2–concentration) are described by a finite number of state variables. This might be state variables characterising the activity levels in a building (e.g. CO_2–production). Such variables might include information about the number of people inside the building. It might be difficult to measure the number of people directly, and in particular their activity level. Instead, we might use measurements of noise level or CO_2 levels. Such measurements are obviously somehow related to the activity level, and the goal could be to use the measurements to identify a finite number of meaningful states for the activity level. Consequently, we would like to be able to formulate a method by which we are able to estimate the actual activity level based on measurement of the CO_2–level.

Indirect methods contain a set of methods and models where the state variables of the system is not directly observed. Instead, we observe a function of the state variables. In a simple setting, this could be a subset of the state variables, or, in the example outlined above, the CO_2–level.

In this chapter we will discuss linear Gaussian state space models as well as generalized linear state space models. We limit the discussion to discrete time state space models, but will refer to some work on continuous-discrete time state space models.

State space models can be seen as a natural extension of time series–models, and in fact the time series models we discussed in Chapter 8 can be formulated as state space models. The state space formulation gives a number of possibilities to deal with initial value estimation and with missing observations in time–series.

Including non-observed states leads to a number of possibilities for formulation of more mechanistically inspired models, and in particular if one takes the step to continuous-discrete time models (inspired by ordinary differential equations) mechanistic models can be formulated more or less directly based on first principles.

10.1 Introduction to filtration in state space models

Often, the states are not observed directly, and thus there are two sources of errors: Process and observation noise. The idea of filtering is to get the best estimate of the underlying process given the observations. Often, only the current state is of interest.

In general, we cannot observe the state of the system directly and have to rely on noisy functions of the state. In some cases, all states of the system are observed, but this is often not the case. The opposite might also be true, namely that we have multiple observations of some of the states, e.g. data from multiple CO_2 sensors. Measurement noise or more generally observation noise originates from the inability to measure without uncertainty. While system noise is perturbations within the system that affects future states, observation noise does not propagate to future states of the system. The situation is illustrated in Figure

DOI: 10.1201/9781003340812-10

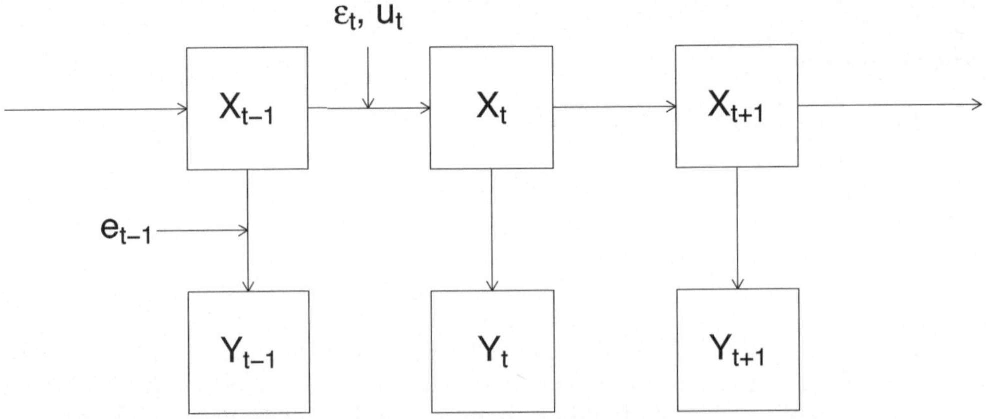

Figure 10.1: Schematic illustration of the stochastic state-space formulation. Y_t is the observation at time t, X_t is a hidden state which is observed under observation noise (e_t), while the state progression is affected by system noise (ϵ_t), and an input (u_t).

10.1. X_t is the true state of the system and Y_t is the observation of the system. The system is affected by the system noise, ϵ_t, and an input, u_t, while the observations are a function of the current state and the observation error e_t.

In general, the stochastic state-space formulation of the model in Figure 10.1 can be written as

$$\boldsymbol{X}_t = \boldsymbol{f}(\boldsymbol{X}_{t-1}, \boldsymbol{u}_t, \boldsymbol{\epsilon}_t; \boldsymbol{\theta}) \tag{10.1a}$$
$$\boldsymbol{Y}_t = \boldsymbol{h}(\boldsymbol{X}_t, \boldsymbol{u}_t, \boldsymbol{e}_t; \boldsymbol{\theta}), \tag{10.1b}$$

where the state space of $\boldsymbol{X}_t$ is a subset of $\mathbb{R}^p$, $\boldsymbol{Y}_t$ is a subset of $\mathbb{R}^l$, and $\boldsymbol{f}(\cdot)$ and $\boldsymbol{h}(\cdot)$ are vector functions of appropriate dimensions. $\boldsymbol{\epsilon}_t$ and $\boldsymbol{e}_t$ are random variables and $\boldsymbol{\theta}$ is a parameter vector to be estimated.

The estimation problem can be divided into two parts; 1) the filtering problem ($\hat{\boldsymbol{X}}_{t|t}$), where the state of the system is estimated given the information provided by the observations $\mathcal{Y}_t = \{\boldsymbol{Y}_1, ..., \boldsymbol{Y}_t\}$ up to time t, and the parameters $\boldsymbol{\theta}$, and 2) the estimation problem, where the best estimate, $\hat{\boldsymbol{\theta}}$, of $\boldsymbol{\theta}$ based on the observations $\mathcal{Y}_n$ is inferred. In the present context, the meaning of the term "best" is that $\hat{\boldsymbol{\theta}}$ maximise the likelihood given the observations ($\mathcal{Y}_n$).

For state space models, conditional distributions and conditional moments are used either as the outcomes of the modelling procedure itself or as a mean to calculate the likelihood. Below we go through the different conditional distributions that are usually used for state space models.

10.1.1 Predictive distributions and moments

As the names suggest, this is the predictive distribution of the state or the observations. When formulated as a distribution, these are

$$\begin{aligned} &p(\boldsymbol{X}_{t+k}|\mathcal{Y}_t) \\ &p(\boldsymbol{Y}_{t+k}|\mathcal{Y}_t), \end{aligned} \tag{10.2}$$

i.e. the distribution of a future (k steps into the future) state or observation given the observations up to present (time t). In the linear and Gaussian case, the distributions are characterised by the first– and second–order moments

$$\begin{aligned}\hat{\boldsymbol{X}}_{t+k|t} &= E[\boldsymbol{X}_{t+k}|\mathcal{Y}_t]; \quad \boldsymbol{\Sigma}^{xx}_{t+k|t} = V[\boldsymbol{X}_{t+k}|\mathcal{Y}_t] \\ \hat{\boldsymbol{Y}}_{t+k|t} &= E[\boldsymbol{Y}_{t+k}|\mathcal{Y}_t]; \quad \boldsymbol{\Sigma}^{yy}_{t+k|t} = V[\boldsymbol{Y}_{t+k}|\mathcal{Y}_t].\end{aligned} \tag{10.3}$$

For likelihood estimation, the 1-step ahead probability density of the observations is needed, and that is calculated based on the 1-step ahead prediction of the state.

10.1.2 Filtered distributions and moments

Filtered distributions and moments refer to the conditional distribution of the state at time t given all observation up to time t

$$p(\boldsymbol{X}_t|\mathcal{Y}_t), \tag{10.4}$$

and the first and second moments are denoted

$$\hat{\boldsymbol{X}}_{t|t} = E[\boldsymbol{X}_t|\mathcal{Y}_t]; \quad \boldsymbol{\Sigma}^{xx}_{t|t} = V[\boldsymbol{X}_t|\mathcal{Y}_t]. \tag{10.5}$$

The primary application of the filtered state is initialising of the calculation of the predictive distribution.

10.1.3 Smoothed distributions and moments

Smoothed state estimates refer to the distribution of the state given all observations

$$p(\boldsymbol{X}_t|\mathcal{Y}_n), \tag{10.6}$$

that is, the state estimate given all available information, and the first and second moments are denoted by

$$\hat{\boldsymbol{X}}_{t|n} = E[\boldsymbol{X}_t|\mathcal{Y}_n]; \quad \boldsymbol{\Sigma}^{xx}_{t|n} = V[\boldsymbol{X}_t|\mathcal{Y}_n]. \tag{10.7}$$

The primary used is for inference on the unobserved state. We will not discuss this further here, but the reader is referred to Simon (2009) for detailed treatment of aspects of smoothing.

We start this chapter by a general account of filtering, and then we account for the linear and Gaussian case, and finally we will present a generalized state space model using the general filtering theory and compare it with generalized mixed effect models (presented in Chapter 6).

10.1.4 Example data

The data that is used throughout this chapter is presented in the example below.

Example 10.1 (Data used in this chapter) *In this chapter, we use CO_2 concentration and the window position (open=0, or closed=1) in 1 dwelling. Both are measured in the living room and the bedroom. The data is given in 10–minute resolution and can be seen in the figure below.*

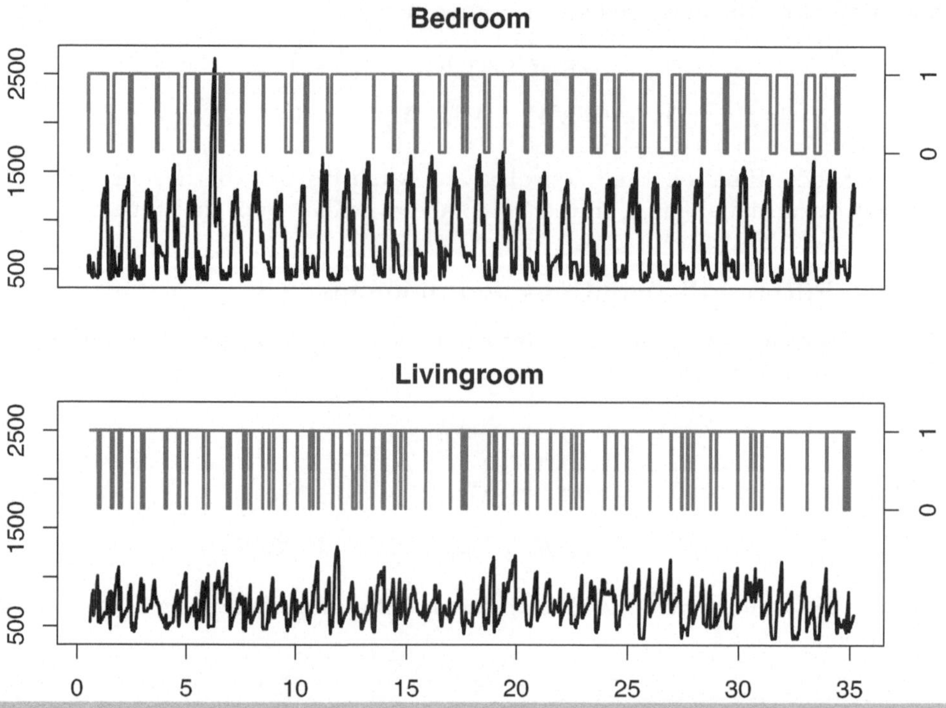

CO_2 or $\log(CO_2)$ concentrations are naturally modelled using Gaussian models as we have seen in Chapter 8, while window position is naturally modelled using Binomial models. We will treat both in the present chapter. Since measurements from the two rooms are taken in the same dwelling, it is also natural to formulate models that are able to model the series simultaneously. The data is presented in more details in Chapters 8 and 9 and in Andersen et al. (2013). The data used in this chapter is contained in the data.frame `win.datSS`*, which we will refer to in some examples.*

10.2 Bayesian recursive filtering

We start by the general setup with the focus on obtaining the likelihood, that we use for parameter estimation.

For likelihood estimation we need the joint distribution of all observations, i.e.

$$L(\boldsymbol{\theta}|\mathcal{Y}_n) = p(\boldsymbol{Y}_n, ..., \boldsymbol{Y}_1; \boldsymbol{\theta}). \tag{10.8}$$

In principle, we could use the same methods as we discussed in Chapter 6 to calculate the marginal likelihood of all observations using model structure (10.1a)–(10.1b) by

$$p(\boldsymbol{Y}_n, ..., \boldsymbol{Y}_1; \boldsymbol{\theta}) = \int p(\boldsymbol{Y}_n, ..., \boldsymbol{Y}_1, \boldsymbol{X}_n, ..., \boldsymbol{X}_1; \boldsymbol{\theta}) d\boldsymbol{X}_1 \cdots d\boldsymbol{X}_n, \tag{10.9}$$

and use the Laplace approximation to approximate the integral or calculate it explicitly in the case of Gaussian linear state space models. However, usually the likelihood is calculated recursively using a filter, that is either exact (linear Gaussian case) or an approximation in the cases of non-linear or non-Gaussian models. This means writing the marginal likelihood of all observations as a product of conditional distributions

$$L(\boldsymbol{\theta}|\mathcal{Y}_n) = p(\boldsymbol{Y}_1)\prod_{t=2}^{n} p(\boldsymbol{Y}_t|\mathcal{Y}_{t-1};\boldsymbol{\theta}). \tag{10.10}$$

In the further derivation, we will omit the parameters ($\boldsymbol{\theta}$). Each term in the recursive formulation of the likelihood can be written as

$$p(\boldsymbol{Y}_t|\mathcal{Y}_{t-1}) = \int p(\boldsymbol{Y}_t|\boldsymbol{X}_t)p(\boldsymbol{X}_t|\mathcal{Y}_{t-1})d\boldsymbol{x}_t. \tag{10.11}$$

The first term in the integral is given directly from the model definition and hence we can concentrate on the second term.

Example 10.2 *Consider the simple state space model*

$$\begin{aligned} X_t &= \phi X_{t-1} + \epsilon_t; \quad \epsilon_t \sim N(0,\sigma^2) \\ Y_t &= X_t + e_t; \quad e_t \sim N(0,s^2), \end{aligned} \tag{10.12}$$

in this case we have

$$Y_t|X_t = x_t \sim N(x_t, s^2) \tag{10.13}$$

and hence

$$p(y_t|X_t = x_t) = \frac{1}{\sqrt{2\pi s^2}}e^{-\frac{1}{2s^2}(x_t - y_t^2)}. \tag{10.14}$$

Now, using the Markov property of the dynamic model, we can write the second term in the integral as

$$\begin{aligned} p(\boldsymbol{X}_t|\mathcal{Y}_{t-1}) &= \int p(\boldsymbol{X}_t, \boldsymbol{X}_{t-1}|\mathcal{Y}_{t-1})d\boldsymbol{X}_{t-1} \\ &= \int p(\boldsymbol{X}_t|\boldsymbol{X}_{t-1})p(\boldsymbol{X}_{t-1}|\mathcal{Y}_{t-1})d\boldsymbol{X}_{t-1}, \end{aligned} \tag{10.15}$$

which is the 1-step ahead predictive distribution of the state (given the observations up to time $t-1$), and again the first term in the integral is given directly from the model definition.

Example 10.3 *Referring to Example 10.2, we have*

$$X_t|X_{t-1} \sim N(\phi X_{t-1}, \sigma^2), \tag{10.16}$$

hence defining the distribution function ($p(\boldsymbol{X}_t|\boldsymbol{X}_{t-1})$).

The second term in the integral is the filtered state distribution and can be written as

$$\begin{aligned} p(\boldsymbol{X}_t|\mathcal{Y}_t) &= \frac{p(\boldsymbol{Y}_t|\boldsymbol{X}_t, \mathcal{Y}_{t-1})p(\boldsymbol{X}_t|\mathcal{Y}_{t-1})}{p(\boldsymbol{Y}_t|\mathcal{Y}_{t-1})} \\ &= \frac{p(\boldsymbol{Y}_t|\boldsymbol{X}_t)p(\boldsymbol{X}_t|\mathcal{Y}_{t-1})}{p(\boldsymbol{Y}_t|\mathcal{Y}_{t-1})}. \end{aligned} \tag{10.17}$$

The first term in the enumerator is given directly from the model definition (see Example 10.2), the second term is given by (10.15) and the enumerator is the likelihood contribution (10.11) of observation $\boldsymbol{Y}_t$. This concludes the recursive filtering and to sum up the recursive algorithm is

1. Initialise with the distribution $p_0 = p(\boldsymbol{X}_0)$, which we can think of as $p(\boldsymbol{X}_0|\boldsymbol{Y}_0)$, and set $t = 1$.

2. Using (10.15), calculate $p(\boldsymbol{X}_t|\mathcal{Y}_{t-1})$.

3. Using (10.11), calculate $p(\boldsymbol{Y}_t|\mathcal{Y}_{t-1})$.

4. Using (10.17), calculate $p(\boldsymbol{X}_t|\mathcal{Y}_t)$.

5. Set $t = t + 1$ and return to step 2 and repeat until $t = n$.

The above constitutes a recursive filter that in principle can be calculated using numerical integration. Numerical integration is however computational demanding for high dimensional systems and some approximations may be used, for example the Laplace approximation (see Chapter 6) or using simulation techniques like MCMC-methods or particle filters, e.g. Givon et al. (2009).

If system and observation equations are linear with additive Gaussian noise, the involved integrals can be solved explicitly and hence the recursive filter will work much faster with exact solutions. This is the linear Kalman filter that is detailed in the next section.

10.3 The linear Kalman filter

When the noise process and the observation noise are both Gaussian and all functions are linear, the Bayesian recursive filter reduces to the Kalman filter, where we only need to keep track of the second–order moment representation.

In the remaining part of this chapter, we consider only linear time-varying systems. This leads, as we will see, to the linear Kalman filter. For non–linear systems, the derivations done here need to be extended and other filters, e.g. the Extended Kalman filter. Both Jazwinski (1970) and Simon (2009) cover filtering in more general settings. The restriction implies that we consider systems described by (which is referred to as a linear time–varying system)

$$\boldsymbol{X}_t = \boldsymbol{A}_t\boldsymbol{X}_{t-1} + \boldsymbol{B}_t\boldsymbol{u}_t + \boldsymbol{G}_t\boldsymbol{\epsilon}_t \tag{10.18a}$$
$$\boldsymbol{Y}_t = \boldsymbol{C}_t\boldsymbol{X}_t + \boldsymbol{D}_t\boldsymbol{u}_t + \boldsymbol{e}_t, \tag{10.18b}$$

where t is time, $\boldsymbol{x}_t \in \mathcal{X} \subset \mathbb{R}^p$ is a state vector, $\boldsymbol{u}_t \in \mathcal{U} \subset \mathbb{R}^m$ is an input vector, $\boldsymbol{y}_t \in \mathcal{Y} \subset \mathbb{R}^l$ is an output vector, $\boldsymbol{A}_t \in \mathbb{R}^{p\times p}$, $\boldsymbol{B}_t \in \mathbb{R}^{p\times m}$, $\boldsymbol{C}_t \in \mathbb{R}^{l\times p}$, $\boldsymbol{D}_t \in \mathbb{R}^{l\times m}$ and $\boldsymbol{G}_t \in \mathbb{R}^{p\times k}$ are matrices, $\{\boldsymbol{\epsilon}_t\}$ is an k-dimensional white noise process with $\boldsymbol{\epsilon}_t \in N(\boldsymbol{0}, \boldsymbol{\Sigma}_t)$, and $\{\boldsymbol{e}_t\}$ is an l-dimensional white noise process with $\boldsymbol{e}_t \in N(\boldsymbol{0}, \boldsymbol{S}_t)$. The matrices are allowed to depend on time (e.g. through some input), but not on the state (nor the observations).

In the case where all matrices are constants

$$\boldsymbol{X}_t = \boldsymbol{A}\boldsymbol{X}_{t-1} + \boldsymbol{B}\boldsymbol{u}_t + \boldsymbol{G}\boldsymbol{\epsilon}_t \tag{10.19a}$$
$$\boldsymbol{Y}_t = \boldsymbol{C}\boldsymbol{X}_t + \boldsymbol{D}\boldsymbol{u}_t + \boldsymbol{e}_t, \tag{10.19b}$$

we refer to the system as a linear time–invariant system.

Equations (10.18a) and (10.19a) are referred to as the system equations, and $\boldsymbol{X}_t$ as the state. Further, equations (10.18b) and (10.19b) are referred to as the measurement equations. The Bayesian recursive filter deals with general distributions of state and observations. In the system (10.18a)–(10.18b) all the conditional distributions are Gaussian and it is therefore enough to keep track of the conditional mean and the conditional variance. We use the following shorthand notation

$$\begin{aligned}\hat{\boldsymbol{X}}_{t|t-k} &= E[\boldsymbol{X}_t|\mathcal{Y}_{t-k}] \\ \boldsymbol{\Sigma}^{xx}_{t|t-k} &= V[\boldsymbol{X}_t|\mathcal{Y}_{t-k}].\end{aligned} \tag{10.20}$$

With this setup, the optimal linear filter can be described by the Kalman filter.

Theorem 10.1 (Kalman filter) *With given initial conditions* $\hat{\boldsymbol{X}}_{1|0} = \boldsymbol{x}_0$, *and* $\boldsymbol{\Sigma}^{xx}_{1|0} = \boldsymbol{\Sigma}^{xx}_0$, *the optimal output* prediction *equation is*

$$\begin{aligned}\hat{\boldsymbol{Y}}_{t|t-1} &= \boldsymbol{C}\hat{\boldsymbol{X}}_{t|t-1} + \boldsymbol{D}_t\boldsymbol{u}_t \\ \boldsymbol{\Sigma}^{yy}_{t|t-1} &= \boldsymbol{C}\boldsymbol{\Sigma}^{xx}_{t|t-1}\boldsymbol{C}^T_t + \boldsymbol{S}_t,\end{aligned} \tag{10.21}$$

*The optimal reconstruction (*update, *i.e. the filtered state estimate) is*

$$\begin{aligned}\hat{\boldsymbol{X}}_{t|t} &= \hat{\boldsymbol{X}}_{t|t-1} + \boldsymbol{K}_t(\boldsymbol{y}_t - \boldsymbol{C}_t\hat{\boldsymbol{X}}_{t|t-1}) \\ \boldsymbol{\Sigma}^{xx}_{t|t} &= \boldsymbol{\Sigma}^{xx}_{t|t-1} - \boldsymbol{K}_t\boldsymbol{\Sigma}^{yy}_{t|t-1}\boldsymbol{K}^T_t,\end{aligned} \tag{10.22}$$

where $\boldsymbol{K}_t$ *is the Kalman* gain*:*

$$\boldsymbol{K}_t = \boldsymbol{\Sigma}^{xx}_{t|t-1}\boldsymbol{C}^T(\boldsymbol{\Sigma}^{yy}_{t|t-1})^{-1}, \tag{10.23}$$

and $\widetilde{\boldsymbol{e}}_k = \boldsymbol{y}_k - \hat{\boldsymbol{y}}_{k|k-1}$ *is the* innovation *error. Finally the optimal state* prediction *is*

$$\hat{\boldsymbol{X}}_{t+1|t} = \boldsymbol{A}_t\hat{\boldsymbol{X}}_{t|t} + \boldsymbol{B}_t\boldsymbol{u}_t \tag{10.24a}$$

$$\boldsymbol{\Sigma}^{xx}_{t+1|t} = \boldsymbol{A}_t\boldsymbol{\Sigma}^{xx}_{t|t}\boldsymbol{A}^T_t + \boldsymbol{\Sigma}_t. \tag{10.24b}$$

The Kalman filter was first introduced by Kalman (1960). Later on, filtering was covered in more detail in the books by Jazwinski (1970), Maybeck (1982), and Meinhold and Singpurwalla (1983).

The matrices ($\boldsymbol{A}_t$, $\boldsymbol{B}_t$, $\boldsymbol{C}_t$, $\boldsymbol{\Sigma}_t$ and $\boldsymbol{S}_t$) are all allowed to depend on time (but not on states or observations). If $\boldsymbol{A}_t$, $\boldsymbol{C}_t$, $\boldsymbol{\Sigma}_t$ and $\boldsymbol{S}_t$ are constant in time, then the conditional variances ($\boldsymbol{\Sigma}^{xx}_{t|t-1}$, $\boldsymbol{\Sigma}^{yy}_{t|t-1}$, $\boldsymbol{\Sigma}^{xx}_{t|t}$) and the Kalman gain ($\boldsymbol{K}_t$) will all converge to stationary values.

The Kalman filter has two components: 1) the state and observation predictions; these can be calculated directly from the model, and 2) the reconstruction step that can be derived from the projection theorem (Theorem 5.3, page 133). The Kalman gain is related to the covariance between the observations and the variance and the inverse of the observation variance. If the covariance ($\boldsymbol{C}_t\boldsymbol{\Sigma}^{xx}_{t|t-1}$) is "large" and prediction variance ($\boldsymbol{\Sigma}^{yy}_{t|t-1}$) is "small", then the update will be strong and of course if either the correlation is weak or the predictive variance large, then the update will be weak (minor).

Example 10.4 *We consider again the simple system in Example 10.2. In this case we have, using the initial mean and variance,* $\hat{X}_{1|0} = \phi x_0$ *and* $\Sigma^{xx}_{1|0} = \phi^2\sigma_0 + \sigma^2$,

$$\begin{aligned}\hat{Y}_{1|0} &= \phi x_0 \\ \Sigma^{yy}_{1|0} &= \phi^2\sigma_0^2 + \sigma^2,\end{aligned} \tag{10.25}$$

the Kalman gain becomes

$$K_1 = \frac{\phi^2\sigma_0^2 + \sigma^2}{\phi^2\sigma_0^2 + \sigma^2 + s^2}, \tag{10.26}$$

and the reconstruction of X_1 *becomes*

$$
\begin{aligned}
\hat{X}_{1|1} =& \phi x_0 + \frac{\phi^2\sigma_0^2 + \sigma^2}{\phi^2\sigma_0^2 + \sigma^2 + s^2}(y_1 - \phi x_0) \\
\Sigma_{1|1}^{xx} =& \phi^2\sigma_0^2 + \sigma^2 - \left(\frac{\phi^2\sigma_0^2 + \sigma^2}{\phi^2\sigma_0^2 + \sigma^2 + s^2}\right)^2 (\phi^2\sigma_0^2 + \sigma^2 + s^2) \\
=& \frac{s^2(\phi^2\sigma_0^2 + \sigma^2)}{\phi^2 + \sigma^2 + s^2}.
\end{aligned}
\tag{10.27}
$$

Hence we see that if $s^2 = 0$*, i.e. we observe the state without noise, we get* $K_1 = 1$*,* $\hat{X}_{1|1} = y_1$ *and* $\Sigma_{1|1}^{xx} = 0$*.*

One strength of the Kalman filter is that we can handle models with unobserved states, an example of that is missing observations or some of the observations in $\boldsymbol{Y}_t$ are missing.

10.3.1 Missing observations

In real–life systems, observations are often incomplete; this can happen for various reasons, for example sensor failures, which is a typical case.

Observations can basically be missing in two different ways, 1) completely missing (all elements of $\boldsymbol{Y}_t$ are missing) or 2) only a subset of the observation is missing. Both situations can be dealt with by the Kalman filter, but will be treated separately below.

Completely missing observation

Let the observation $\boldsymbol{Y}_t$ be completely missing, which corresponds to no information about the state at time t. We can express this as $V[\boldsymbol{e}_t] = \boldsymbol{S}_t = \infty\boldsymbol{I}$ (in practice a large number rather than ∞), which gives $\boldsymbol{\Sigma}_{k|k-1}^{yy} \approx \infty\boldsymbol{I}$, $(\boldsymbol{\Sigma}_{t|t-1}^{yy})^{-1} = \boldsymbol{0}$ and $\boldsymbol{K}_t = \boldsymbol{0}$. The reconstruction is calculated by

$$
\begin{aligned}
\hat{\boldsymbol{X}}_{t|t} =& \hat{\boldsymbol{X}}_{t|t-1} + \boldsymbol{K}_t\tilde{\boldsymbol{e}}_t \\
=& \hat{\boldsymbol{X}}_{t|t-1} \\
\boldsymbol{\Sigma}_{t|t}^{xx} =& \boldsymbol{\Sigma}_{t|t-1}^{xx} - \boldsymbol{K}_t\boldsymbol{\Sigma}_{t|t-1}^{yy}\boldsymbol{K}_t^T \\
=& \boldsymbol{\Sigma}_{t|t-1}^{xx},
\end{aligned}
\tag{10.28}
$$

and it is seen that completely missing observations correspond to no update in the state estimate.

Partly missing observation

Assume that observation $\boldsymbol{Y}_t$ is partly missing, meaning that only $l_t < l$ values of the observation vector are recorded. Denote the index set for the full observation vector by $I = \{1, 2, .., l\}$, the index of the missing observations by I_m and the index of the non-missing observations by I_O. The reconstruction could be calculated by setting the observation variance $\boldsymbol{S}_t = \boldsymbol{S} + \boldsymbol{I}_{missing}$, where $(\boldsymbol{I}_{missing})_{I_m(i),I_m(i)} = \infty, i = 1, .., l_t$. We will however need to evaluate $(\boldsymbol{\Sigma}_{t|t-1}^{yy})^{-1}$, when the likelihood is evaluated in the next section. Another way is to have a time dependent observation equation, with $\boldsymbol{C}_t^O = \boldsymbol{C}_t(I_O, :)$, i.e

$$
\boldsymbol{Y}_t(I_O) = \boldsymbol{C}_t^O\boldsymbol{X}_t + \boldsymbol{e}_t; \quad \boldsymbol{e}_t \sim N(\boldsymbol{0}, \boldsymbol{S}(I_O, I_O)),
\tag{10.29}
$$

and then use the usual filter equations.

Example 10.5 *Consider a model with the observation equation*

$$\begin{bmatrix} Y_{1,t} \\ Y_{2,t} \end{bmatrix} = \begin{bmatrix} c_{1,1} & \cdots & c_{1,p} \\ c_{2,1} & \cdots & c_{2,p} \end{bmatrix} \boldsymbol{X}_t + \boldsymbol{e}_t; \quad \boldsymbol{e}_t \sim N(\boldsymbol{0}, \boldsymbol{S}). \tag{10.30}$$

Now assume that at a specific time-point t^* only y_{1,t^*} is available, i.e. y_{2,t^*} is missing, in that case we simply set the observation equation to

$$\begin{aligned} Y_{1,t^*} =& \boldsymbol{C}_{t^*} \boldsymbol{X}_{t^*} + e_{t^*} \\ =& \begin{bmatrix} c_{1,1} & \cdots & c_{1,p} \end{bmatrix} \boldsymbol{X}_{t^*} + e_{t^*}; \quad e_{t^*} \sim N(0, S_{11}), \end{aligned} \tag{10.31}$$

and use the usual filter for updating the state and state variance-covariance, i.e. we can get an estimate of the missing observation.

10.4 Likelihood estimation

In Sections 10.2 and 10.3 and in general in this chapter, the underlying process is a Markov–process (see Figure 10.1). Recall that the fundamental property of a Markov process is

$$P(\boldsymbol{X}_t < x | \boldsymbol{X}_{t-1}, ..., \boldsymbol{X}_1) = P(\boldsymbol{X}_t < x | \boldsymbol{X}_{t-1}), \tag{10.32}$$

that is, all information of the future is given by $\boldsymbol{X}_{t-1}$.

We are interested in the joint density of all observations given a set of parameters. To emphasise the dependence on the parameters, the system equation of the linear model is written as

$$\boldsymbol{X}_t = \boldsymbol{A}_t(\boldsymbol{\theta}) \boldsymbol{X}_{t-1} + \boldsymbol{B}(\boldsymbol{\theta}) \boldsymbol{u}_t + \boldsymbol{\epsilon}_t, \tag{10.33}$$

with the *observation equation*

$$\boldsymbol{Y}_t = \boldsymbol{C}_t(\boldsymbol{\theta}) \boldsymbol{X}_t + \boldsymbol{e}_t, \tag{10.34}$$

with $\boldsymbol{\epsilon}_t \sim N(\boldsymbol{0}, \boldsymbol{\Sigma}_t(\boldsymbol{\theta}))$ and $\boldsymbol{e}_t \sim N(\boldsymbol{0}, \boldsymbol{S}_t(\boldsymbol{\theta}))$. The joint density of the observations is (we denote the collection of all observations up to time t by $\mathcal{Y}_t = \{\boldsymbol{Y}_t, .., \boldsymbol{Y}_1\}$)

$$\begin{aligned} f(\mathcal{Y}_n | \boldsymbol{\theta}) =& f(\boldsymbol{Y}_n | \mathcal{Y}_{n-1}, \boldsymbol{\theta}) f(\mathcal{Y}_{n-1} | \boldsymbol{\theta}) \\ =& \prod_{i=1}^{n} f(\boldsymbol{Y}_t | \mathcal{Y}_{t-1}, \boldsymbol{\theta}). \end{aligned} \tag{10.35}$$

Note that $f(\boldsymbol{Y}_1 | \boldsymbol{Y}_0, \boldsymbol{\theta}) = f(\boldsymbol{Y}_1 | \boldsymbol{x}_0, \boldsymbol{\theta})$ has a special role since it depends on the initial conditions (the distribution of $\boldsymbol{x}_0$), which may be given or estimated as part of the parameters in $\boldsymbol{\theta}$. Let $\mathcal{Y}_{n^*}$ denote the set of all partly or fully observed observations, then the likelihood of $\boldsymbol{\theta}$ is (in the Gaussian case)

$$L(\boldsymbol{\theta} | \mathcal{Y}_{n^*}) = \prod_{t=1}^{n^*} \frac{1}{(2\pi)^{l_t/2} \sqrt{\det \boldsymbol{\Sigma}_{t|t-1}^{yy}}} \exp\left[-\frac{1}{2} \widetilde{\boldsymbol{e}}_t^T (\boldsymbol{\Sigma}_{t|t-1}^{yy})^{-1} \widetilde{\boldsymbol{e}}_t \right], \tag{10.36}$$

where $l_t \leq l$ is the number of non-missing observations at time t. The log-likelihood $l(\boldsymbol{\theta} | \mathcal{Y}_{n^*})$ becomes

$$l(\boldsymbol{\theta} | \mathcal{Y}_{n^*}) = -\frac{1}{2} \sum_{t=1}^{n^*} \left[\log \det \boldsymbol{\Sigma}_{t|t-1}^{yy} + \widetilde{\boldsymbol{e}}_t^T (\boldsymbol{\Sigma}_{t|t-1}^{yy})^{-1} \widetilde{\boldsymbol{e}}_t \right] + \texttt{const}, \tag{10.37}$$

where const is a constant not depending on the parameters.

As usual, the maximum likelihood estimate (MLE) of $\boldsymbol{\theta}$ is

$$\hat{\boldsymbol{\theta}} = \arg\max_{\boldsymbol{\theta}} l(\boldsymbol{\theta}|\mathcal{Y}_{n^*}), \tag{10.38}$$

and the variance-covariance matrix of $\hat{\boldsymbol{\theta}}$ is approximated by the negative inverse Hessian of $l(\boldsymbol{\theta}|\mathcal{Y}_{n^*})$ evaluated at $\boldsymbol{\theta} = \hat{\boldsymbol{\theta}}$ (the observed Fisher Information).

As a special example, we treat the state space formulation of ARMA models in the next section.

10.5 State space formulation of ARMA models

Stochastic systems are often given in state space formulation, which has quite general applicability. One is that higher order systems can be written as first–order system, but in a higher dimension. Here we refer to a first–order system as a Markov process, the definition implies that for a Markov process, the current state of the system contains all needed information about the past. Or phrased differently, if we know the current state of the system, we cannot get more information by looking further into the past. The definition implies that the AR(1) process is a Markov process, while the AR(2) process is not. The definition also applies when X_t is a vector, and it is, as we will see, possible to write an ARMA(p, q) process as a vector valued Markov process. The main application within time series analysis is to deal with missing values and estimating initial values (i.e. X_t for $t \leq 0$) of the process.

In this section we consider systems defined by

$$\begin{aligned} \boldsymbol{X}_t &= \boldsymbol{A}\boldsymbol{X}_{t-1} + \boldsymbol{B}\boldsymbol{u}_t + \boldsymbol{G}\boldsymbol{\epsilon}_t; \quad \boldsymbol{\epsilon}_t \sim N(\boldsymbol{0}, \boldsymbol{\Sigma}) \\ \boldsymbol{Y}_t &= \boldsymbol{C}\boldsymbol{X}_t + \boldsymbol{e}_t; \quad \boldsymbol{e}_t \sim N(\boldsymbol{0}, \boldsymbol{S}), \end{aligned} \tag{10.39}$$

where $\boldsymbol{X}_t \in \mathbb{R}^p$, $\boldsymbol{A} \in \mathbb{R}^{p\times p}$, $\boldsymbol{B} \in \mathbb{R}^{p\times m}$, $\boldsymbol{u}_t \in \mathbb{R}^m$, $\boldsymbol{G} \in \mathbb{R}^{p\times l}$, $\boldsymbol{\epsilon} \in \mathbb{R}^l$, $\boldsymbol{\Sigma} \in \mathbb{R}^{l\times l}$, $\boldsymbol{Y}_t \in \mathbb{R}^k$, $\boldsymbol{C} \in \mathbb{R}^{k\times n}$ and $\boldsymbol{S} \in \mathbb{R}^{k\times k}$.

The univariate ARMA(p, q) process can be written as a stochastic state space model with

$$\begin{aligned} \boldsymbol{X}_t &= [X_{1,t}, X_{2,t}, \ldots, X_{d,t}]^T; \quad d = \max(p, q+1) \\ \boldsymbol{A} &= \begin{bmatrix} -\phi_1 & 1 & 0 & \cdots & 0 \\ -\phi_2 & 0 & 1 & \cdots & 0 \\ \vdots & \vdots & \vdots & \ddots & \vdots \\ -\phi_{d-1} & 0 & 0 & \cdots & 1 \\ -\phi_d & 0 & 0 & \cdots & 0 \end{bmatrix} \\ \boldsymbol{G} &= [1, \theta_1, \cdots, \theta_{d-1}]^T \\ \boldsymbol{C} &= [1 \quad 0 \quad \cdots \quad 0] \\ \boldsymbol{e}_t &= 0. \end{aligned} \tag{10.40}$$

The formulation above is not unique and other representations could be formulated, the representation is exemplified below.

Example 10.6 (A state space model of an $ARMA(2,1)$ process) *Recall that the $ARMA(2,1)$ process is given by*

$$Y_t = -\phi_1 Y_{t-1} - \phi_2 Y_{t-2} + \epsilon_t + \theta_1 \epsilon_{t-1}, \tag{10.41}$$

where $\{\epsilon_t\}$ is white noise with variance σ^2.

The process can be written in state space form as

$$\begin{aligned} \begin{pmatrix} X_{1,t} \\ X_{2,t} \end{pmatrix} &= \begin{bmatrix} -\phi_1 & 1 \\ -\phi_2 & 0 \end{bmatrix} \begin{pmatrix} X_{1,t-1} \\ X_{2,t-1} \end{pmatrix} + \begin{pmatrix} 1 \\ \theta_1 \end{pmatrix} \epsilon_t, \\ Y_t &= \begin{bmatrix} 1 & 0 \end{bmatrix} \begin{pmatrix} X_{1,t} \\ X_{2,t} \end{pmatrix}. \end{aligned} \tag{10.42}$$

The variance of $\boldsymbol{X}_t$ can be found from the variance of $\boldsymbol{X}_{t-1}$ by

$$V[\boldsymbol{X}_t] = \boldsymbol{A} V[\boldsymbol{X}_{t-1}] \boldsymbol{A}^T + \boldsymbol{G \Sigma G}^T. \tag{10.43}$$

For the stationary distribution we have $V[\boldsymbol{X}_t] = V[\boldsymbol{X}_{t-1}]$ and with $\boldsymbol{\Sigma} = \sigma^2$, the stationary distribution is given by the solution to

$$V[\boldsymbol{X}_t] = \boldsymbol{A} V[\boldsymbol{X}_t] \boldsymbol{A}^T + \sigma^2 \boldsymbol{G G}^T, \tag{10.44}$$

where $V[\boldsymbol{X}_t]$ contains the stationary variance-covariance matrix of $\boldsymbol{X}_t$. The solution can be given in terms of vectorizations and Kronecker products using the relation (see Durbin and Koopmann (2001), p. 112)

$$\text{vec}(\boldsymbol{A} V[\boldsymbol{X}_t] \boldsymbol{A}^T) = (\boldsymbol{A} \otimes \boldsymbol{A}) \text{vec}(V[\boldsymbol{X}_t]), \tag{10.45}$$

as

$$\text{vec}(V[\boldsymbol{X}_t]) = \sigma^2 (\boldsymbol{I} - \boldsymbol{A} \otimes \boldsymbol{A})^{-1} \text{vec}(\boldsymbol{G G}^T). \tag{10.46}$$

As we show in the next example, this can be used for evaluating the likelihood function. We will use the notation

$$\boldsymbol{\Sigma}^{xx} = V[\boldsymbol{X}_t], \tag{10.47}$$

for the stationary (or unconditional) variance-covariance of the state vector $\boldsymbol{X}_t$.

Example 10.7 *We consider again the ARMA(2,1) process. The likelihood is given by*

$$L(\boldsymbol{\theta}; \boldsymbol{Y}) = f(Y_1; \boldsymbol{\theta}) \cdot f(Y_2 | Y_1 \boldsymbol{\theta}) \cdots f(Y_n | Y_{n-1}, \ldots, Y_1; \boldsymbol{\theta}). \tag{10.48}$$

In Chapter 8, we discussed maximum likelihood estimation, simply disregarding the first observations (i.e. conditioning on it). Using the state space formulation, we can write

$$Y_1 \sim N(0, \boldsymbol{C \Sigma}^{xx} \boldsymbol{C}^T) = N(0, (\boldsymbol{\Sigma}^{xx})_{11}), \tag{10.49}$$

which gives us the first term in the likelihood. After having observed Y_1, we can calculate

$$\boldsymbol{\Sigma}^{xx}_{1|1} = V[\boldsymbol{X}_1 | Y_1], \tag{10.50}$$

to that end we need the variance of $[\boldsymbol{X}_1^T, Y_1]^T$

$$V\left[\begin{pmatrix} \boldsymbol{X}_1^T \\ Y_1 \end{pmatrix}\right] = \begin{bmatrix} \boldsymbol{\Sigma}^{xx} & \boldsymbol{\Sigma}^{xx} \boldsymbol{C}^T \\ \boldsymbol{C \Sigma}^{xx} & \boldsymbol{C \Sigma}^{xx} \boldsymbol{C}^T \end{bmatrix}, \tag{10.51}$$

and hence, using the Projection theorem (Theorem 5.3, page 133), we get

$$\begin{aligned}\hat{X}_{1|1} &= \Sigma^{xx} C^T (C \Sigma^{xx} C^T)^{-1} Y_1 \\ \Sigma^{xx}_{1|1} &= \Sigma^{xx} - \Sigma^{xx} C^T (C \Sigma^{xx} C^T)^{-1} C \Sigma^{xx},\end{aligned} \tag{10.52}$$

and now the predictive mean variance of $X_{2|1}$ can be written as

$$\begin{aligned}\hat{X}_{2|1} &= A \hat{X}_{2|1} \\ \Sigma^{xx}_{2|1} &= A \Sigma^{xx}_{1|1} A^T + \sigma^2 G G^T,\end{aligned} \tag{10.53}$$

the distribution of $Y_2|Y_1$ is now given as

$$Y_2|Y_1 \sim N(C \hat{X}_{2|1}, C \Sigma^{xx}_{2|1} C^T), \tag{10.54}$$

the likelihood contribution of Y_2 can be calculated using the above. Also, we see that the algorithm described above is iterative so we can calculate all terms in the likelihood.

We illustrate this with a small example, using the parameters;

```
phi1 <- 0.9; phi2 <- -0.3; theta <- 0.5; sigma <- 1
```

and a few observations;

```
y <- c(1.73, 2.82,  4.12, 2.69, -1.24, -2.29, -0.31)
```

*We can set up the matrices **A**, **B** and **C** by;*

```
A <- matrix(0,ncol=2,nrow=2)
A[1,2] <- 1
A[ ,1] <- c(phi1, phi2)
G <- c(1, theta)
C <- c(1,0)
```

The initial (stationary) variance-covariance matrix Σ^{xx} can be found by;

```
X.hat <- c(0,1)
Sig.x <- matrix(solve(diag(4) - kronecker(A, A)) %*%
                as.vector(G %*% t(G)), ncol = 2)
```

From which the variance of the first observation is found by;

```
(Sig.y <- C %*% Sig.x %*% C)

##      [,1]
## [1,]  4.1
```

which is much larger than the usual 1-step ahead error (which in this case is 1). In the first step, the prediction is 0. The state and state variance are now updated by;

```
X.hat <- X.hat + Sig.x %*% C %*% solve(Sig.y) %*%
    (y[1] - C %*% X.hat)
Sig.x <- Sig.x - Sig.x %*% C %*% solve(Sig.y) %*% C %*% Sig.x
```

A new prediction for the state and state covariance is obtained by;

```
## prediction
X.hat <- A %*% X.hat
Sig.x <- A %*% Sig.x %*% t(A)+ sigma * G %*% t(G)
```

which is used for the next observation prediction;

```
(Sig.y <- C %*% Sig.x %*% C)

##      [,1]
## [1,]  1.6
```

As can be seen in this case, the variance seems to converge fairly fast towards 1, which it eventually will be. The rate of the convergence depends on the value of the parameters (in particular on θ). The state and state variance-covariance can now be updated with the next observation.

The example above generalises and we can set up the general algorithm as follows:

1. Set up matrices $\boldsymbol{A}$, $\boldsymbol{G}$ and $\boldsymbol{C}$ using models definition, and set $t = 1$.
2. Calculate $\boldsymbol{\Sigma}^{xx}_{1|0} = \boldsymbol{\Sigma}^{xx}$ using (10.46), and set $\boldsymbol{X}_{1|0} = \boldsymbol{0}$.
3. Calculate the likelihood contribution of $\boldsymbol{Y}_t$ using

$$\begin{aligned}\boldsymbol{\Sigma}^{yy}_{t|t-1} &= \boldsymbol{C}\boldsymbol{\Sigma}^{xx}_{t|t-1}\boldsymbol{C}^T \\ \hat{\boldsymbol{Y}}_{t|t-1} &= \boldsymbol{C}\hat{\boldsymbol{X}}_{t|t-1}.\end{aligned} \tag{10.55}$$

4. Update state and state covariance by

$$\begin{aligned}\hat{\boldsymbol{X}}_{t|t} &= \boldsymbol{\Sigma}^{xx}\boldsymbol{C}^T(\boldsymbol{C}\boldsymbol{\Sigma}^{xx}\boldsymbol{C}^T)^{-1}(\boldsymbol{y}_t - \boldsymbol{C}\hat{\boldsymbol{X}}_{t|t}) \\ \boldsymbol{\Sigma}^{xx}_{t|t} &= \boldsymbol{\Sigma}^{xx} - \boldsymbol{\Sigma}^{xx}_{t|t-1}\boldsymbol{C}^T(\boldsymbol{C}\boldsymbol{\Sigma}^{xx}_{t|t-1}\boldsymbol{C}^T)^{-1}\boldsymbol{C}\boldsymbol{\Sigma}^{xx}_{t|t-1}.\end{aligned} \tag{10.56}$$

5. Propagate state and state covariance forward using the prediction step

$$\begin{aligned}\hat{\boldsymbol{X}}_{t+1|t} &= \boldsymbol{A}\hat{\boldsymbol{X}}_{t|t} \\ \boldsymbol{\Sigma}^{xx}_{t+1|t} &= \boldsymbol{A}\boldsymbol{\Sigma}^{xx}_{t|t}\boldsymbol{A}^T + \sigma^2\boldsymbol{G}\boldsymbol{G}^T.\end{aligned} \tag{10.57}$$

6. Set $t := t + 1$ and go back to step 3 and repeat until $t = n$.

The procedure above is the Kalman filter presented in Theorem 10.1. In addition to determining initial estimates, the algorithm can be used in a similar way to deal with missing values in time series.

As a final remark, the method covered here also includes ARIMAX models (see Chapter 8, here given in the `arima` set up) as we can consider the process

$$e_t = y_t - \boldsymbol{u}_t\boldsymbol{\beta}, \tag{10.58}$$

where $\boldsymbol{u}_t$ is an input variable and apply the discussed methods to e_t.

10.6 Parameters and state estimation

For estimating parameters in state space models, one could of course simply implement the methods discussed above. Here we will instead use the R-package `KFAS` (Helske, 2017). The model formulation in `KFAS` is based on the state space formulation

$$\begin{aligned} \boldsymbol{X}_t =& \boldsymbol{T}_t \boldsymbol{X}_{t-1} + \boldsymbol{R}_t \boldsymbol{\epsilon}_t \\ \boldsymbol{Y}_t =& \boldsymbol{Z}_t \boldsymbol{X}_t + \boldsymbol{e}_t, \end{aligned} \tag{10.59}$$

where $\boldsymbol{e}_t \sim N(\boldsymbol{0}, \boldsymbol{H}_t)$, and $\boldsymbol{\epsilon}_t \sim N(\boldsymbol{0}, \boldsymbol{Q}_t)$ are white noise processes independent of each other. We see that the formulation differs from the formulation given in (10.18a)-(10.18b) (there is no input). The formulation can however be given in an equivalent formulation by treating some of the parameters as constant states.

Example 10.8 *In Chapter 8 we studied the ARX(2)-model*

$$(Y_t - u_t) + \phi_1(Y_{t-1} - u_{t-1}) + \phi_2(Y_{t-2} - u_{t-2}) = \epsilon_t, \tag{10.60}$$

where u_t is a known input and $\epsilon_t \sim N(0, \sigma^2)$, and $u_t = \boldsymbol{x}_t \boldsymbol{\beta}$. The model can be written in state space formulation as

$$\begin{aligned} \begin{bmatrix} X_{1,t} \\ X_{2,t} \end{bmatrix} =& \begin{bmatrix} -\phi_1 & 1 \\ -\phi_2 & 0 \end{bmatrix} \begin{bmatrix} X_{1,t-1} \\ X_{2,t-1} \end{bmatrix} + \begin{bmatrix} 1 \\ 0 \end{bmatrix} \epsilon_t \\ Y_t =& u_t + \begin{bmatrix} 1 & 0 \end{bmatrix} \begin{bmatrix} X_{1,t} \\ X_{2,t} \end{bmatrix}, \end{aligned} \tag{10.61}$$

which is not in the form (10.59). However, by extending the state space, the model can be written in the form

$$\begin{aligned} \begin{bmatrix} \alpha_t \\ X_{1,t} \\ X_{2,t} \end{bmatrix} =& \begin{bmatrix} 0 & 0 & 0 \\ 0 & -\phi_1 & 1 \\ 0 & -\phi_2 & 0 \end{bmatrix} \begin{bmatrix} \alpha_{t-1} \\ X_{1,t-1} \\ X_{2,t-1} \end{bmatrix} + \begin{bmatrix} 0 & 0 \\ 1 & 0 \\ 0 & 0 \end{bmatrix} \boldsymbol{\epsilon}_t \\ Y_t =& \begin{bmatrix} u_t & 1 & 0 \end{bmatrix} \begin{bmatrix} \alpha_t \\ X_{1,t} \\ X_{2,t} \end{bmatrix}, \end{aligned} \tag{10.62}$$

with $\boldsymbol{\epsilon}_t \sim N(\boldsymbol{0}, \boldsymbol{Q})$, and $(\boldsymbol{Q})_{22} = \sigma^2$ and $(\boldsymbol{Q})_{ij}$ zero otherwise, which is in the form (10.59). If further $V[\alpha_0] = 0$ then α_t will be constant and hence $\alpha_t = \alpha_0$, which is equivalent to the model (10.61) when $\alpha_0 = 1$.

State space models are quite flexible and therefore the model set up is also more tricky than many of the other models that we have studied in this book so far. It is however helpful to set up the model with emphasis on the parameters that need to be estimated

$$\begin{aligned} \boldsymbol{X}_t =& \boldsymbol{T}(\boldsymbol{\theta}, t) \boldsymbol{X}_{t-1} + \boldsymbol{R}(\boldsymbol{\theta}, t) \boldsymbol{\epsilon}_t \\ \boldsymbol{Y}_t =& \boldsymbol{Z}(\boldsymbol{\theta}, t) \boldsymbol{X}_t + \boldsymbol{e}_t \\ \boldsymbol{\epsilon}_t \sim& N(\boldsymbol{0}, \boldsymbol{Q}(\boldsymbol{\theta}, t)) \\ \boldsymbol{e}_t \sim& N(\boldsymbol{0}, \boldsymbol{H}(\boldsymbol{\theta}, t)), \end{aligned} \tag{10.63}$$

where the dependence of t may be through some input and with initial conditions

$$\begin{aligned} \hat{\boldsymbol{X}}_0 &= \boldsymbol{X}_0(\boldsymbol{\theta}) \\ \boldsymbol{\Sigma}_0^{xx} &= \boldsymbol{\Sigma}_0(\boldsymbol{\theta}), \end{aligned} \tag{10.64}$$

which implies that we would have to set up all the matrix functions for the state space model, while the filtering part is taken care of by the statistical package. This is also the way that KFAS works, as we will show in the example below.

Example 10.9 *Using the example above, we need to initialise the matrices and the initial conditions. In our case, $\boldsymbol{Z}_t$ is varying over time and should therefore be implemented as an array;*

```
n <- dim(win.datSS)[1]
Zt <- array(0, dim=c(1,3,n))
Zt[1,2, ] <- 1
```

Implying that $\boldsymbol{Z}_t$ is initialised by

$$\boldsymbol{Z}_t = \begin{bmatrix} 0 & 1 & 0 \end{bmatrix}, \tag{10.65}$$

in the likelihood set up below, $(\boldsymbol{Z}_t)_{11}$ will be a function of time including the input/regression.

The remaining matrices are constants and are initialised by;

```
Qt <- diag(2); Qt[1,1] <- 0;Qt[2,2] <- 1
Tt <- matrix(c(1, 0, 0, 0, 0.9, -0.1, 0, 1, 0), 3, 3)
Rt <- matrix(c(0,0,0,0,1, 0), 3, 2)
Ht <- matrix(0, 1, 1)
```

Finally, the initial values for the optimisation set are set-up by;

```
a1 <- matrix( 1, 3 , 1)
P1 <- diag(3) * 0.05
P1[1,1] <-0
```

Here, a1 *are the initial values of the states, and the important part for our purpose is that the first element of* a1 *is set to 1, while the rest will be estimated during the likelihood estimation. The initial value of the state variance is quite arbitrarily set as 0.05. The final step before the likelihood set up is to define the* KFAS *model object;*

```
model_gaussianReg <- SSModel(log(win.datSS$CO2CB) ~ -1 +
                        SSMcustom(Z = Zt, T = Tt,
                                  R = Rt, Q = Qt,
                                  a1 = a1, P1 = P1), H = Ht)
```

The KFAS *package includes a number of different default methods for setting up different model structures. To keep the discussion simple, we only consider the* SSMcustom, *where the matrices are defined directly by the user, but see Helske (2017) for automated structures.*

Referring to the previous example we set

$$u_t = \boldsymbol{x}_t\boldsymbol{\beta}, \tag{10.66}$$

where $\boldsymbol{x}_t$ is a row in the design matrix (here a 1 and an indicator for window position), and

$$\boldsymbol{Z}_t = \begin{bmatrix} u_t & 1 & 0 \end{bmatrix}, \tag{10.67}$$

We can set up the likelihood as;

```
KFASnll1 <- function(pars, model, X, full=FALSE) {
  model$T[2, 2, ] <- pars[1]
  model$T[3, 2, ] <- pars[2]
  model$Q[2, 2, 1] <- exp(pars[3])
  model$a1[-1, 1] <- pars[4:5]
  model$Z[1, 1, ] <- X %*% pars[6:7]
  if(full){model}
  -logLik(model)
}
```

and estimate the parameters using a general purpose optimiser;

```
X <- cbind(1,win.datSS$WindowClosedB)
optKFAS <- nlminb(c(0.9, -0.4, 0, 0, 0, 0, 0), KFASnll1,
                  model = model_gaussianReg, X = X)
```

We analysed the model and extensions to the model in Chapter 8, and hence we will not go into details with the properties here, but just compare with the ARIMA model

```
fitArx <- arima(log(win.datSS$CO2CB), order = c(2, 0, 0),
                include.mean = TRUE,
                xreg = win.datSS$WindowClosedB)
c(-optKFAS$objective,logLik(fitArx))

## [1] 7181 7181
```

We see that the models have almost the exact same likelihood, further the parameters (not shown) are also (almost) identical. There is a small difference in the way that the models treat initial values of the state, and that explain differences in the parameter values.

10.6.1 Generalisations of the ARIMA model

In the classical ARIMA models, parameters are assumed to be constant. The `arima` method in R allows for a time-varying input for the mean, but not for the AR- or MA-parameters. The state space formulation in `KFAS` gives the flexibility to define models with time varying parameters. Such a model would emphasize that the dynamics of the system is a function of the input, e.g. we would expect that time constants for CO_2 concentration in a room differs depending on the window being open or closed, and airtightness of the space. In a general state space formulation that is expressed in (10.63), the dependence on time is expressed in form of some input. As an example we use an ARMAX(2,1), which we may write as (using the formalism of KFAS)

$$\boldsymbol{X}_t = \begin{bmatrix} 1 & 0 & 0 \\ 0 & \phi_1(u_t,\boldsymbol{\theta}) & 1 \\ 0 & \phi_2(u_t,\boldsymbol{\theta}) & 0 \end{bmatrix} \boldsymbol{X}_{t-1} + \begin{bmatrix} 0 & 0 \\ 0 & 1 \\ 0 & \theta(u_t,\boldsymbol{\theta}) \end{bmatrix} \boldsymbol{\epsilon}_t \tag{10.68}$$

$$Y_t = \begin{bmatrix} \mu(u_t,\boldsymbol{\theta}) & 1 & 0 \end{bmatrix} \boldsymbol{X}_t, \tag{10.69}$$

where $\boldsymbol{\epsilon}_t \sim N(\boldsymbol{0}, \boldsymbol{Q}_t)$ and

$$\boldsymbol{Q}_t = \begin{bmatrix} 0 & 0 \\ 0 & \sigma^2(u_t,\boldsymbol{\theta}) \end{bmatrix}. \tag{10.70}$$

The initial values of the system should be set such that $\boldsymbol{X}_{1,t} = 1$ for all t, we will return to the formulation of $\phi_i(u_t,\boldsymbol{\theta})$, $\theta(u_t,\boldsymbol{\theta})$, and $\sigma^2(u_t,\boldsymbol{\theta})$. In addition, KFAS will find the maximum likelihood estimates disregarding stationarity and invertability conditions, and hence if such considerations are important one has to take care of those in the likelihood or model formulation (a simple example is given below).

Example 10.10 *For an ARMAX(2,1) model with time–varying parameters all matrices have to be initialized as arrays;*

```
n <- dim(win.datSS)[1]
Zt <- array(0, dim=c(1,3,n))
Zt[1,2, ] <- 1

Qt <- array(0, dim=c(2,2,n)); Qt[2,2, ] <- 1
Tt <- array(0, dim=c(3,3,n))
Tt[1,1, ] <- 1
Tt[2,3, ] <- 1
Rt <- array(0, dim=c(3,2,n))
Rt[2,2, ] <- 1
Ht <- matrix(0, 1, 1)

a1 <- matrix( 1, 3 , 1)
P1 <- diag(3) * 0.05
P1[1,1] <-0
```

Note here that we set $V[\boldsymbol{X}_{1,0}] = 0$, and $E[\boldsymbol{X}_{1,0}] = 1$, while the other values are more arbitrary. The model is defined in KFAS *by;*

```
model_gaussianRegT <- SSModel(log(win.datSS$CO2CB) ~ -1 +
                              SSMcustom(Z = Zt, T = Tt,
                                        R = Rt, Q = Qt,
                                        a1 = a1, P1 = P1), H = Ht)
```

and the likelihood function is defined as;

```
KFASnll21T <- function(pars, model, X, full=FALSE) {
    model$T[2,2, ] <- X %*% pars[1:2]
    model$T[3,2, ] <- X %*% pars[3:4]
    eta <- X %*% pars[5:6]
    model$R[3,2, ] <- 2 * exp(eta) / (1 + exp(eta)) - 1
    model$Q[2,2, ] <- exp(X %*% pars[7:8])
    model$a1[-1,1] <- pars[9:10]
    model$Z[1,1, ] <- X %*% pars[11:12]
    if(full){return(model)}
    -logLik(model)
}
```

Here, X *is the input variables (first column equal 1), and the second column is the window position (0 open and 1 closed). In order to ensure that the system is invertible, we estimate $\theta(\boldsymbol{x}_t)$ by*

$$\theta(\boldsymbol{x}_t) = 2\frac{e^{\boldsymbol{x}_t\boldsymbol{\beta}_\theta}}{1+e^{\boldsymbol{x}_t\boldsymbol{\beta}_\theta}} - 1 \tag{10.71}$$

such that $\theta(\boldsymbol{x}_t) \in (-1, 1)$. Further, it is ensured that $\sigma_{22}(\boldsymbol{x}_t) > 0$ (by setting $\sigma_{22}(\boldsymbol{x}_t) = e^{\boldsymbol{x}_t\boldsymbol{\beta}_\sigma}$). The parameters of the model are estimated by;

```
pars <- c(0.9,0, -0.2, 0 , 0, 0, 0, 0, 0, 0, 0, 0)
optKFAS21Time <- nlminb(pars, KFASnll21T,
                        model = model_gaussianRegT, X = X)
```

The first question is of course if the model can be reduced (does the parameter depend on the window position). Such a question could be answered by comparing the likelihood of different models (likelihood ratio test), or we may simply consider the Wald statistics (using the observed Fisher Information matrix);

```
H <- hessian(KFASnll21T, optKFAS21Time$par,
             model = model_gaussianRegT, X = X)
VarTheta <- solve(H)
```

Here, the parameters of primary interest are index numbers 2, 4, 6, 8 and 12, which describe the difference between the two window positions

```
index <- c(2,4,6,8,12)
optKFAS21Time$par[index] / sqrt(diag(VarTheta)[index])

## [1]  12.3  -1.5  21.1 -18.6  22.5
```

The conclusion here (using the usual 5% significance level) is that all parameters, except ϕ_2, dependent on the window position, and as such we could consider to reduce the system (i.e. remove the dependence of window position for ϕ_2). Here we will rather discuss the interpretation of parameters. The model at the MLE is extracted by;

```
mod21T <- KFASnll21T(optKFAS21Time$par, model_gaussianRegT, X,
                     full=TRUE)
```

The stationary expectation in the two situations (window open/closed) is given by

```
c(mod21T$Z[1,1,X[ ,2]==0][1], mod21T$Z[1,1,X[ ,2]==1][1])

## [1] 6.1 6.2
```

In the log-domain. In the natural domain, we have

```
exp(c(mod21T$Z[1,1,X[ ,2]==0][1], mod21T$Z[1,1,X[ ,2]==1][1]))

## [1] 439 505
```

hence 439 ppm and 505 ppm when window is open and closed, respectively. The AR parameters are

```
mod21T$T[2:3,2,X[ ,2]==0][ ,1]

## custom2 custom3
##    0.86   -0.07
```

when the window is open and

```
mod21T$T[2:3,2,X[ ,2]==1][ ,1]

## custom2 custom3
##    1.08   -0.09
```

when the window is closed. The MA–parameters are given by

```
c(mod21T$R[2:3, 2, X[ , 2] == 0][2, 1],
  mod21T$R[2:3, 2, X[ ,2] == 1][2,1])

## custom3 custom3
##   0.095   0.779
```

This implies that the autocorrelation is stronger when the window is closed, and the error variance is

```
c(mod21T$Q[ , , X[ ,2] == 0][2 ,2 ,1],
  mod21T$Q[ , , X[ ,2] == 1][2 ,2 ,1])

## [1] 0.0060 0.0021
```

So the error variance is larger when the window is open than when it is closed. These variances should however be seen relative to the AR and MA parameters. This comparison is most effectively done by examining the stationary variance under the two conditions. Using (10.46)*, the stationary variance can be calculated as (when the window is open);*

```
A <- mod21T$T[2:3,2:3,X[ ,2]==0][ , ,1]
G <- mod21T$R[2:3, 2, X[ , 2] == 0][ , 1]
sigma.sq <- mod21T$Q[ , , X[ ,2] == 0][2 ,2 ,1]
stat.var.open <- matrix(
    sigma.sq * solve(diag(4) - kronecker(A, A)) %*%
    as.vector(G %*% t(G)), 2, 2)
```

Using the same calculation, the stationary variance when the window is closed can also be calculated.

The stationary standard deviations in the two cases are;

```
sqrt(c(stat.var.open[1,1],stat.var.closed[1,1]))

## [1] 0.14 0.74
```

and hence a huge difference in the stationary variance/standard deviation in the two situations can be seen. A large proportion of the difference can probably be attributed to the occupant behaviour, i.e. when the window is open, the CO_2 concentration should be around 400 ppm (log(400)=6), while with the window closed it depends on the number of occupants in the room (and their activity level).

10.6.2 Multivariate time–series models

In Chapter 8, we discussed univariate linear time–series models. Such models can also be generalised to multivariate time–series models

$$\boldsymbol{y}_t = -\boldsymbol{\phi}_1 \boldsymbol{y}_{t-1} - \cdots - \boldsymbol{\phi}_p \boldsymbol{y}_{t-p} + \boldsymbol{\epsilon}_t + \boldsymbol{\theta}_1 \boldsymbol{\epsilon}_{t-1} + \cdots + \boldsymbol{\theta}_q \boldsymbol{\epsilon}_{t-q}, \tag{10.72}$$

where $\boldsymbol{y}_t \in \mathbb{R}^r$ and $\boldsymbol{\epsilon}_t \sim N_r(\boldsymbol{0}, \boldsymbol{Q})$. While there are dedicated packages for such models, e.g. `marima` (Spliid, 2017), these models can also be written in state space form. The coefficients in the AR and MA part are all matrices and hence the number of parameters grows fast with the order of the model. Stationarity is related to eigen-values in the state space formulation of the problem, see Hamilton (1994) (all eigenvalues should be less than 1).

Example 10.11 *The CO_2 data we have considereds so far contain observations from different rooms in the same dwelling, and it is reasonable to assume that there is a flow of air between the rooms in the dwelling and hence that there is a correlation between CO_2 observations in different rooms. Such an assumption would lead to a multivariate time series model. To keep it simple, we formulate a vector ARX(1) model*

$$\begin{bmatrix} y_{1,t} - \mu_{1,t} \\ y_{2,t} - \mu_{2,t} \end{bmatrix} = \begin{bmatrix} t_{11} & t_{12} \\ t_{21} & t_{22} \end{bmatrix} \begin{bmatrix} y_{1,t-1} - \mu_{1,t-1} \\ y_{2,t-1} - \mu_{2,t-1} \end{bmatrix} + \begin{bmatrix} \epsilon_{1,t} \\ \epsilon_{2,t} \end{bmatrix} \tag{10.73}$$

with $\boldsymbol{\epsilon}_t = \begin{bmatrix} \epsilon_{1t} & \epsilon_{2t} \end{bmatrix}^T \sim N(\boldsymbol{0}, \boldsymbol{Q})$. Or given in the state space formulation

$$\begin{aligned} \begin{bmatrix} x_{1,t} \\ x_{2,t} \\ x_{3,t} \\ x_{4,t} \end{bmatrix} &= \begin{bmatrix} 1 & 0 & 0 & 0 \\ 0 & 1 & 0 & 0 \\ 0 & 0 & t_{11} & t_{12} \\ 0 & 0 & t_{21} & t_{22} \end{bmatrix} \begin{bmatrix} x_{1,t-1} \\ x_{2,t-1} \\ x_{3,t-1} \\ x_{4,t-1} \end{bmatrix} + \begin{bmatrix} \epsilon_{1t} \\ \epsilon_{2t} \\ \epsilon_{3t} \\ \epsilon_{4t} \end{bmatrix} \\ \begin{bmatrix} y_{1,t} \\ y_{2,t} \end{bmatrix} &= \begin{bmatrix} \mu_{1t} & 0 & 1 & 0 \\ 0 & \mu_{2t} & 0 & 1 \end{bmatrix} \boldsymbol{x}_t, \end{aligned} \tag{10.74}$$

with appropriate definitions of initial values and initial variances. The output here is log–concentration of CO_2 in the two rooms (bedroom and living room). The model may be implemented in `KFAS` *with*

```
Zt <- array(0, dim=c(2, 4, n))
Zt[1, 3, ] <- Zt[2, 4, ] <- 1
Qt <- matrix(0, 4, 4)
Tt <- matrix(0, 4, 4)
Tt[1, 1] <- Tt[2, 2] <- 1
Rt <- diag(4); Rt[1, 1] <- 0; Rt[2, 2] <- 0
a1 <- matrix(1, 4, 1)
P1 <- diag(4) * 0.4
P1[1, 1] <- P1[2, 2] <-0
Ht <- diag(2) * 0
```

and the model definition;

```
model_multiVar <- SSModel(log(resp) ~ -1 +
                     SSMcustom(Z = Zt, T = Tt, R = Rt, Q = Qt,
                               a1 = a1, P1 = P1), H = Ht)
```

Here, `resp` *is a two–column matrix with the columns being the CO_2–concentration in the bedroom room and the living room, respectively. In order to have a benchmark, we start by estimating a model with $t_{12} = t_{21} = 0$, and $\boldsymbol{Q}_{12} = \boldsymbol{Q}_{21} = 0$, that is independent time–series models. The negative log-likelihood of such a model can be defined by;*

```
KFASnllMultivar0 <- function(pars, model, X, full=FALSE) {
    model$Z[1,1, ] <- X[ ,c(1:2)] %*% pars[1:2]
    model$Z[2,2, ] <- X[ ,c(1,3)] %*% pars[3:4]
    model$T[3,3, ] <- pars[5]
    model$T[4,4, ] <- pars[6]
    model$Q[3,3, ] <- exp(pars[7])
    model$Q[4,4, ] <- exp(pars[8])
    model$a1[-(1:2),1] <- pars[9:10]
    if(full){return(model)}
    -logLik(model)
}
```

where the columns of `X` *contain a vector of ones (intercept) and indicators for the window position in the two rooms (bedroom and living room respectively). Finally, the model can be estimated by*

```
pars <- c(6, 0, 6, 0, 0.9, 0.9, -3, -3, 0, 0)
optMultiKFAS0 <- nlminb(pars, KFASnllMultivar0,
                        model = model_multiVar, X = X)
```

The full model can be formulated as (some lines omitted as definition of `Z` *and* `T` *is unchanged);*

```
KFASnllMultivar1 <- function(pars, model, X, full=FALSE) {
    ## ...
    rho <- 2 * exp( pars[11]) / (1 + exp(pars[11])) - 1
    model$Q[3, 3, ] <- sigma1^2
    model$Q[4, 4, ] <- sigma2^2
    model$Q[4, 3, ] <- model$Q[3, 4, ] <- sigma1 * sigma2 * rho
    model$a1[-(1:2), 1] <- pars[12:13]
    if(full){return(model)}
    -logLik(model)
}
```

where we see (`Q[4, 3, ]`*) that rather than estimating the covariance between* ϵ_{1t} *and* ϵ_{2t} *we estimate the correlation (which is between -1 and 1) between the two. The MLEs are calculated by;*

```
pars <- c(6, 0, 6, 0, 0.9, 0.9, 0, 0, -3, -3, 0, 0, 0)
optMultiKFAS1 <- nlminb(pars, KFASnllMultivar1,
                        model = model_multiVar, X = X)
```

The difference in likelihood can be obtained by

```
-optMultiKFAS1$objective + optMultiKFAS0$objective

## [1] 119
```

With only 3 extra parameters, this is highly significant. The matrix $\boldsymbol{T}$ *can be extracted by*

```
modMult1 <- KFASnllMultivar1(optMultiKFAS1$par,
                        model = model_multiVar, X, full = TRUE)
(T <- modMult1$T[3:4 , 3:4, 1])

##         custom3 custom4
## custom3  0.9894  0.0023
## custom4  0.0018  0.9773

eigen(T)$values

## [1] 0.99 0.98
```

hence we see a weak interaction and an eigen-value very close to 1, so we may consider using a vector ARIMAX model. The other difference between the two models is the correlation between errors

```
cov2cor(modMult1$Q[3:4 ,3:4 ,1])[1,2]

## [1] -0.22
```

which is negative, implying that if there is a positive error in one room we may expect that there is a negative error in the other room.

The example above illustrates how correlation between time series can be modelled using multivariate time series, and this may give some insight into interactions between the series. The comments on time–varying parameters discussed above still apply and such models can in principle be applied here also. However the number of parameters grows fast and considerations on the available data should be made, in particular there should be data for different type of events (e.g. window position). In our particular example, there is a very limited number of observations where the window is open in both rooms. In many applications, (multivariate) time–series models may be considered as an initial analysis before formulating models based on mechanistic understanding of the system. We consider such models in the next section.

10.7 Models based on first principles

The Examples 10.8-10.11 above shows how ARIMAX models can be formulated as state space models. Another application is to use a more mechanistic description of the system. A mechanistic system will often be formulated in continuous time as a set of ODE's (Ordinary Differential Equations)

$$\frac{d\boldsymbol{x}_t}{dt} = f(\boldsymbol{x}). \tag{10.75}$$

The stochastic counterpart of an ODE is a so-called stochastic differential equation (SDE), which is defined by

$$d\boldsymbol{x}_t = f(\boldsymbol{x}_t)dt + \boldsymbol{\sigma} d\boldsymbol{w}_t, \tag{10.76}$$

where $\boldsymbol{\sigma}$ is a matrix of appropriate dimension and $d\boldsymbol{w}_t$ is differential white noise, meaning that $\boldsymbol{w}_t - \boldsymbol{w}_s \sim N(0, t-s)$ $(t > s)$. In a continuous-discrete time stochastic system, there will be in addition an observation equation

$$\boldsymbol{y}_k = h(\boldsymbol{x}_{t_k}) + \boldsymbol{e}_k; \quad \boldsymbol{e}_k \sim N(0, \boldsymbol{S}). \tag{10.77}$$

We will not treat continuous time stochastic systems in this book, but note that the R-package `ctsmr` (Juhl et al., 2016), does allow for estimation in quite general non-linear systems of SDEs, and Wolf et al. (2019b) use an SDE based approach (and `ctsmr`) to model CO_2 concentration. Rather, we will treat the discretised system

$$\boldsymbol{x}_{t+1} = \boldsymbol{x}_{t+1} + f(\boldsymbol{x}_t)\Delta + \boldsymbol{\sigma}\boldsymbol{\epsilon}_{t+1}. \tag{10.78}$$

If f is linear, the system is a linear state space model that can be handled by `KFAS`.

Example 10.12 *Laussmann and Helm (2011) suggest the following model for gas concentration using simple mass-balance considerations*

$$\frac{dC_t}{dt} = -C_t\frac{Q}{V} + \frac{E}{V}, \tag{10.79}$$

where C is the concentration above the outside concentration, V is the volume of the room, Q is the air change rate and E is the production/emitted gas in the room. In the case of CO_2 concentration we may assume that the outside concentration is 400 ppm. In the data we have available, the volume of the room is not known and hence it will only be possible to estimate the relative air change $(Q_r = Q/V)$ and the relative *production ($E_r = E/V$). The production is not observed in our case, and we cannot assume that it is constant. But, we can formulate the production as a random walk and with that an SDE description of the system is*

$$\begin{aligned} dC_t =& (-Q_r(t)C_t + E_{r,t})dt + \sigma_1 dw_{1,t} \\ dE_{r,t} =& \sigma_2 dw_{2,t}. \end{aligned} \tag{10.80}$$

The filter can now be used to estimate the production as a function of time. It is natural to assume that the air change is a function of the window position

$$Q_r(t) = Q_r(WinPos_t). \tag{10.81}$$

With this, we can set up the discretized system as

$$\begin{aligned} \begin{bmatrix} C_{t+1} \\ E_{r,t+1} \end{bmatrix} &= \begin{bmatrix} 1 - Q_r(t)\Delta & \Delta \\ 0 & 1 \end{bmatrix} \begin{bmatrix} C_t \\ E_{r,t} \end{bmatrix} + \boldsymbol{R}\epsilon_{t+1} \\ Y_t &= \begin{bmatrix} 1 & 0 \end{bmatrix} \begin{bmatrix} C_t \\ E_t \end{bmatrix} + e_t. \end{aligned} \tag{10.82}$$

It is reasonable to assume that the primary source of variation is the production, leading to

$$\boldsymbol{R} = \begin{bmatrix} 0 & 0 \\ 0 & 1 \end{bmatrix}; \quad \boldsymbol{Q} = \begin{bmatrix} 0 & 0 \\ 0 & q_{22} \end{bmatrix}, \tag{10.83}$$

and we set $\boldsymbol{H} = 25^2$, the value is quite arbitrary but reflects that the CO_2 concentration is measured with some error (i.e. a standard deviation of 25 ppm). The air change is modelled by

$$Q_r(t)\Delta = \beta_0 + \beta_1 \texttt{WinPos}_t, \tag{10.84}$$

and hence the unit of $Q_r(t)$ will be $1/h$, and $\delta = 10min = \frac{1}{6}h$.
The model is defined by;

```
n <- dim(win.datSS)[1]
Zt <- matrix(0, 1, 2)
Zt[1, 1] <- 1
Qt <- diag(2); Qt[1, 1] <- 0;
Tt <- array(0, dim=c(2, 2, n))
Tt[2, 2, ] <- 1
Tt[1, 2, ] <- 1 / 6
Rt <- diag(2); Rt[1, 1] <- 0
Ht <- matrix(25^2, 1, 1)
```

the initial values for the optimisation if set by;

```
a1 <- matrix(1, 2, 1)
P1 <- diag(2) * 2200
```

and the model object is defined by

```
model_gaussianRegPhys <- SSModel((win.datSS$CO2CB-400) ~ -1 +
                                 SSMcustom(Z = Zt, T = Tt,
                                           R = Rt, Q = Qt,
                                           a1 = a1, P1 = P1), H = Ht)
```

and the likelihood is defined as;

```
KFASnllPh <- function(pars, model, X, full=FALSE) {
    model$T[1,1, ] <- 1 - X %*% pars[1:2] / 6
    model$Q[2,2,1] <- exp(pars[3])
    model$a1[,1] <- pars[4:5]
    if(full){return(model)}
    -logLik(model)
}
```

The parameter estimation is done by;

```
X <- cbind(1, win.datSS$WindowClosedB)
optKFASPh <- nlminb(c(0, 0, 0, 0, 0), KFASnllPh,
                    model = model_gaussianRegPhys, X = X)
H <- hessian(KFASnllPh, optKFASPh$par,
             model = model_gaussianRegPhys, X = X)
```

the parameters are given below

```
tab <- cbind(theta = optKFASPh$par, se = sqrt(diag(solve(H))))
rownames(tab) <- c("beta0","beta1","q22","C0","E0")
tab

##         theta      se
## beta0    3.13   0.137
## beta1   -0.93   0.112
## q22     10.86   0.033
## C0     136.88  52.684
## E0      13.48 141.744
```

The parameters tell us that when the window is open, $Q_r \approx 3/h$, and when the window is closed, $Q_r \approx 2/h$. While the first value seems reasonable, the second one seems way too high, indicating that the model is probably too simple and more/other structures should be included. Finally the smoothed state of respiration (CO_2 production) can be extracted by

```
modPh <- KFASnllPh(optKFASPh$par, model_gaussianRegPhys, X, full=TRUE)
States <- KFS(modPh)
```

The object `State` *contains the 1-step predictions of state, filtered states and smoothed state as well as variances. The figure below shows the smoothed state of the relative CO_2–production.*

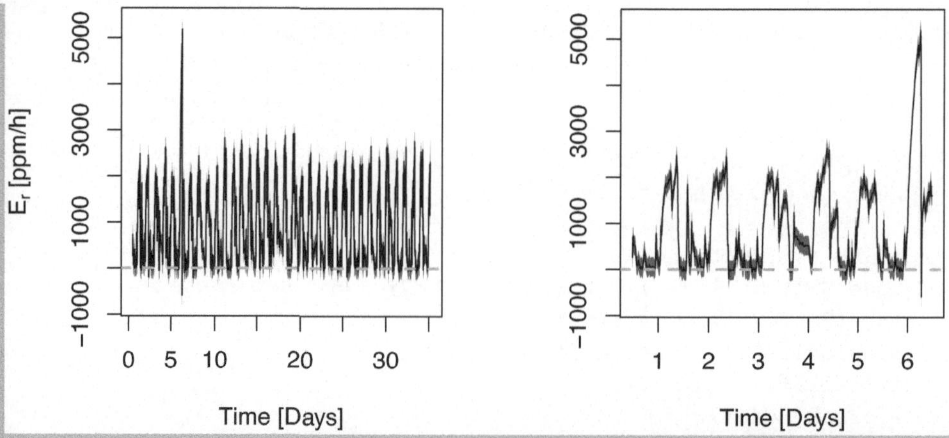

Which could be converted into CO_2 production using the volume of the room had it been available. A qualitative interpretation can however be done using the presented results, namely that high relative CO_2 production would imply that there are occupants present in the room.

10.8 Diagnostics

Just as we saw for linear time–series models, residual analysis is important for model development and validation, and basically all the methods presented in Section 8.6 carry over to the state space models, as long as we work on the appropriate residuals. In the case of time varying state space models these are the standardized residuals, that is the raw residuals (innovation error) divided by the standard deviation

$$r_{i,t} = \frac{y_{i,y} - \hat{y}_{i,t|t-1}}{\sqrt{\left(\Sigma^{yy}_{t|t-1}\right)_{ii}}}. \tag{10.85}$$

If the model is appropriate, then each of the residual series should be white noise, when considered marginally. However when considered in a multivariate setting, there will be some crosscorrelation between the series. Before considering crosscorrelation, the autocorrelation should be dealt with by the model, as crosscorrelation may show artificial patterns if the autocorrelation of the individual series is not zero (see Madsen (2008), p. 150). We illustrate this with a small example.

Example 10.13 *The figure below shows the main residual plot for the model in Example 10.12. There are clearly a number of problems with the model, for example the tails are way too heavy, indicating that the model is not appropriate. Further, both the autocorrelation and the partial autocorrelation have many lags that are different from zero, indicating that the dynamic model is not correctly specified, and more structure should be included. One possibility is to add AR or MA part in the CO_2 production part of the model (which is for now a random walk), or patterns related to the time of day (as we have also seen in the examples of Chapter 8). There is also a clear peak in the partial autocorrelation at lag 144, corresponding to a diurnal variation.*

Standardized residuals can be extracted by

```
res <- rstandard(KFS(modPh))
```

and the figures below are based on the standardized residuals.

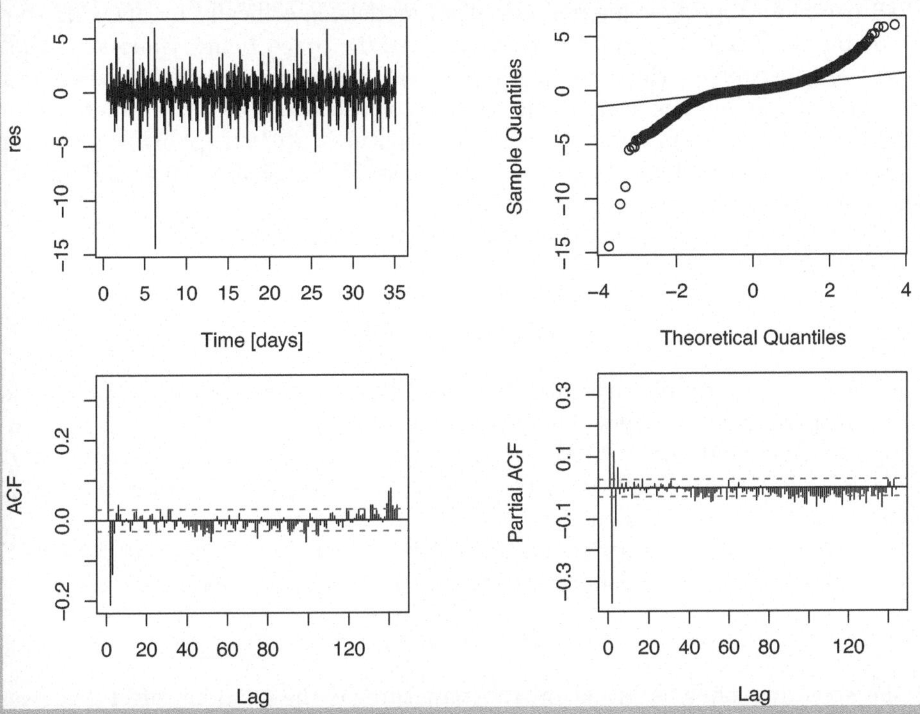

To summarise the example more/other structures is needed in order to capture the variation the the presented data.

We will not go further into the concrete model development here, but rather point to some further possibilities in the next section.

10.8.1 Some comments on model development

We saw in Example 10.13 above that the model had a number of issues. One is related to the autocorrelation and partial autocorrelation. We have covered ways to handle this throughout this chapter and Chapter 8 (i.e. add AR/MA terms in the model).

The other part was related to the distribution of the residuals. Here, the issue might be that the variance actually depends on the level of the CO_2–production/level. This would imply that the variance is state dependent, and that is actually the way we have modelled CO_2 so far (by modelling log-CO_2). For the physical model, introducing such dependence implies non-linear models, which we do not cover here. However, state dependent diffusion models based on SDE–description are possible using the so-called Lamperti transform (see e.g. Møller and Madsen (2010)). Further, see Thygesen (2023) for a more general introduction to SDEs.

10.9 Generalized state space models

The KFAS package supports some non-Gaussian models, specifically by allowing the observations to be non-Gaussian, but the underlying state space being Gaussian. The model structure is comparable to the structure supported by the generalized mixed effect models (see Section 6.2). While generalized mixed effect models focus on the difference between subjects, they also support some correlation structures. KFAS supports any linear state space model, and hence a large family of autocorrelation structures, but random effects for different subjects are not supported.

The model structure for generalized state space models in KFAS is

$$\begin{aligned} \boldsymbol{X}_t =& \boldsymbol{T}_t \boldsymbol{X}_t + \boldsymbol{R}_t \boldsymbol{\epsilon}_t \\ Y_t \sim& ED(g^{-1}(\boldsymbol{Z}_t \boldsymbol{X}_t), \phi_t), \end{aligned} \tag{10.86}$$

where ED is a probability density/mass function (from the exponential dispersion family), and ϕ_t is a dispersion parameters. The observations follow an exponential dispersion model (the supported distributions in KFAS are Gaussian, Poisson, binomial, gamma and negative binomial). The approximation of the likelihood is based on filtering and the Laplace approximation, that we also discussed in Chapter 6.

When estimating parameters in the generalized linear mixed effect model (see Section 6.2), we wrote the model as a random effect model, with the second stage model

$$\boldsymbol{X} \sim N(\boldsymbol{0}, \boldsymbol{\Psi}), \tag{10.87}$$

where the structure of $\boldsymbol{\Psi}$ is based on the structure of the random effect (e.g. variation between subjects and autocorrelation), and the likelihood was calculated by (10.9) (using the Laplace approximation). KFAS basically uses the recursive filter (see Section 10.2) approximating the observation equation by a Gaussian distribution applying the Laplace approximation. We illustrate the estimation set up with a small example.

Example 10.14 *Here we consider the window position as an outcome of a Bernoulli random variable, but with an underlying AR(1) process to describe the variation, i.e.*

$$Y_t \sim Binom(p(x_t, \boldsymbol{\theta}), n_t). \tag{10.88}$$

In our, case we have $n_t = 1$ for all t, and further

$$\begin{aligned} p(x_t; \boldsymbol{\theta}) =& \frac{e^{\beta_0 + x_t}}{1 + e^{\beta_0 + x_t}} \\ x_t =& \phi x_{t-1} + \epsilon_t; \quad \epsilon_t \sim N(0, \sigma^2). \end{aligned} \tag{10.89}$$

The model is defined in KFAS *by;*

```
model_window <- SSModel(win.datSS$WindowClosedB ~ -1 +
                        SSMcustom(Z = Zt, T = Tt, R = Rt,
                                  Q = Qt, a1 = a1, P1 = P1),
                        distribution = "binomial")
```

where the matrices are set up as in the AR(1) model, and the likelihood is defined by;

```
KFASnllBinom <- function(pars, model, full=FALSE) {
    model$Z[1,1, ] <- pars[1]
    model$T[2,2, ] <- pars[2]
    model$Q[2,2, ] <- exp(pars[3])
    model$a1[-1,1] <- pars[4]
    if(full){return(model)}
    -logLik(model)
}
```

The parameters are found with the usual optimisation;

```
optWindowKFAS1 <- nlminb(rep(0, 4), KFASnllBinom,
                         model = model_window,
                         upper = c(100, 0.999, 100, 100),
                         lower = c(-100, -0.999, -100, -100))
mod <- KFASnllBinom(optWindowKFAS1$par, model_window, full = TRUE)
```

To illustrate the difference/similarities, the same model is implemented in **glmmTMB** *by (we need to define observation number and dwelling number for* **glmmTMB** *to work, but these are basically dummy variables);*

```
win.datSS$obs.no <- 1:length(win.datSS$WindowClosedB)
win.datSS$dwel <- 1
ar1.windowGlmmTMB <- glmmTMB(WindowClosedB ~ 1 +
                             ar1(as.factor(obs.no) - 1 | dwel),
                             data = win.datSS, family = binomial)
```

The likelihood of the two methods is compared by;

```
c(-optWindowKFAS1$objective, logLik(ar1.windowGlmmTMB))

## [1] -397 -397
```

Hence, we obtain the same likelihood, which also implies that the models are equivalent. For comparing the parameters, the summary table of the **glmmTMB** *model is given below;*

```
## [...]
## Conditional model:
##  Groups Name               Variance Std.Dev. Corr
##  dwel   as.factor(obs.no)1 1322     36.4     0.96 (ar1)
## Number of obs: 4996, groups:  dwel, 1
##
## Conditional model:
##             Estimate Std. Error z value   Pr(>|z|)
## (Intercept)    25.60       5.15    4.97 0.00000066 ***
## ---
## [...]
```

The parameters based on KFAS *are given below;*

```
##      beta0        phi logsigma2        x0
##      26.06       0.96      4.67    -17.65
```

We see that the intercept and the AR(1) coefficient are similar for the two methods, while the variance seems different. This is because glmmTMB *uses the parametrisation*

$$\boldsymbol{X} \sim N(\boldsymbol{0}, \sigma_s^2 \boldsymbol{R}), \tag{10.90}$$

where $\boldsymbol{R}_{ij} = \phi^{|i-j|}$, *and* σ_s^2 *is that stationary variance of the AR(1) process, which is*

$$\sigma_s^2 = \frac{\sigma^2}{1-\phi^2}, \tag{10.91}$$

and hence we can find the stationary variance of the KFAS *implementation by*

```
exp(pars["logsigma2"]) / (1 - pars["phi"] ^ 2)

## logsigma2
##      1358
```

which is close to 1322 as found by glmmTMB. *Using the function* KFS, *smoothed state estimates and other features related to the state estimates can be extracted for a* KFAS *model object, by*

```
mod <- KFASnllBinom(optWindowKFAS1$par, model_window, full = TRUE)
filtered <- KFS(mod)
```

The figure below shows the smoothed states and the observed window position as a function of time.

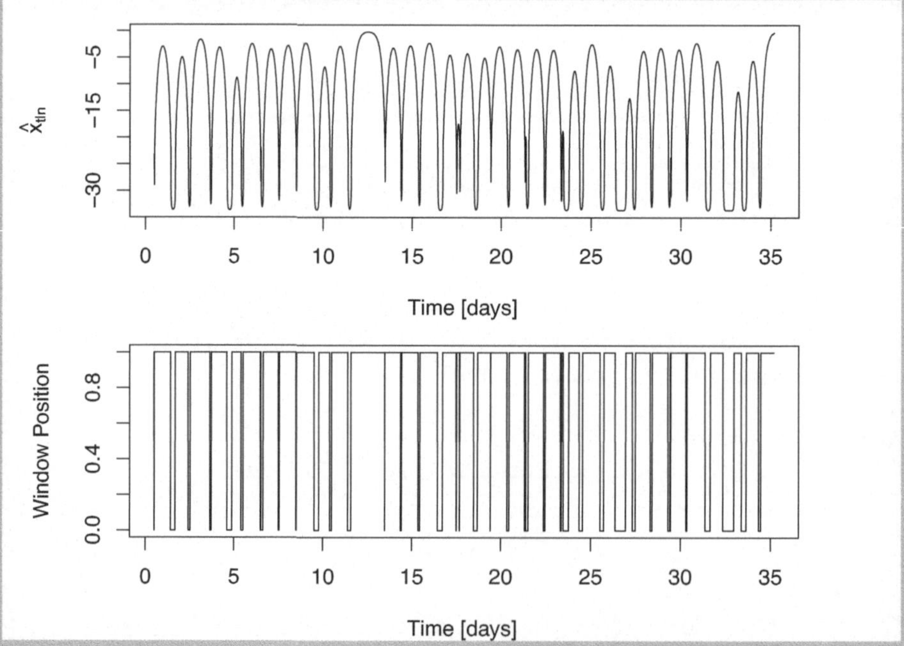

The figure indicates that there is a clear diurnal pattern that could be modelled either using a model that takes hour of day into account or using a seasonal ARMA model for the state.

While `glmmTMB` allows easy implementation on different dwellings (as random effects), the state space method allows for easy implementations of more general correlation structures, like AR(2) structures (see the example below) or even structures based on a mechanistic understanding of the system.

Example 10.15 *To exemplify the type of additional structure that could be put into the model we extend the model to an AR(2) model, using the structures of the matrices presented in Example 10.8, and a model definition as in Example 10.14. The likelihood function is calculated by;*

```
KFASnllBinom2 <- function(pars, model, full=FALSE) {
    model$Z[1,1, ] <- pars[1]
    model$T[2,2, ] <- pars[2]
    model$T[3,2, ] <- pars[3]
    model$Q[2,2, ] <- exp(pars[4])
    model$a1[-1,1] <- pars[5:6]
    if(full){return(model)}
    -logLik(model)
}
```

The parameters can be estimated by;

```
optWindowKFAS2 <- nlminb(pars,KFASnllBinom2, model = model_window21,
                         upper = c(100, 2, 1, 10, 100, 100),
                         lower = c(-100, -1, -1, -10, -100, -100))
```

The log-likelihoods can be compared by;

```
-c(optWindowKFAS2$objective, optWindowKFAS1$objective)

## [1] -318 -397
```

and hence the extra structure gives a highly significant improvement on the model so that we would expect a better predictive skill. The AR parameters are;

```
## [1] "1.690"  "-0.704"
```

i.e. both parameters are far from 0. Using Example 8.7 the roots can be found at 0.94 and 0.75, and hence the system is stable.

10.10 Summary and concluding remarks

In this chapter, we have discussed linear Gaussian state space models and shown how to build different kind of structures using these models. These structures include ARMA and ARMAX type models, and models based on a mechanistic understanding of the systems.

The models can be compared to the Linear Mixed Effect model treated in Chapter 5, but the time varying parameters allow for more general dynamic structures of the models, while including random effects related to subjects is not directly accessible using the presented methods. In addition to the linear Gaussian state space models, we have treated generalized linear state space models that can be compared to the models presented in the first part of Chapter 6. However, a richer class of autocorrelation structure is directly available using the state space formulation, but random effects related to subjects are more challenging to include.

The focus on linear models has the limitation that non-linear models are completely left out of the treatment. The natural generalisation of the presented models is non-linear state space model using the extended Kalman filter, with a classical reference being Jazwinski (1970) and a more recent reference focusing on the filtering aspects being Simon (2009).

A further generalisation is the class of continuous-discrete time stochastic models, where filtering may be based on the continuous-discrete time extended Kalman filter. Such models are ideal for mechanistically based models since it builds on stochastic differential equations (a generalisation of ordinary differential equations). The R-package `ctsmr` (Juhl et al., 2016) deals with these type of models in a quite general setup, and Thygesen (2023) gives a more broad introduction to stochastic differential equations.

11

Hidden Markov models

Hidden Markov Models (HMMs) is a very broad class of models, where an underlying state, which is not observed, influences the distribution of the outcomes, as shown schematically in Figure 10.1. The construction of HMMs is quite general and includes, for example the models presented in Chapter 10. In the present book, we refer to HMMs as models with a discrete state space, implying that the Markov part of the HMM is defined by the models presented in Chapter 9.

In occupancy modelling, HMMs can be used for estimation of a hidden state, that e.g. represents presence or activity based on quantitative measurements of some variable. Liisberg et al. (2016) use electricity load to identify patterns in presence. Wolf et al. (2019b), Wolf et al. (2019a), and Wolf et al. (2019c) use CO_2 concentration to estimate presence and activity in buildings.

In an ideal setting, the state represents some feature of the system e.g. number of occupants in a room, state of the window or similar. In practice of course the interpretation might be more complicated or challenging.

Formally, we define HHMs as a class of models where the state (X_t) can take a finite number of values (i.e. $X_t \in \{1, 2, ..., m\}$), and the observations follow a state dependent distribution ($p_i(y_t) = p(y_t|X_t = i)$). Hence, while the models in Chapter 10 are based on Gaussian models for the innovation, the models in this chapter are based on Markov chain models as presented in Chapter 9. The basic model structure considered in this chapter is

$$P(X_t = i) = \boldsymbol{u}_t^i, \tag{11.1}$$

where $\boldsymbol{u}_t$ is a probability mass function, i.e. $\sum \boldsymbol{u}_t^i = 1$, and further

$$\boldsymbol{u}_t = \boldsymbol{u}_{t-1}\boldsymbol{\Gamma}, \tag{11.2}$$

where $\boldsymbol{\Gamma}$ is the transition probability matrix (see Chapter 9), the observations are described by

$$p(Y_t = y_t|X_t = i) = p_i(y_t), \tag{11.3}$$

where $p_i(y_t)$ is a parametrised probability density or probability mass function. In order to set up the likelihood, we need a filter in line with the formulation in Chapter 10. The formulations given in (11.1)-(11.3) will be generalised throughout the chapter, and rely heavily on the development in Chapter 9.

> **Example 11.1** *Throughout Chapter 8, we modelled CO_2 concentration as ARIMAX models using the window position as an input. This implies that we assume that the window position is known. Had the window not been known (which is natural for future observations) a natural extension is to assume a two state HMM where the states are interpreted as window open or closed.*

As mentioned above, the development in this chapter relies heavily on the Markov models presented in Chapter 9. From a practical point of view, we reuse a lot of the R-functions that were developed there, and further properties of the underlying Markov Models are also covered in Chapter 9.

DOI: 10.1201/9781003340812-11

11.1 Recursive filter

In Chapter 10, we presented the general filtering problem (Section 10.2, formulation with univariate state) with three components (recall that $\mathcal{Y}_t = [Y_1, \ldots, Y_t]$, i.e. the observed information up to time t);

1. State filter/reconstruction

$$p(X_t|\mathcal{Y}_t) = \frac{p(\mathbf{Y}_t|X_t)p(X_t|\mathcal{Y}_{t-1})}{p(\mathbf{Y}_t|\mathcal{Y}_{t-1})}, \tag{11.4}$$

   which is the distribution of the state given the observation up to time t.

2. State prediction

$$p(X_t|\mathcal{Y}_{t-1}) = \int p(X_t|X_{t-1})p(X_{t-1}|\mathcal{Y}_{t-1})dX_{t-1}, \tag{11.5}$$

   which is the 1-step ahead prediction of the state given the observations up to time t.

3. Observation equation (1-step prediction)

$$p(\mathbf{Y}_t|\mathcal{Y}_{t-1}) = \int p(\mathbf{Y}_t|X_t)p(X_t|\mathcal{Y}_{t-1})dX_t, \tag{11.6}$$

   this is the likelihood contribution from observation $\mathbf{Y}_t$.

In HMMs the state distribution is given as a finite dimensional vector, which we will denote $\boldsymbol{u}_t$. In addition, we will use the following notation

$$\begin{aligned} \hat{\boldsymbol{u}}_{t|t} &= [p(X_t = 1|\mathcal{Y}_t), \ldots, p(X_t = m|\mathcal{Y}_t)] \\ \hat{\boldsymbol{u}}_{t|t-1} &= [p(X_t = 1|\mathcal{Y}_{t-1}), \ldots, p(X_t = m|\mathcal{Y}_{t-1})] \\ f_{t|t-1} &= p(\mathbf{Y}_t = \boldsymbol{y}_t|\mathcal{Y}_{t-1}) \\ g_{i,t} &= p(\mathbf{Y}_t = \boldsymbol{y}_t|X_t = i). \end{aligned} \tag{11.7}$$

The form here is kept general such that $\mathbf{Y}_t$ may be multivariate. The Markov chain is finite dimensional and hence the integral involved translates to summations. The elements of the vectors $\hat{\boldsymbol{u}}_{t|t}$ and $\hat{\boldsymbol{u}}_{t|t-1}$ are

$$\begin{aligned} \hat{\boldsymbol{u}}^i_{t|t} &= p(X_t = i|\mathcal{Y}_t) \\ \hat{\boldsymbol{u}}^i_{t|t-1} &= p(X_t = i|\mathcal{Y}_{t-1}). \end{aligned} \tag{11.8}$$

Using this notation we can write (referring to (11.4) and (11.6))

$$\hat{\boldsymbol{u}}_{t|t-1} = \hat{\boldsymbol{u}}_{t|t}\boldsymbol{\Gamma} \tag{11.9a}$$

$$f_{t|t-1} = \sum_{i=1}^{m} g_{i,t}\hat{\boldsymbol{u}}^i_{t|t-1}. \tag{11.9b}$$

$$\hat{\boldsymbol{u}}^i_{t|t} = \frac{g_{i,t}\hat{\boldsymbol{u}}^i_{t|t-1}}{f_{t|t-1}} \tag{11.9c}$$

Defining the matrix $\boldsymbol{g}_t$ such that diagonal elements equal $g_{i,t}$, and zero otherwise, (11.9c) and (11.9b) can be written in matrix-vector notation as

$$f_{t|t-1} = \hat{\boldsymbol{u}}_{t|t-1}\boldsymbol{g}_t\mathbf{1} \tag{11.10a}$$

$$\hat{\boldsymbol{u}}_{t|t} = \frac{\hat{\boldsymbol{u}}_{t|t-1}\boldsymbol{g}_t}{f_{t|t-1}}, \tag{11.10b}$$

which defines the needed recursions. The initial state distribution ($\hat{\boldsymbol{u}}_{0|0}$) has to be defined, either as a known vector or as the stationary distribution (see Theorem 9.1) defined by $\boldsymbol{\Gamma}$.

Example 11.2 (Simple Gaussian HMM) *Consider the HMM*

$$P(\boldsymbol{X}_t = i|\boldsymbol{u}_t) = \boldsymbol{u}_t^i \tag{11.11a}$$

$$\boldsymbol{u}_t = \boldsymbol{u}_{t-1}\boldsymbol{\Gamma} \tag{11.11b}$$

$$Y_t|X_t = i \sim N(\mu_i, \sigma_i^2). \tag{11.11c}$$

In this case, we have $g_{i,t} = \phi_i(y_t)$*, where* ϕ_i *is a normal probability density function with mean* μ_i *and variance* σ_i^2*, and*

$$\boldsymbol{g}_t = \begin{bmatrix} \phi_1(y_t) & \cdots & 0 \\ \vdots & \ddots & \vdots \\ 0 & \cdots & \phi_m(y_t) \end{bmatrix}. \tag{11.12}$$

Using the results stated above, the recursive filter can be set up using e.g. the stationary distribution for $\hat{\boldsymbol{u}}_{0|0}$.

To summarise, the algorithm can be formulated as the following steps:

1: Set $t = 0$ and initiallise $\hat{\boldsymbol{u}}_{0|0}$, e.g. to $\hat{\boldsymbol{u}}_{0|0} = \mathbf{1}^T(\boldsymbol{I} - \boldsymbol{\Gamma} + \boldsymbol{E})^{-1}$ (see Theorem 9.1).

2: Set $t := t + 1$ and calculate $\hat{\boldsymbol{u}}_{t|t-1}$ using (11.9a).

3: Calculate $f_{t|t-1}$ using (11.9b).

4: Calculate $\hat{\boldsymbol{u}}_{t|t}$ using (11.9c).

5: Repeat step 2–4 until $t = n$ (end of time series).

Even though we generalise the filter throughout this chapter, it constitutes the basis of all the development in this chapter, including the likelihood that can be calculated directly from $f_{1|0}, \ldots, f_{n|n-1}$.

The function below implements the filter for time invariant ($\boldsymbol{\Gamma}$ is a constant matrix) HMMs as defined above, using the stationary distribution (see Chapter 9) to initialise the filter. The matrices `Gamma` and `g` are defined by the model definitions with the rows of `g` being the diagonal elements of $\boldsymbol{g}_t$. The output of the function is all values of 1-step ahead state predictive distribution (`Ut`), reconstructed state probabilities (`Utt`), and 1-step ahead density of the observations (`ft`).

```
HMM.filter <- function(Gamma, g){
    ## Initialize matrices and vectors
    m <- dim(Gamma)[1]
    n <- dim(g)[1]
```

```
    Ut <- matrix(ncol = m, nrow = n)
    Utt <- matrix(ncol = m, nrow = n)
    ft <-numeric(n)

    ## Inital state and distribution
    Ut[1, ] <- stat.dist(Gamma)
    ft[1] <- sum(Ut[1, ] %*% diag(g[1, ]))
    Utt[1, ] <- Ut[1, ] %*% diag(g[1, ]) / ft[1]

    ## Recursive filtering
    for(i in 2:n){
        Ut[i, ] <- Utt[i-1, ] %*% Gamma
        v <- Ut[i, ] %*% diag(g[i, ])
        ft[i] <- sum(v)
        Utt[i, ] <- v / ft[i]
    }
    ## Return state predictions and reconstruction
    return(list(Ut = Ut, Utt = Utt, ft = ft))
}
```

Before introducing likelihood estimation, we present the data used for this section and apply the filter directly without parameter estimation.

Example 11.3 (Data and filter) *In this chapter we use the same data as we used in Example 8.1, and other examples of Chapter 8. But rather than formulating an ARIMAX model for* CO_2*, we will use a HMM to distinguish between different situations. In Chapter 8 the window position was used as an input to the model, implying that the model is valid if the future window position is known. Using a HMM, we formulate a state dependent model where the probability distribution is state dependent. A simple model is the one given in Example 11.2. The data is shown in the figure below.*

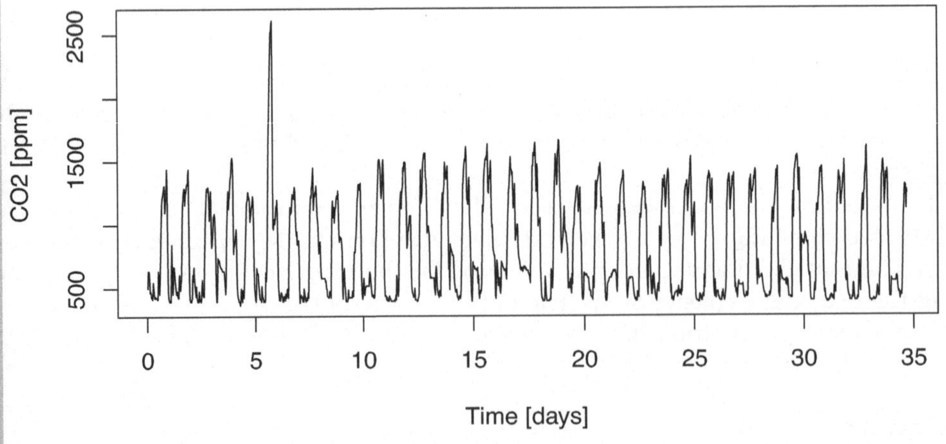

As a simple model, we consider

$$\mathbf{\Gamma} = \begin{bmatrix} 0.95 & 0.05 & 0 \\ 0.05 & 0.9 & 0.05 \\ 0 & 0.05 & 0.95 \end{bmatrix}. \tag{11.13}$$

The transition probability matrix is quite arbitrary, but reflects that X_t can transition from state 1 to 2, from 2 to either 1 or 3, and from 3 to 2 in 1 time-step. This is specified in R by;

```
Gamma <- matrix(0, 3, 3)
Gamma[1, ] <- c(0.95, 0.05, 0)
Gamma[2, ] <- c(0.05, 0.9, 0.05)
Gamma[3, ] <- c(0, 0.05, 0.95)
```

Further, we need the matrix of observed condition densities (g_{it}). This is implemented in the function `normal.obs` *given below*

```
normal.obs <- function(mu, sigma, y){
    n <-  length(y)
    m <- length(mu)
    g <- matrix(NA, ncol = m, nrow = n)
    for(i in 1:m){
        g[ ,i] <- dnorm(y, mean = mu[i], sd = sigma[i])
    }
    g
}
```

The mean is (again quite arbitrary) set as quartiles of the observed CO_2 concentration and standard deviations is set at 150ppm;

```
mu <- quantile(y, probs = c(0.25, 0.5, 0.75))
sigma <- rep(150, 3)
```

With the above, we can find the state dependent distribution and the state estimates;

```
g <- normal.obs(mu,sigma,y)
filt <- HMM.filter(Gamma, g)
```

The figure below shows the filtered probability mass function (black indicates probability of being in state 1, dark grey indicates probability of being in state 2 and light grey indicates probability of being in state 3).

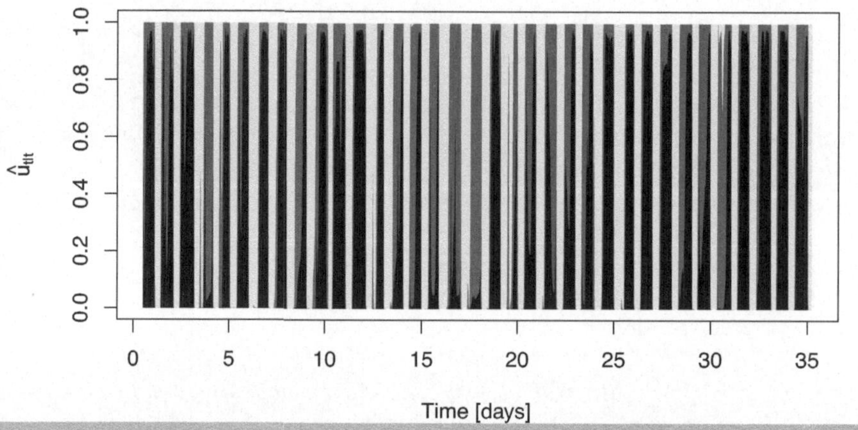

We see a clear diurnal variation in the filtered probabilities. This is the same pattern as seen in the data itself. Of course, we should note that the parameters are not inferred from data, but rather chosen as some quite arbitrary values.

The example above illustrates how the state can be inferred based on a model definition, and hence can be used for identifying the most probable states over time (given the model). In our example, this would refer to periods of high or low CO_2 concentration (i.e. high or low activity). For estimation of parameters, we also need the likelihood, as will be discussed in the next section.

11.2 Maximum likelihood estimation

The likelihood function can be constructed directly from the results in the previous section as

$$L(\boldsymbol{\theta}; \mathcal{Y}_n) = \prod_{t=1}^{n} f_{t|t-1}(\boldsymbol{\theta}), \tag{11.14}$$

where we have emphasised that $f_{t|t-1}$ is a function of the parameters $\boldsymbol{\theta}$. This dependence goes through the state dependent distribution $\boldsymbol{g}_t$ and the transition probability matrix $\boldsymbol{\Gamma}$. Hence, the densities are obtained directly as a result of the filtering algorithm. As usual, we use the log-likelihood function

$$l(\boldsymbol{\theta}) = \sum_{t=1}^{n} \log(f_{t|t-1}(\boldsymbol{\theta})), \tag{11.15}$$

for parameter estimation, and the maximum likelihood estimates (MLEs) are given by

$$\hat{\boldsymbol{\theta}} = \arg\min_{\boldsymbol{\theta}} -l(\boldsymbol{\theta}). \tag{11.16}$$

The parameter vector $\boldsymbol{\theta}$ contains parameters related to the transition probability matrix, $\boldsymbol{\Gamma}$, and parameters related to the state dependent distributions. The number of parameters for the transition probability matrix is $m(m-1)$, in the general case where all elements of the transition probability matrix are estimated. The number of parameters for the state dependent distribution is dependent on the chosen distribution, but if we choose the normal distribution, as in Example 11.2, it will be $2m$ parameter.

The likelihood is implemented in the R-function below. Besides the parameters, the function takes the observations (`y`), the number of states (`m`) and a function **`g.fun`** as input, and further it is possible to get the filtered values and parameter list as output (by setting **`full = TRUE`**);

```
HMM.lli <- function(pars, y, ms, g.fun, full = FALSE){
    Gamma <- Markov.Inv.Link(pars[1 : (ms * (ms - 1))], ms)
    g.list <- g.fun(pars[- (1 : (ms * (ms - 1)))], y, ms)
    filt <- HMM.filter(Gamma, g.list$g)
    if(full){
        return(list(Gamma = Gamma, g.list = g.list, filt = filt))
    }
    - sum(log(filt$ft))
}
```

The R-function **`g.fun`** should return a list where one element is a matrix (**`g`**) with all observed state dependent densities/probabilities. Further, other outputs from the function are allowed (e.g. state dependent parameters). The function can be passed to a general purpose optimizer (e.g. **`nlminb`**).

Parameter estimation is usually done in a working domain, e.g. variances are usually estimated in the log-domain and for the transition probability matrix the logit transformation defined in Chapter 9 (Section 9.4) will be used (included in the function **`Markov.Inv.Link`**, see page 265).

Example 11.4 (Example 11.3 continued) *We continue with the model we defined in Example 11.3, but now estimate the parameters, using the model*

$$Y_t | X_t = i \sim N(\mu_i, \sigma_i^2). \tag{11.17}$$

The state-dependent parameters are μ_i and σ_i^2, except we define one state as the background CO_2. Implying that we fix the mean value given state 1 to the background level ($\mu_1 = 400ppm$). With a 3 state model (and 1 mean fixed) we get 11 parameters (6 $\boldsymbol{\Gamma}$-parameters + 2 mean value parameters+ 3 variance parameters).

The state-dependent distribution is implemented below;

```
g.fun1 <- function(pars, y, ms){
    ## Initialization of state depedent density and distribution
    n <- length(y)
    g <- matrix(NA, ncol = ms, nrow = n)
    G <- matrix(NA, ncol = ms, nrow = n)

    ## Parameters for state depedent distributions
    mu <- c(400, pars[1 : (ms - 1)])
    sigma2 <- exp(pars[ -c(1 : (ms - 1))])
    sigma <- sqrt(sigma2)

    ## Calculation of state depedent density and distribution
    for(i in 1 : ms){
        g[ ,i] <- dnorm(y, mean = mu[i] , sd = sigma[i])
        G[ ,i] <- pnorm(y, mean = mu[i] , sd = sigma[i])
    }
    ## Return results
    return(list(g = g, G = G, mu = mu, sigma = sigma))
}
```

`pars` *are the parameters related to the state dependent distribution and variances are estimated in the log-domain. In addition to the state-dependent densities (*`g`*), the function also returns the parameters for the state dependent distribution (*`mu` *and* `sigma`*) and the state dependent distribution function (*`G`*), which will be used for the residuals analysis later on in this chapter (Section 11.5).*

In order to set up the estimation, initial values of the parameters have to be defined for the transition probability matrix. Zucchini and MacDonald (2009) suggest to initialise with diagonal elements close to 1, and then distribute the probabilities equally among the other element. Here we use

```
gamma.pars <- rep(-3,6)
```

corresponding to an initial guess of the transition probability matrix;

```
Markov.Inv.Link(gamma.pars, 3)

##       [,1]  [,2]  [,3]
## [1,] 0.909 0.045 0.045
## [2,] 0.045 0.909 0.045
## [3,] 0.045 0.045 0.909
```

Initial values for the state-dependent variables are chosen based on observed quantiles of the data, and the variances are chosen close to the observed variance in data. The parameters can now be estimated by;

```
pars <- c(gamma.pars, 800, 1000, rep(10, 3))
opt3s <- nlminb(pars, HMM.lll, y = y, ms = 3, g.fun = g.fun1)
```

And the results are extracted by;

```
res3s <- HMM.ll1(opt3s$par, y, ms = 3, g.fun1, full = TRUE)
```

From which we can extract the state-dependent parameters;

```
## Mean values
  res3s$g.list$mu

## [1]  400  547 1188

## standard deviation
  res3s$g.list$sigma

## [1]  24  85 252

## Coefficient of variation
  res3s$g.list$sigma / res3s$g.list$mu

## [1] 0.059 0.155 0.212
```

Hence, the three mean value parameters are the background CO_2 *concentration (which is fixed), a medium* CO_2 *concentration and a high* CO_2 *concentration. Further, we observe that both the standard deviation and the coefficient of variation are increasing with the mean value. Based on the examples in Chapter 8 we might have expected that the coefficient of variation is constant (since we used the log-normal distribution). We will explore this a bit further later on in this chapter using the gamma distribution.*

An interpretation of state 1 is that either the window is open or the room has been vacated for a long time, while state 3 is the state where there are occupants in the room (and there have been for some time). Finally, state 2 would be a transition state between the other two states.

The transition probability matrix is extracted, and the stationary distribution calculated by;

```
res3s$Gamma

##            [,1] [,2]     [,3]
## [1,] 8.5e-01 0.15 4.8e-51
## [2,] 8.2e-02 0.86 6.3e-02
## [3,] 3.4e-09 0.05 9.5e-01

stat.dist(res3s$Gamma)

##      [,1] [,2] [,3]
## [1,]  0.2 0.35 0.45
```

The function `stat.dist` *is defined in Chapter 9 (page 263). The results support the interpretation above, namely that transition from state 1 to state 3 and from state 3 to state 1 is not possible. With this interpretation, the stationary distribution tells us the room is close to the background* CO_2 *concentration about 20% of the time.*

The natural extension of the 3 state model would be a 4 state model. The estimation procedure is the same, and the number of parameters is now 19 rather than 11, so the models are quite parameter intensive. The log-likelihood of the 3 and 4 state models are compared below;

```
- c(opt3s$objective, opt4s$objective)

## [1] -10602 -10389
```

Hence, we observe an increase in the log-likelihood of around 200 which is significant on any reasonable level (considering the difference in degrees of freedom being 8).
The results from the 4 state model is extracted by ;

```
res4s <- HMM.ll1(opt4s$par,y,m=4,g.fun1,full=TRUE)
```

and state-dependent mean values, standard deviation and coefficient of variation are;

```
##            mu  sd    cv
## State 1  400  22 0.054
## State 2  536  77 0.143
## State 3  890 111 0.125
## State 4 1307 197 0.151
```

From those values, it seems that the transition state is split into two, and further we see that the high concentration state is now even higher. The standard deviation is still increasing with the mean value of the state, but the coefficient of variation is relatively stable across state 2-4.
The transition probability matrix and stationary distribution are given by;

```
res4s$Gamma

##          [,1]    [,2]   [,3]    [,4]
## [1,] 8.5e-01 1.5e-01 0.0028 9.7e-09
## [2,] 8.3e-02 8.5e-01 0.0679 6.8e-09
## [3,] 4.3e-04 1.6e-01 0.6842 1.5e-01
## [4,] 2.6e-08 6.1e-09 0.0723 9.3e-01

stat.dist(res4s$Gamma)

##      [,1] [,2] [,3] [,4]
## [1,] 0.19 0.35 0.15 0.32
```

and indeed we see that transitions can effectively be done is from 1 to 2, from 2 to 1 or 3, from 3 to 2 or 4, and from 4 to 3. Hence, the transitions are made to "neighbouring" states.

11.2.1 Optimisation

In the above we have used direct likelihood optimisation, and we will not cover other ways to optimize the likelihood here. But the Expectation Maximization–algorithm (EM–algorithm) which is often used for HMMs should be mentioned. One of the difficulties of HMMs is that the number of parameters is often quite high and therefore difficult to solve, implying that robust methods may have an advantage. The basic idea of the EM algorithm is to write the full likelihood of all (missing and non-missing) observations, and then alternate between

E-step: For a chosen set of parameters calculate the expected value of the missing observations

M-step: Using the estimated missing values, optimise the parameters and go back to the E-step

The algorithm alternates between the E-step and the M-step until convergence (e.g. small changes in the parameters).

Generally EM–algorithms are slower than the usual gradient method (Cappé et al., 2005) we use here but also more robust as it usually guarantees that the likelihood increase in each step, hence it will converge to some maximum (which may be local). For HMM–models, the E-step is what we refer to as a state estimation and the transition probability matrix is inferred from that. Finally, the parameters are the parameters of the state dependent distribution. In the case of the Normal HMM, we consider here, there is a closed form solution to the M-step, but that relies on the distribution assumption. For further details on implementation, see Zucchini and MacDonald (2009).

Even with the quite simple models that we consider in this chapter, care should be taken in choosing the initial values of the parameter, and hence the EM–algorithm may be helpful here also. However, HMMs have quite complicated likelihood functions and multiple maxima might occur. This is not solved by using the EM–algorithm. One practical solution to that is to use a range of initial values for the estimation and choose the one that gives the best likelihood. For models with a continuous outcome (like the ones we use in our examples), the likelihood might also be unbounded for specific parameter values. The problem can occur because densities can in principle take any value from the real axis. A solution is to use a discretized density function (i.e. translate the probability density into a probability mass function), see Zucchini and MacDonald (2009) for further discussion.

11.3 Statistical tests and parameter uncertainties

The statistical methods we have treated so far carry over to HMMs. Hence, statistical testing using likelihood ratio tests can be applied for nested models, and likelihood–based information criteria (AIC, BIC) can be applied for non-nested models. Further, tests on individual parameters can be carried out using either profile likelihood–based confidence intervals or Wald intervals using the observed Fisher information. For the state dependent distribution parameters (in our case μ_i and σ_i^2), uncertainties can be considered marginally, while for the transition probability matrix, natural parameters are functions of several working parameters (see Section 9.4) and that have to take into account when calculating uncertainties of the transition probability matrix. Some bootstrap methods for that are discussed in Chapter 9.

In this chapter we will rely only on likelihood–based information criteria to compare non-nested models, and likelihood ratio test to compare nested models. For parameter uncertainty we use methods related directly to the observed Fisher Information (i.e. Wald statistics). This implies that the uncertainty of transition probabilities are difficult to interpret (we can say something based on magnitudes), and there is no direct access to uncertainty of, for example the stationary distribution. If such assessments are important, this can be obtained either using bootstrap based in the observed Fisher Information (see Chapter 9), parametric bootstrap (also covered in Chapter 9), or profile likelihood. The latter methods are either computationally intensive (parametric bootstrap) or both computational intensive and complicated to implement (profile likelihood).

11.4 Adding structures to the models

Many additional structures on the HMMs can be added. We have divided those structures into three different overall structures. Each of which we will discuss below and give some examples either worked through including estimation or by discussing the resulting structures.

11.4.1 Multivariate observations

Multivariate observation is not so much a model structure as it is a feature of the data. Of course, such a feature will give additional parameters in the model. In general, the model can be written as

$$P(\boldsymbol{Y}_t = \boldsymbol{y}_t | X_t = i) = p_i(\boldsymbol{y}_t), \tag{11.18}$$

where $p_i()$ is a probability density/mass function. The formulation is quite general as it includes both multiple measurements of the same hidden state and combinations of different HMMs that are assumed to be connected somehow. We briefly discuss examples of each below.

Example 11.5 (Multiple sensors) *In the analysis so far, we have used one CO_2 sensor, but we could of course imagine that we had more than one sensor or measure different environmental variables (e.g. temperature) as well. In this case we may have a state-dependent distribution given by*

$$\boldsymbol{Y}_t | X_t = i \sim N(\boldsymbol{\mu}_i, \boldsymbol{\Sigma}_i), \tag{11.19}$$

where the structure of $\boldsymbol{\mu}_i$ and $\boldsymbol{\Sigma}_i$ would be part of the model identification, and e.g. if we have two sensors measuring the same thing (e.g. CO_2) it would be reasonable to have $\boldsymbol{\mu}_i = \boldsymbol{1}\mu_i$. But, if we measure CO_2 and temperature it should have different values. Further it would be part of the identification to establish if $\boldsymbol{\Sigma}_i$ should be a diagonal matrix (corresponding to independent measurements) or not, for example if two CO_2 sensors are installed, we would probably expect correlation between the measurements.

Example 11.6 (Multiple rooms) *In Section 10.6.2, we considered models involving two rooms in the same dwelling. A similar model in the framework of HMM models would be*

$$\boldsymbol{Y}_t | X_t = i \sim N(\boldsymbol{\mu}_i, \boldsymbol{\Sigma}_i), \tag{11.20}$$

but where the state vector now refers to the state in each room i.e. (1,1), (1,2), etc. If both rooms have 4 states as we saw above, this would amount to 16 states. For example, state 1 could be low activity in both room 1 and 2, state 2 low activity in room 1 and higher activity in room 2 and so on.

As illustrated by the two examples, the way that data have been obtained could directly suggest model structures, and hence structures of e.g. the transition probability matrix. Of course, the usual model development process still applies and e.g. the number of states may be difficult to infer directly from the way data is obtained.

11.4.2 Structure in the state-dependent distribution

Structures in the state-dependent distribution may involve combinations of different assumptions. This includes consideration regarding the distribution type (e.g. Gaussian, gamma, log-normal), and, for example some sort of autocorrelation structure, and in fact this may also be combined with multivariate observations such that the state-dependent distribution is different for different elements of the multivariate distribution. For the choice of state dependent distribution, we may include basically any distribution we can think of or any transformation of the data, for example the Box-Cox transformation (see Chapter 3), and further the distribution or transformation may even dependent on the state.

Here, we will give two examples of how to generalise the distributions related to the example we have used so far in this chapter.

Example 11.7 (Changing state-dependent distribution) *So far we have assumed that the state dependent distribution is Gaussian. But as CO_2 concentrations are positive, it may be reasonable to use a gamma distribution. The gamma distribution is usually parametrised in terms of a shape parameter (k) and a scale parameter (θ) or a shape and a rate parameter (this is also what is implemented in R). Following Chapter 4, we can parametrise in terms of the mean value and a dispersion parameter. In order to be able to compare directly with Example 11.4, we parametrise by*

$$Y_i|X_t = i \sim G(\mu_i, cv_i^2), \tag{11.21}$$

meaning that the expectation is $E[Y_i|X_t] = \mu_i$ and the coefficient of variation is $cv_i = \frac{\mu_i}{\sqrt{V[Y_i|X_t]}}$. When transforming to shape and scale parameter's we have

$$k_i = \frac{1}{cv_i^2}; \quad \theta_i = \mu_i cv_i^2, \tag{11.22}$$

and hence the state-dependent distribution can be implemented using the standard R implementation of the gamma distribution. With the R-functions we have considered so far, we just have to change the function (`g.fun1`) to include the gamma distribution (only lines that change are included);

```
g.funGamma <- function(pars, y, ms){
## ...
    cv <- exp(pars[(ms : (2 * ms - 1))])
    k <- 1 / cv^2
    theta <- mu  * cv^2

    for(i in 1 : ms){
        g[ ,i] <- dgamma(y, shape = k[i], scale = theta[i])
        G[ ,i] <- pgamma(y, shape = k[i], scale = theta[i])
    }
    return(list(g = g, G = G, mu = mu, cv = cv))
}
```

The optimisation of the likelihood is done in a similar way and also the results in terms of the parameter (mean values, variances and the transition probability matrix), are almost identical (not shown). The log-likelihood;

```
c(Gamma = -opt4sGam$objective, Gaus = -opt4s$objective)

##  Gamma   Gaus
## -10332 -10389
```

does however increase by about 60 units, and as the number of parameters are the same, we should of course choose the gamma distribution.

The state-dependent mean and variance only change slightly between the two models, and the difference is mainly attributed to other features of the state-dependent distribution. We will not present all parameters, but as the distribution is now parametrised through the coefficient of variation, we can directly asses the uncertainty of that using the observed information;

The table below shows estimates of $\log(cv)$ *with standard errors, estimated coefficients of variation, and 95% confidence intervals of coefficient of variation;*

```
##    log.cv se.log.cv   cv cv.lower cv.upper
## 1    -3.0     0.069 0.05    0.044    0.058
## 2    -1.9     0.035 0.15    0.136    0.156
## 3    -2.1     0.064 0.12    0.109    0.141
## 4    -2.0     0.034 0.14    0.128    0.147
```

the confidence intervals indicate that the coefficient of variation could indeed be the same across state 2-4. The figure below shows the state-dependent distribution for the two different models.

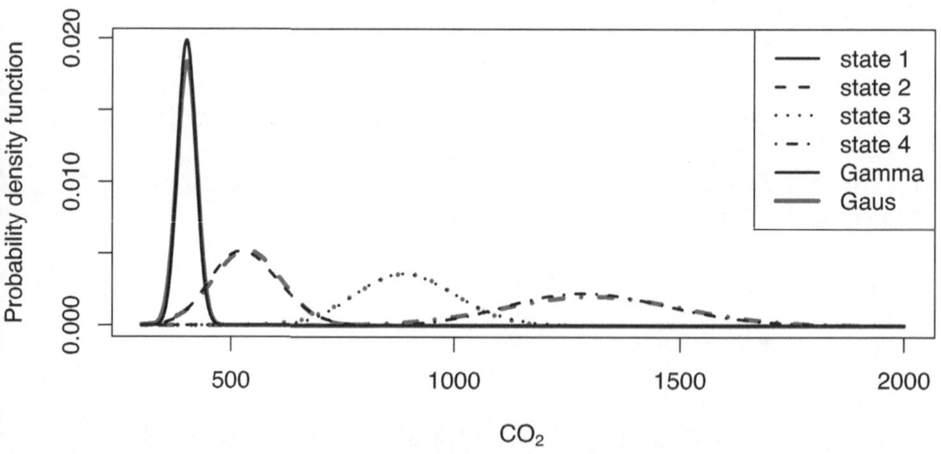

It is quite hard to distinguish the Gaussian and gamma distributions. In fact, a large part (app. 50%) of the improvement in the log-likelihood can be traced back to the high CO_2 *concentrations around day 6 (see Example 11.3), and is a consequence of the gamma distribution having heavier tails than the normal distribution. We omit further*

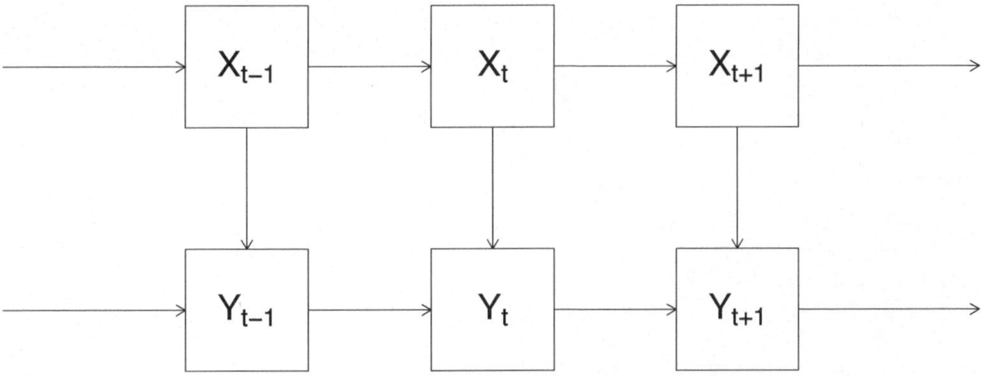

Figure 11.1: Schematic illustration HMM with autocorrelation in the observations.

analysis here, but it can be done by considering the vector **filt$ft**, *which is an output from* `HMM.ll1`.

The example above illustrates that it is easy to include different types of distribution. Further, we do not need to have the same state dependent distribution or transformation across states, so the framework is very flexible. Different transformations correspond to different distributions, as the most obvious example assuming a log-normal distribution corresponds to assuming a normal distribution for log-transformed data.

In Chapter 8 we discussed ARIMAX type of models using the same data as we use in this chapter. These models take care of autocorrelation in data. And in much of the arguments we have done so far this is the same arguments we have used for Markov model (see Example 9.7), but the Markov models and ARIMA-type models can also be combined using HMMs. As discussed in Section 10.6.1 the time constant (AR–parameters) of the system may depend on the state of the system (in Section 10.6.1 this was the window position). Hence we can adapt Figure 10.1 to a situation where we have an underlying discrete state and dependence between observations (Figure 11.1).

Example 11.8 (HMM and AR–models) *We consider the state-dependent distribution*

$$Y_t|(X_t = i, Y_{t-1} = y_{t-1}) = \mu_i + \phi_i(\mu_i - y_{t-1}) + \epsilon_t^i, \tag{11.23}$$

with $\epsilon_t^i \sim N(0, \sigma_i^2)$ white noise and independent of each other.

Again, the density function for the state dependent distribution needs to be updated (only lines that change, compared to `g.fun1` *are included);*

```
g.funAR1 <- function(pars, y, ms){
    ## ...
    phi <- pars[-c(1 : (2 * ms - 1))]
    for(i in 1 : ms){
        mu.t <- mu[i] + phi[i] * (y[-n] - mu[i])
        g[ ,i] <- dnorm(y[-1], mean = mu.t, sd = sigma[i])
        G[ ,i] <- pnorm(y[-1], mean = mu.t, sd = sigma[i])
    }
    return(list(g = g, G = G, mu = mu, sigma = sigma, phi = phi))
}
```

Parameter estimation (finding the MLE) is done exactly as we have seen in Example 11.4. We use the previous model as a starting point for the estimation (last four parameters are $\phi_1, \ldots, \phi_4$);

```
pars <- c(opt4s$par, rep(0, 4))
opt4sAR1 <- nlminb(pars, HMM.ll1, y = y, ms = 4, g.fun = g.funAR1)
```

The log-likelihood of the models considered so far are compared with an AR(2) model (based on findings in Chapter 8) below;

```
## HMM-Gaus  HMM-Gam  HMM-AR1        AR2
##   -10389   -10332    -9354   -10207
```

Here, we see a huge improvement using state dependent AR(1) models. It should be noted that formally the likelihoods are not directly comparable, as first two likelihoods are calculated using all data, while `HMM-AR1` *condition on the first data point and* `arima` *again use all data by back-casting y_{-1} and y_0. The differences are however so large that we can disregard this difference in the formal methods. There is also a large difference in number of parameters between the models:* `HMM-Gaus` *and* `HMM-Gam` *each have 19 parameter,* `HMM-AR1` *have 23 parameters, and* `AR2` *have 3 parameters. However, again the difference in likelihood is so large that this will not influence the statistical conclusions.*

Parameters for the state dependent distributions are given in the table below;

```
##          mu sigma  phi
## [1,]   400    13 0.97
## [2,]   453    42 0.30
## [3,]   760   108 0.62
## [4,]  1590    72 0.85
```

Here we see some notable changes: there are now 2 states with very low mean values (one is fixed at 400), but with very different AR(1) parameters (0.97 and 0.3), that is one state (state 1) with a high time constant and one (state 2) with a low time constant. An interpretation is that state 1 corresponds to an empty room with closed windows and state 2 corresponds to a room with open windows. In addition. there is a state with medium CO_2 concentration and quite low time constant and finally a state with high CO_2-concentration and a higher time constant.

The table also shows that there are quite different variances in the 4 states. These variances are the 1-step ahead variance. Another way to look at the variance is the stationary variance in each state. These are (given as standard deviations);

```
sqrt(res4sAR1$g.list$sigma^2 / (1 - res4sAR1$g.list$phi^2))

## [1]  58  44 138 135
```

and as we can see, these are much more comparable between the states (low stationary variance when mean value low and high variance when mean value is high).

Finally, the transition probability matrix and stationary distribution is given as;

```
res4sAR1$Gamma

##          [,1]    [,2]  [,3]    [,4]
## [1,] 8.2e-01 0.0946 0.087 9.3e-09
## [2,] 1.8e-01 0.6875 0.136 5.5e-09
## [3,] 9.6e-02 0.2104 0.414 2.8e-01
## [4,] 2.3e-08 0.0049 0.133 8.6e-01

stat.dist(res4sAR1$Gamma)

##      [,1] [,2] [,3] [,4]
## [1,] 0.29 0.21 0.17 0.34
```

Also here, we see some differences. Qualitatively we see that more transitions are allowed. Transitions from 1 can now be made to both states 2 and 3, and also from state 3 to any other state. Transitions between 1 and 2 (both directions) are now much more likely (probably indicating that it is hard to distinguish the two), and the probability of transition from state 3 to 3 is now very low (0.41) (hence the system will leave the state fast after a visit). The stationary probability distribution shows that most of the time is spend in state 4, and further, that about 50% of the time is spend in state 3 or 4. If the interpretation holds, these are the states where the window is closed and there is presence in the room. Finally 50% of the time is spend in state 1 or 2 (i.e. the room is empty and/or the window open).

Here we have seen two examples that aim at illustrating that we can both work with the distribution assumption and the innovation between observations. These two cases can of course be combined in basically endless ways to describe the structures present in data. Of course, such structures should be supported by data and the appropriate evaluation schemes to support the analysis. Such considerations may be important and many models are supported by the setup here. The last part of the model is the transition probability matrix which we discuss in the next section.

11.4.3 Structures in the transition probability matrix

The last element in the HMMs is the transition probability matrix (TPM), that was covered in detail in Chapter 9. The structures that were covered in Chapter 9 are also relevant here, e.g. higher order Markov models where the transitions do not only dependent on the present but also some lags into the past (see Section 9.5).

Also time varying transition probabilities may be relevant here. Such time dependence may be through some regression variables (see Section 9.6).

The additional structures are relatively easy to include in the likelihood functions we have formulated so far. An example of including covariates is given below.

Example 11.9 (Including covariates) *We continue with the previous example, and include time of day in the TPM. As we have so far interpreted the states as activity in the room, it seems reasonable to include time of day here.*

In order to do this, we use sine and cosine of time of day as covariate;

```
time <- win1.agr$HalfHour
X <- cbind(cos(2 * pi * time / 48), sin(2 * pi * time / 48))
```

Assuming that the TPM is given as an array, we can define the function for filtering as (only lines that differ from `HMM.filter` *are shown);*

```
HMM.filterGamma.t <- function(Gamma, g){
    ## ...
    Ut[1, ] <- stat.dist(Gamma[1, , ])
    ## ...

    for(i in 2:n){
        Ut[i, ] <- Utt[i-1, ] %*% Gamma[i, , ]
        ## ...
    }
    return(list(Ut = Ut, Utt = Utt, ft = ft))
}
```

In order to include a time-dependent TMP, we use the function `Markov.Inv.Link.Reg` *defined in Chapter 9 (page 276), and further update the negative log-likelihood function*

```
HMM.llGamReg <- function(pars, y, ms, g.fun,  X.reg,
                         full = FALSE){
    p <- dim(X.reg)[2]
    n.parGam <- (ms * (ms - 1) + 4 * p)
    Gamma <- Markov.Inv.Link.Reg(pars[1 : n.parGam], ms, X.reg)
    g.list <- g.fun(pars[- (1 : n.parGam)], y, ms)
    filt <- HMM.filterGamma.t(Gamma, g = g.list$g)
    if(full){
        return(list(Gamma=Gamma, g.list = g.list, filt = filt))
    }
    - sum(log(filt$ft))
}
```

The likelihood can now be optimized to find the MLE by (first 8 parameters are related to regression parameters in the TPM);

```
pars <- c(rep(0, 8), opt4sAR1$par)
opt4sAR1.Gx <- nlminb(pars, HMM.llGamReg, y = y, ms = 4,
                      X.reg = X, g.fun = g.funAR1)
```

In order to verify that there is a significant improvement of the model, the log-likelihood of the models can be compared in a likelihood table

```
##   df   ll Dll    p.v
## 1 31 9293  62 7e-23
## 2 23 9354  NA     NA
```

Hence a highly significant improvement in the likelihood is observed.

The figure below shows the first row of the TMP as a function of hour of day. There are a few things to note on this. First, the transition from state 1 to 4 is never possible, second the probability of staying in state 1 is small during night and high during day.

The interpretation is that state 1 is an empty room (with window closed), and since it is a bedroom, this interpretation seems reasonable.

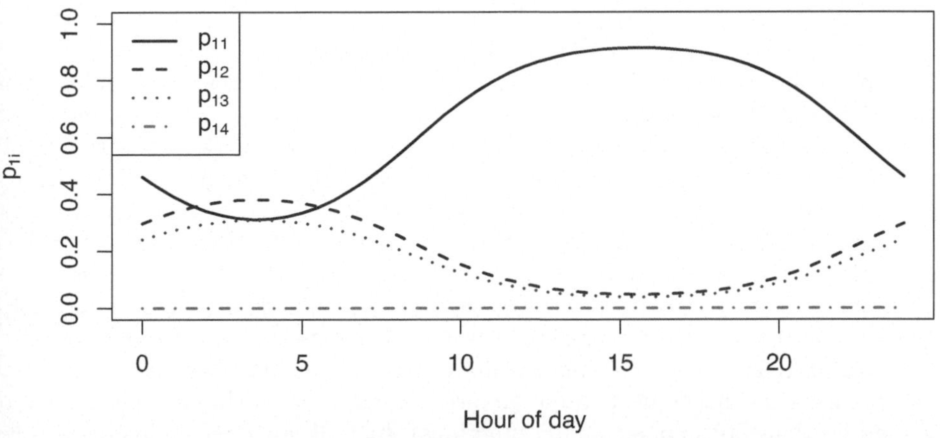

We can also note that the transition from state 1 to 2 is always higher than the transition from state 1 to 3. This may or may not be reasonable, but the model formulation actually means that those probabilities cannot cross, and hence it is a feature of the model. A more general model that allows such crossings would include covariates in more of the diagonal elements (hence more parameters).

In addition to analysing the resulting model (the figure above), we can also look at the filtered states. The figure below shows average filtered states as a function of time of day, and e.g. on average, the probability of being in state 4 is higher during night and morning, while in the morning there is a shift towards state 3 and then 2, and finally state 1, which is the dominant state during the day. Also it is interesting to see the resemblance between state 2 and the window opening. In the figure below, the average of the window position at every hour of the day is given. Even though these are not perfectly aligned, the shift towards state 2 follows the window opening quite well, in particular in the morning.

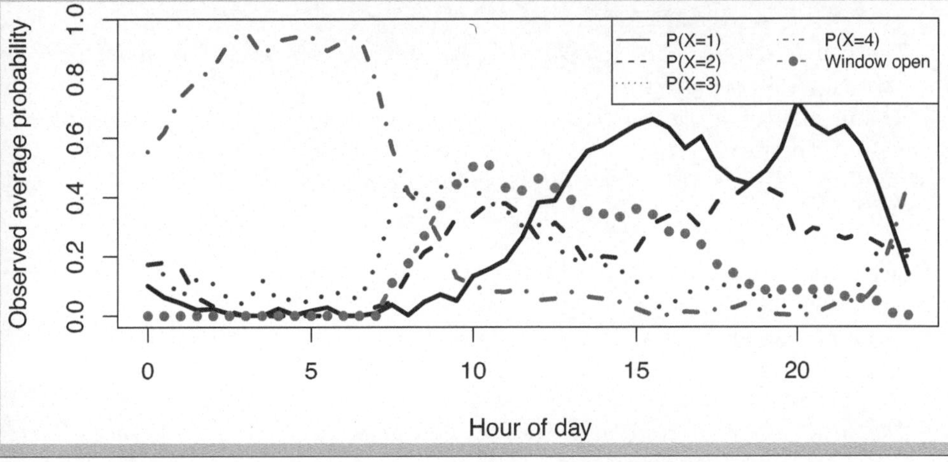

The example above illustrates how we can work with the TPM and have it as part of the modelling. The example focussed on modelling the transition probabilities as a function of time of day, but of course other effects (like indoor or outdoor temperature could be included). Also, the model is in a way restrictive as only the diagonal elements of the working domain parameters are a function of time (see Section 9.6). In a more general setting, we could follow the formalism in (9.46) and set

$$\gamma_{ij}(\boldsymbol{x}_t) = \begin{cases} \frac{e^{\tau_{ij}(\boldsymbol{x}_t)}}{1+\sum_{j\neq i} e^{\tau_{ij}(\boldsymbol{x}_t)}} & j \neq i \\ \frac{1}{1+\sum_{j\neq i} e^{\tau_{ij}(\boldsymbol{x}_t)}} & j = i, \end{cases} \tag{11.24}$$

with $\tau_{ij}(\boldsymbol{x}_t) = \boldsymbol{x}_t\boldsymbol{\beta}_{ij}$. This will give a much more flexible model that allows different ordering of transitions out of the state as a function of time. For example, in the example above the transition probability from state 1 to 2 is always higher than the transition probability from 1 to 3. Further following the formalism of Section 9.5, higher order Markov chains could be included in the HMM. More states can of course also be included in the Markov Chains. The cost is that the number of parameters grow quite fast; for example the 4 state model with no covariates requires 12 parameters for the TPM, the model in the example above requires 20 parameters, and using the model outlined in (11.24) would require 36 parameters for the TPM. Of course, the usual comments on model reduction also apply here and again referring to the example above we saw that the transition from state 1 to 4 is always zero and hence no parameters are needed for that (implementing this requires some modifications of the presented R-functions). In addition, specific hypotheses could, of course, be tested, e.g. in our example we interpret state 1 as an empty room and state 2 as the window being open, and hence (at least in an infinitesimal setting) there can be no direct transition between the two, which could again be included in the model.

In the discussion of the state-dependent distributions we saw examples of changing the distribution assumption and including AR-terms in state dependent models. Here, one could also include regression parameters, again with the cost of increasing the number of parameters, and probably also changing the interpretation of the model.

We finalise this section by a statistical analysis of the parameter uncertainties of the model parameters of the most complex model that we analysed.

Example 11.10 (Parameter Uncertainty) *The most complex model we have analysed here has 31 parameters, and the full Hessian matrix can be obtained by;*

```
H <- hessian(HMM.llGamReg, opt4sAR1.Gx$par,  y = y,
             ms = 4, X.reg = X, g.fun = g.funAR1)
```

The standard errors are calculated by the square root on the diagonal elements of the inverse Hessian matrix.

As the number of parameters is very high, we will divide them into groups when discussing the uncertainties. The first part are the mean values of the state dependent distribution. The estimates and uncertainties are given by;

```
##        MLE   se
## mu1   400   NA
## mu2   459  5.9
## mu3   778 31.9
## mu4  1566 39.9
```

The mean value in state 1 is fixed at 400 ppm, and hence no standard error on that. For the mean value parameters all are well defined with small standard errors. The variance of the state dependent distribution is calculated in the log-domain. The results are given in the table below;

```
##           MLE    se
## lsig.sq1 5.2 0.108
## lsig.sq2 7.6 0.143
## lsig.sq3 9.3 0.115
## lsig.sq4 8.7 0.095
```

which are again well defined and separated. The final part of the state-dependent distribution are the AR(1) parameters;

```
##       MLE     se
## phi1 0.97 0.0047
## phi2 0.30 0.0141
## phi3 0.61 0.0292
## phi4 0.84 0.0135
```

Again we see a very clear separation. For the TPM there are two groups of parameters, those related to the variation in time ($\boldsymbol{\beta}$, see (9.66)*, page 276) and those related to the level (τ_{ij}, see* (9.67)*, page 276). The parameters related to variation in time are;*

```
##             par.beta se.beta
## beta11.cos     -0.93    0.20
## beta11.sin     -1.28    0.37
## beta22.cos     -0.65    0.24
## beta22.sin     -0.29    0.28
## beta33.cos     -2.63    0.99
## beta33.sin      1.72    0.85
## beta44.cos      2.52    0.44
## beta44.sin      1.68    0.57
```

Even though the standard errors are in general larger, most are still well defined and away from 0. The exception are the parameters related to state 3, where only the sine part is significantly different from 0. For the τ parameters, related to the level, the MLE and standard error is;

```
##          MLE      se
## tau12  -1.38    0.29
## tau13  -1.59    0.36
## tau14 -18.29     NaN
## tau21  -1.09    0.22
## tau23  -1.39    0.25
## tau24 -18.63     NaN
## tau31   0.76    1.07
## tau32   1.61    1.05
## tau34   2.00    1.08
## tau41 -17.42 1428.60
## tau42  -3.56    0.83
## tau43  -0.31    0.48
```

Here we notice two NA's and one very large standard error. These are all related to transitions that are close to 0, and hence the likelihood is (almost) flat around these parameter values, and a good option would probably be to fix the related transition probabilities to 0. Finally the parameters related to transition out of state 3 are quite uncertain, even though it seems that these may all be 0. That is not appealing as a null hypothesis, and it would probably be more relevant to consider other simplifications (e.g. removing state 3).

As a final remark on the Fisher information matrix, this also holds information about which parameters have a large correlation. Based on the above, parameters related to the transition from state 3 seems to be candidates for simplifications of the model. The correlation matrix related to those is;

```
##            beta33.1 beta33.2 tau31 tau32 tau34
## beta33.1       1.00    -0.61 -0.87 -0.89 -0.88
## beta33.2      -0.61     1.00  0.82  0.83  0.84
## tau31         -0.87     0.82  1.00  0.96  0.96
## tau32         -0.89     0.83  0.96  1.00  0.98
## tau34         -0.88     0.84  0.96  0.98  1.00
```

In general, the correlations are strong and some are very high (above 0.95), which again points to over-parametrisation of this part of the model.

As an overall conclusion, the model should probably be simplified in some way. In particular, there are transitions that are estimated at 0, and further there are parts of the TPM that could be simplified. On the other hand, the discussion also points to other parts of the TPM that are too simplistic. We will not go further with this development here, but merely leave it as comments to the model.

In this section we have discussed the model and parameters using likelihood based methods (quadratic approximation) for identifying over-parametrisation and pointed at examples of model development based on those. We have omitted more advanced methods for evaluating the uncertainty (e.g. profile likelihood), primarily because of the computational complexity that makes it difficult to use these in practice. However, as we have seen in previous chapters, the residuals also hold valuable information of model deficiencies, and we will discuss that in the next section.

11.5 Diagnostics

We saw in Chapter 4 that the response residuals are not very useful for model diagnostic in generalized linear models, and the situation in HMMs is basically the same. Namely, the residuals will follow different distributions depending on which state the system is in.

In HMMs, pseudo–residuals are often used to analyse the residuals. The general idea of pseudo residuals is that if a random variable Y has the distribution function F then the random variable $F(Y)$ follows a uniform distribution, see for example Zucchini and MacDonald (2009). For a continuous random variable, pseudo–residuals may be defined as

$$r^{ps} = \Phi^{-1}(F(y)), \tag{11.25}$$

where $\Phi(x)$ is the distribution function of a standard normal distribution.

In the case of HMMs, the distribution function for a given state vector $\boldsymbol{u}_t$ can be calculated as

$$F(y_t) = \sum_{i=1}^{m} \boldsymbol{u}_t^i F_i(y_t), \tag{11.26}$$

and the pseudo residuals are constructed as explained above. The state vector may be replaced by $\hat{\boldsymbol{u}}_{t|t}$ or $\hat{\boldsymbol{u}}_{t|t-1}$, the latter is called the forecast pseudo-residuals. And further $\boldsymbol{u}_t$ may also be replaced by any smoothed version of the state. Here we will use the forecast pseudo residuals, as that implies that the residuals should be independent given that the model is correct.

The pseudo residuals of the models we have considered in this chapter are compared in Figure 11.2 and for reference the standardized residuals of an AR(2) model is included. The development through the models is quite clear. In the first two models, the state space is discrete and there is no autoregressive part in the state-dependent model. In both cases there are some extreme residuals (related to day 6) and a strong autocorrelation. Including an AR(1) term in the state-dependent model reduces the lag-1 autocorrelation and also the extreme residuals are taken care of. At the same time, the seasonal part of the autocorrelation function remains. Furthermore, including dependence in the TPM reduces the peak in the auto correlation function around 24 hours. Finally, the AR(2) model basically removes the lower lag autocorrelation, but the peak around 24 hours remains and further a few extreme values in the residuals also remain.

Overall, the residual analysis in Figure 11.2 suggests that the developed model should actually include an additional lag in the AR model or further structures in the TPM. We will not pursue this further here, but refer to the discussion above, and further refer to more in depth discussion of residual analysis in Chapter 8, which can also be applied to the pseudo residual used in this section.

11.6 Prediction

Having identified and estimated a model, we may want to use it for predictions. Such predictions should include the predictive distribution somehow. This implies that we should be able to predict the state distribution and the distribution of the outcome. In order to do so, we need the distribution of the current state; fortunately, this part is already developed during the likelihood estimation.

Prediction of future outcomes is done using the current state estimate $\hat{\boldsymbol{u}}_{t|t}$ and then doing a forward recursion using the TMP. Hence, the k-step prediction of the state distribution is

$$\hat{\boldsymbol{u}}_{t+k|t} = \hat{\boldsymbol{u}}_{t|t} \prod_{i=1}^{k} \boldsymbol{\Gamma}_{t+i}, \tag{11.27}$$

in the case of a homogeneous Markov chain, this can be reduced to

$$\hat{\boldsymbol{u}}_{t+k|t} = \hat{\boldsymbol{u}}_{t|t} \boldsymbol{\Gamma}^k, \tag{11.28}$$

and in the case where the state-dependent distribution is independent of the past, the predictive distribution of the outcome is

$$f_{t+k|t}(y) = \sum_{i=1}^{m} \hat{\boldsymbol{u}}_{t+k|t}^i f_i(y|X_{t+k} = i). \tag{11.29}$$

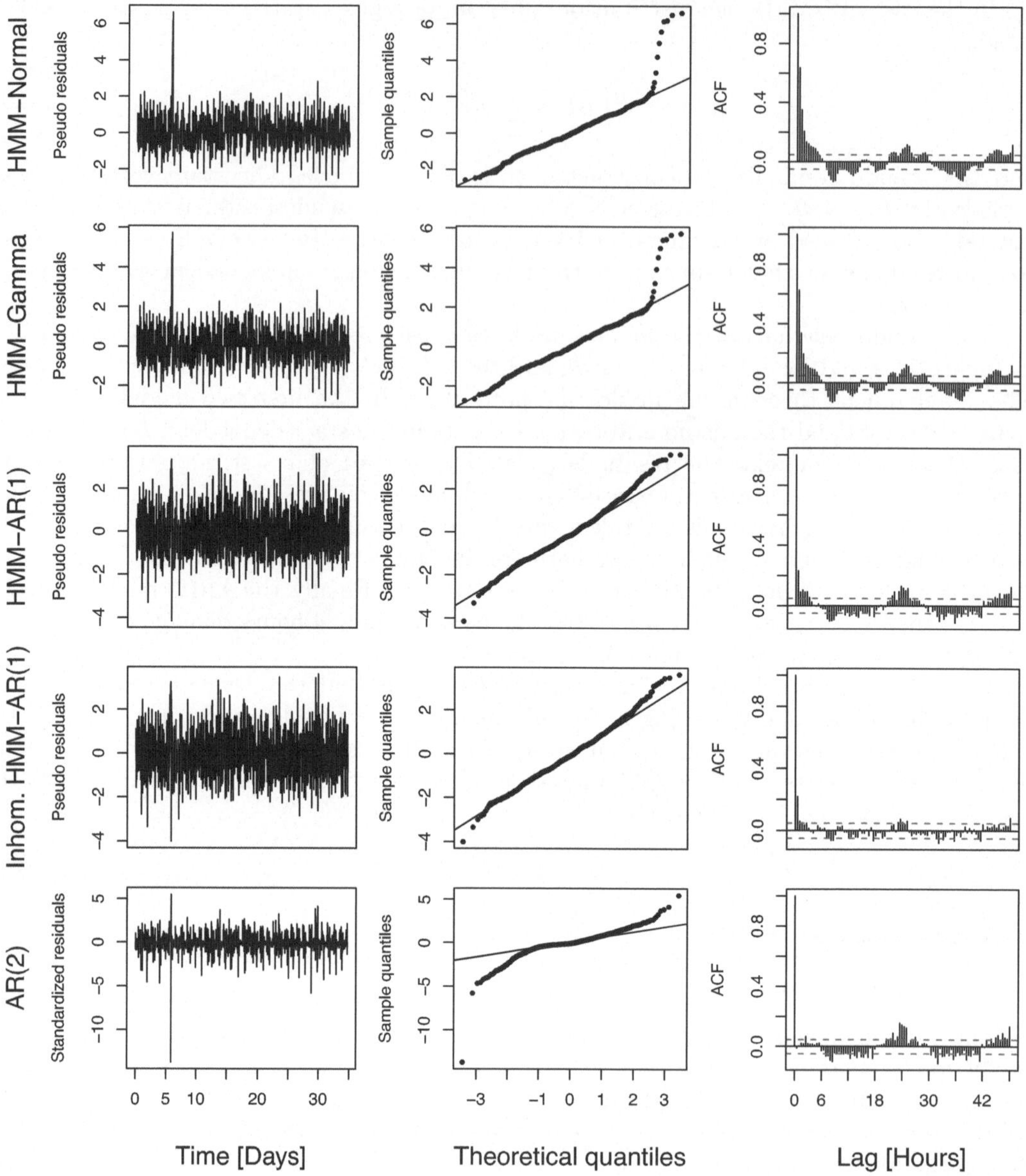

Figure 11.2: Pseudo residuals and assessment of their distribution, and autocorrelation of the residual of models considered in this chapter, for reference the residuals for an AR(2) model is also included.

For more complicated structures, it might be more challenging. However, it is always possible to obtain the distribution by simulation. This implies

1. Simulate outcomes from the filtered state distribution $(\hat{\boldsymbol{u}}_{t|t})$, a large number of times
2. Simulate each outcome forward k steps using the Markov chain model
3. Draw random numbers from the state-dependent distribution

The outcome will constitute the predictive distribution.

Example 11.11 (Prediction) *As an example, we simulate the most complex of the models we have discussed. The simulation is done 24 hours ahead of the last observation. The last observed CO_2 concentration, which is taken at 4 AM in the morning is;*

```
(y0 <- rev(y)[1])

## [1] 1240
```

and the corresponding state (filtered probability distribution) is;

```
n <- dim(res4sAR1Gx$filt$Utt)[1]
(u0 <- res4sAR1Gx$filt$Utt[n, ])

## [1] 2.9e-03 1.1e-34 1.7e-02 9.8e-01
```

Hence, the probability that the system is in state 4 is very close to 1. In order to predict CO_2 concentration over the next hours, we need the TPM, which in our case only dependents on time of day, and can hence be calculated by (`t0` *is the time of day at the last observation, that is the start of the prediction);*

```
t <- t0 + 1:49
X <- cbind(cos(2 * pi * t / 48), sin(2 * pi * t / 48))
Gamma <- Markov.Inv.Link.Reg(opt4sAR1.Gx$par[1 : 20], 4, X)
```

one realisation of state can be calculated by

```
x0 <- sample(1:4,prob=u0,size=1)
x <- sim.markovReg(Gamma,x0)
```

where `sim.markovReg` *is presented in Chapter 9 (Example 9.16, page 280) and finally the CO_2 levels can be simulated by*

```
T <- length(t)
y.pred <- numeric(T + 1)
y.pred[1] <- y0
e <- rnorm(T,0,sd = res4sAR1Gx$g.list$sigma[x])
mu <- res4sAR1Gx$g.list$mu
phi <- res4sAR1Gx$g.list$phi
for(i in 2:(T+1)){
    y.pred[i] <- mu[x[i - 1]] + phi[x[i - 1]] *
        (y.pred[i - 1] - mu[x[i - 1]]) + e[i - 1]
}
```

the vector `y.pred` *now holds one realization of the model given the initial distribution. The distribution can be found by simulating the outcome multiple times. The figure below shows histograms of the forward distribution based on 100,000 realisations (the grey line segments indicate mean values for each of the AR(1) models). The first figure (starting in the top-left corner), shows that for short lead time the distribution is* *quite narrow with a centre close to the last observed value of CO_2 concentration. Moving further 8 hours ahead, there is a large shift towards the lower CO_2 concentrations, which continues with a slower drift, and finally is getting back to the original time of day, the distribution is bimodal with a lot of variability.*

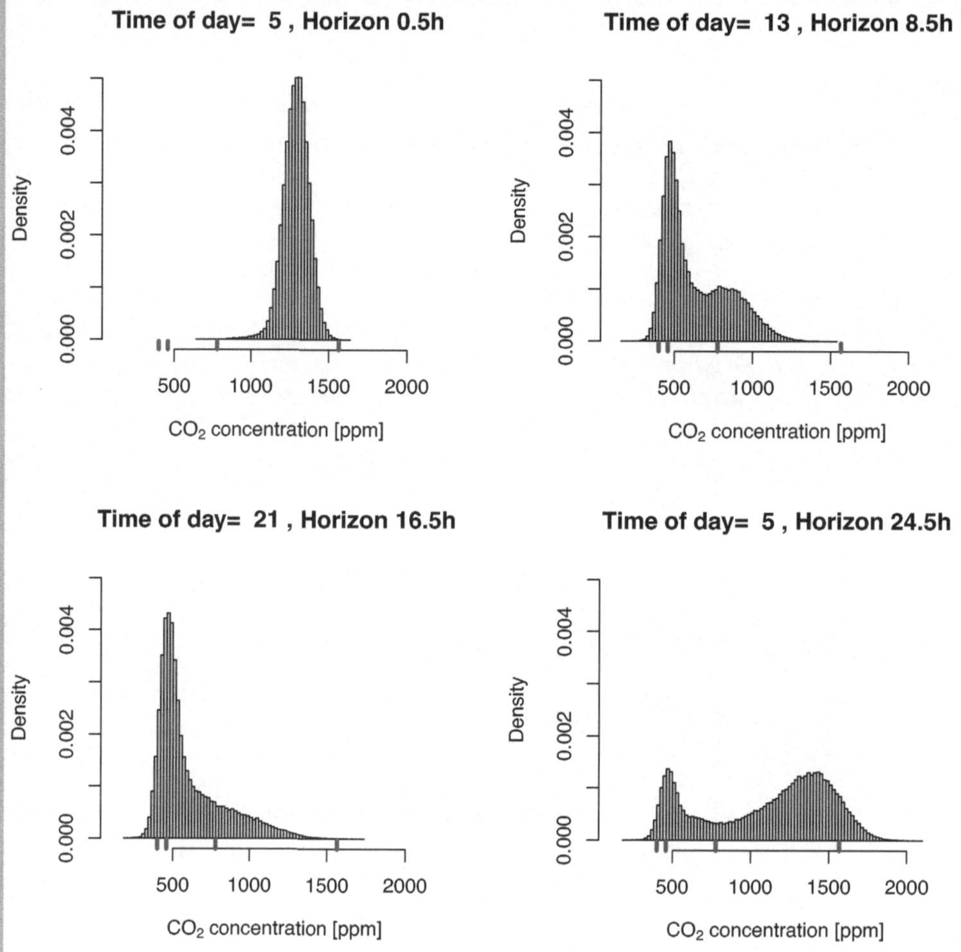

The example shows how to find predictive densities. It does not in any way assess if those predictive distribution are in good agreement with the data. Such assessment should rely on some kind of predictive score, see e.g. Bjerregård et al. (2021) for practical introduction to such scores, that can be used for comparing models.

11.7 Summary and concluding remarks

In this chapter we have discussed discrete time HMMs and some structures that might be considered in the development of such models. Further, we have emphasised the interpretation of the hidden states. The development is largely based on the Markov models discussed in Chapter 9. All examples here are based on continuous observations, but in general, the theory applies for discrete observations as well (e.g. observations of number of occupants in a room).

This presentation being limited to one chapter also implies that there are important parts of HMM theory that are not covered here, for example we base everything on direct likelihood estimation, and neglect the important EM algorithm. Also, smoothing or state decoding is limited to the filtered state and 1-step ahead prediction. If the model should be used for identifying the state of the system, rather than prediction, state decoding/smoothing is important, but the reader is referred to for example Zucchini and MacDonald (2009).

We have relied on custom made R-functions to illustrate the theory of HMMs, but of course there exist some R-packages that deal with HMMs. Examples of such packages are `depMixS4` (Visser and Speekenbrink, 2010), and `msm`, the latter was discussed at the end of Chapter 9, and there are probably other packages out there. However, in order to keep focus on the element of HHM models, we have chosen to base the development on the custom made R-functions presented in this chapter. This also implies that the functions are written to be transparent rather than effective in terms of for example speed of calculation.

Bibliography

Andersen P, Iversen A, Madsen H, Rode C (2014) Dynamic modeling of presence of occupants using inhomogeneous markov chains. *Energy and Buildings* 69:213–223.

Andersen R, Fabi V, Toftum J, Corgnati S, Olesen B (2013) Window opening behaviour modelled from measurements in danish dwellings. *Building and Environment* 69:101–113.

Andersen R (2009) Occupant Behaviour with regard to Control of the Indoor Environment Ph.D. diss., Technical University of Denmark.

Auguie B (2017) *gridExtra: Miscellaneous Functions for "Grid" Graphics* R package version 2.3.

Bates D, Mächler M, Bolker B, Walker S (2015) Fitting linear mixed-effects models using lme4. *Journal of Statistical Software* 67:1–48.

Bingham NH, Fry JM (2010) *Regression - Linear Models in Statistics* Springer-Verlag London.

Bjerregård M, Møller J, Madsen H (2021) An introduction to multivariate probabilistic forecast evaluation. *Energy and AI* 4:1–17.

Box GEP, Cox DR (1964) An analysis of transformations. *Journal of Royal Statistical Society, Series B* 26:211–246.

Box G, Jenkins J (1976) *Time Series Analysis: Forecasting and Control* Holden-Day, San Francisco.

Boyce PR, Veitch JA, Newsham GR, Jones C, Heerwagen J, Myer M, Hunter C (2006) Occupant use of switching and dimming controls in offices. *Lighting Research & Technology* 38:358–376.

Brockhoff P, Møller J, Andersen E, Bacher P, Christiansen L (2015) *Introduction to Statistics - eNotes* DTU-COMPUTE.

Brockwell P, Davis R (1991) *Time Series: Theory and Methods* Springer, New York, 2 edition.

Brooks ME, Kristensen K, van Benthem KJ, Magnusson A, Berg CW, Nielsen A, Skaug HJ, Maechler M, Bolker BM (2017) glmmTMB balances speed and flexibility among packages for zero-inflated generalized linear mixed modeling. *The R Journal* 9:378–400.

Calì D, Osterhage T, Streblow R, Müller D (2016) Energy performance gap in refurbished german dwellings: Lesson learned from a field test. *Energy and buildings* 127:1146–1158.

Cappé O, Moulines E, Rydén T (2005) *Inference in Hidden Markob Models* Springer, New York.

Christensen RHB (2022) `ordinal`—regression models for ordinal data R package version 2022.11-16. https://CRAN.R-project.org/package=ordinal.

De Carli M, Olesen BW, Zarrella A, Zecchin R (2007) People's clothing behaviour according to external weather and indoor environment. *Building and Environment* 42:3965–3973.

Demidenko E (2004) *Mixed Models - Theory and Applications* John Wiley & Sons Inc., New Jersey.

Dobson J A, Barnett G A (2008) *An introduction to generalized linear models* CRC Press, Taylor & Francis Group, 3 edition.

Dowle M, Srinivasan A (2020) *data.table: Extension of 'data.frame'* R package version 1.13.2.

Durbin J, Koopmann SJ (2001) *Time series analysis by state space methods* Oxford University Press Inc, New York.

Fabi V, Andersen RV, Corgnati SP, Olesen BW (2013) A methodology for modelling energy-related human behaviour: Application to window opening behaviour in residential buildings. *Building Simulation* 6:415–427.

Fanger PO (1970) *Thermal Comfort Analysis and Applications in Environmental Engineering* McGraw-Hill, New York.

Favero M, Kloppenborg Møller J, Calì D, Carlucci S (2022) Human-in-the-loop methods for occupant-centric building design and operation. *Applied Energy* 325:1–21.

Ferrari SL, Cribari-Neto F (2004) Beta regression for modelling rates and proportions. *Journal of Applied Statistics* 31:799–815.

Fox J, Weisberg S (2019) *An R Companion to Applied Regression* Sage, Thousand Oaks CA, third edition.

Genz A, Bretz F (2009) *Computation of Multivariate Normal and t Probabilities* Lecture Notes in Statistics. Springer-Verlag, Heidelberg.

Genz A, Bretz F, Miwa T, Mi X, Leisch F, Scheipl F, Hothorn T (2021) *mvtnorm: Multivariate Normal and t Distributions* R package version 1.1-3.

Gilbert P, Varadhan R (2019) numDeriv: Accurate Numerical Derivatives R package version 2016.8-1.1, https://cran.r-project.org/package=numDeriv.

Givon D, Stinis P, Weare J (2009) Variance reduction for particle filters of systems with time scale separation. *Transactions on Signal Processing* 57(2):424–435.

Gram-Hanssen K (2010) Residential heat comfort practices: understanding users. *Building Research & Information* 38:175–186.

Grimmit GR, Stirzaker DR (2001) *Probability and Random Processes, third edition* Oxford University Press, New York.

Grün B, Leisch F (2012) Extended beta regression in r: Shaken, stirred, mixed, and partitioned. *Journal of Statistical Software* 48:1–25.

Gunay HB, O'Brien W, Beausoleil-Morrison I, Gilani S (2016) Modeling plug-in equipment load patterns in private office spaces. *Energy and Buildings* 121:234–249.

Gunay HB, O'Brien W, Beausoleil-Morrison I, Gilani S (2017) Development and implementation of an adaptive lighting and blinds control algorithm. *Building and Environment* 113:185–199.

Haldi F (2010) Towards a unified model of occupants' behaviour and comfort for building energy simulation Technical report, EPFL.

Haldi F, Calì D, Andersen RK, Wesseling M, Müller D (2017) Modelling diversity in building occupant behaviour: a novel statistical approach. *Journal of Building Performance Simulation* 10:527–544.

Haldi F, Robinson D (2008) On the behaviour and adaptation of office occupants. *Building and environment* 43:2163–2177.

Haldi F, Robinson D (2009) Interactions with window openings by office occupants. *Building and Environment* 44:2378–2395.

Haldi F, Robinson D (2010) Adaptive actions on shading devices in response to local visual stimuli. *Journal of Building Performance Simulation* 3:135–153.

Haldi F, Robinson D (2011) The impact of occupants' behaviour on building energy demand. *Journal of Building Performance Simulation* 4:323–338.

Hamilton J D (1994) *Time Series Analysis* Princeton University Press, Princeton.

Hastie T, Tibshirani R, Friedman J (2001) *The Elements of Statistical Learning; Data Mining, Inference and Prediction* Springer, New York.

Helske J (2017) KFAS: Exponential family state space models in R. *Journal of Statistical Software* 78:1–39.

Hosmer D, Lemeshow S, May S (2008) *Applied survival analysis, regression modelling of time-to-event data* Wiley-Interscience, New Jersey.

Hothorn T, M L, Bühlmann P (2018) Most likely transformation. *Scandinavian Journal of Statistics* 45:110–134.

Hothorn T (2022) Transformation models: The tram package. *R package vignette version 0.7-2* .

Hunt D (1979) The use of artificial lighting in relation to daylight levels and occupancy. *Building and Environment* 14:21–33.

Hyndman R, Athanasopoulos G, Bergmeir C, Caceres G, Chhay L, O'Hara-Wild M, Petropoulos F, Razbash S, Wang E, Yasmeen F (2023) *forecast: Forecasting functions for time series and linear models* R package version 8.21.

Hyndman RJ, Khandakar Y (2008) Automatic time series forecasting: the forecast package for R. *Journal of Statistical Software* 26:1–22.

Jackson CH (2011) Multi-state models for panel data: The msm package for r. *Journal of Statistical Software* 38(8):1–28.

Jazwinski AH (1970) *Stochastic Processes and Filtering Theory* Academic Press, New York.

Juhl R, Møller J, Jørgensen J, Madsen H (2016) Modeling and prediction using stochastic differential equations In Kirchsteiger H, Bagterp Jørgensen J, Renard E, del Re L, editors, *Prediction Methods for Blood Glucose Concentration*, Lecture Notes in Bioengineering, pp. 183–209. Springer.

Kalman RE (1960) A new approach to linear filtering and prediction problems. *Trans. ASME Journal of Basic Engineering* 82:35–45.

Karjalainen S (2012) Thermal comfort and gender: a literature review. *Indoor air* 22:96–109.

Kim J, Zhou Y, Schiavon S, Raftery P, Brager G (2018) Personal comfort models: Predicting individuals' thermal preference using occupant heating and cooling behavior and machine learning. *Building and Environment* 129:96–106.

Kristensen K, Nielsen A, Berg C, Skaug H, Bell B (2016) TMB: Automatic differentiation and laplace approximation. *Journal of Statistical Software* 70:1–21.

Lassen N, Møller J, Goia F (2021) Information match between continuous occupant data streams and one-time manual surveys on indoor climate. *Building and Environment* 204:1–19.

Lattin J, Carroll J, Green P (2003) *Analyzing multivariate data* Duxbury Applied Series. Thomson Brooks/Cole, Pacific Grove, CA.

Laussmann D, Helm D (2011) Air change measurements using tracer gases In Mazzeo NA, editor, *Chemistry, Emission Control, Radioactive Pollution and Indoor Air Quality*, pp. 365–406. IintchOpen, Rijeka.

Lee S, Karava P, Tzempelikos A, Bilionis I (2019) Inference of thermal preference profiles for personalized thermal environments with actual building occupants. *Building and Environment* 148:714–729.

Lee Y, Nelder J (1996) Hierarchical generalized linear models. *Journal of the Royal Statistical Society* B,58:619–678.

Lee Y, Nelder J (2001) Hierarchical generalized linear models: a synthesis of generalised linear models, random-effect models and structured dispersions. *Biometrika* 88:987–1006.

Lehmann E, Casella G (1998) *Theory of Point Estimators* Springer, New York.

Liisberg J, Møller J, Bloem H, Cipriano J, Mor G, Madsen H (2016) Hidden markov models for indirect classification of occupant behaviour. *Sustainable Cities and Society* 27:83–98.

Love JA (1998) Manual switching patterns in private offices. *International Journal of Lighting Research and Technology* 30:45–50.

Madsen H (2008) *Time Series Analysis* Chapman and Hall, New York.

Madsen H, Thyregod P (2011) *Introduction to General and Generalized Linear Models* Texts in Statistical Science. CRC Press, Boca Raton.

Maechler M, Dutang C, Goulet V (2023) *expm: Matrix Exponential, Log, 'etc'* R package version 0.999-7.

Mahdavi A, Berger C, Amin H, Ampatzi E, Andersen R, Azar E, Barthelmes V, Favero M, Hahn J, Khovalyg D, Knudsen H, Luna-Navarro A, Roetzel A, Sangogboye F, Schweiker M, Taheri M, Teli D, Touchie M, Verbruggen S (2021) The role of occupants in buildings' energy performance gap: Myth or reality? *Sustainability* 13:1–44.

Mahdavi A, Mohammadi A, Kabir E, Lambeva L (2008) Occupants' operation of lighting and shading systems in office buildings. *Journal of Building Performance Simulation* 1:57–65.

Maniccia D, Rutledge B, Rea MS, Morrow W (1999) Occupant use of manual lighting controls in private offices. *Journal of the Illuminating Engineering Society* 28:42–56.

Markovic R, Grintal E, Wölki D, Frisch J, van Treeck C (2018) Window opening model using deep learning methods. *Building and Environment* 145:319–329.

Maybeck PS (1982) *Stochastic Models, Estimation and Control; Vol 1,2,3* Academic Press, New York.

Meinhold RJ, Singpurwalla ND (1983) Understanding the kalman filter. *American Statistician* 70:123–127.

Møller J, Madsen H (2010) *From State Dependent Diffusion to Constant Diffusion in Stochastic Differential Equations by the Lamperti Transform* IMM-Technical Report-2010-16. Technical University of Denmark, DTU Informatics, Building 321.

Newsham G (1994) Manual control of window blinds and electric lighting: implications for comfort and energy consumption. *Indoor Environment* 3:135–144.

Nicol JF (2001) Characterising occupant behaviour in buildings: towards a stochastic model of occupant use of windows, lights, blinds, heaters and fans In *Proceedings of the Seventh International IBPSA Conference, Rio*, Vol. 2, pp. 1073–1078.

O'Brien W, Gunay HB, Tahmasebi F, Mahdavi A (2017) A preliminary study of representing the inter-occupant diversity in occupant modelling. *Journal of Building Performance Simulation* 10:509–526.

O'Brien W, Kapsis K, Athienitis AK (2013) Manually-operated window shade patterns in office buildings: A critical review. *Building and Environment* 60:319–338.

O'Brien W, Wagner A, Schweiker M, Mahdavi A, Day J, Kjærgaard MB, Carlucci S, Dong B, Tahmasebi F, Yan D et al. (2020) Introducing iea ebc annex 79: Key challenges and opportunities in the field of occupant-centric building design and operation. *Building and Environment* 178:106738.

Page J, Robinson D, Morel N, Scartezzini JL (2008) A generalised stochastic model for the simulation of occupant presence. *Energy and Buildings* 40:83–98.

Parys W, Saelens D, Hens H (2011) Coupling of dynamic building simulation with stochastic modelling of occupant behaviour in offices – a review-based integrated methodology. *Journal of Building Performance Simulation* 4:339–358.

Pawitan Y (2013) *In All Likelihood – Statistical Modelling and Inference using Likelihood* Oxford University Press, Oxford.

Petersen KB, Pedersen MS (2012) *The Matrix Cookbook* Technical University of Denmark, Lyngby Version 20121115.

Pigg S, Eilers M, Reed J (1996) Behavioral aspects of lighting and occupancy sensors in private offices: a case study of a university office building In *Summer Study on Energy Efficiency in Buildings.* American Council for an Energy-Efficient Economy, Washington.

R Core Team (2022) *R: A Language and Environment for Statistical Computing* R Foundation for Statistical Computing, Vienna, Austria.

Rea MS (1984) Window blind occlusion: a pilot study. *Building and Environment* 19:133–137.

Reinhart CF (2004) Lightswitch-2002: a model for manual and automated control of electric lighting and blinds. *Solar Energy* 77:15–28.

Rijal HB, Humphreys MA, Nicol JF (2009) Understanding occupant behaviour: the use of controls in mixed-mode office buildings. *Building Research & Information* 37:381–396.

Rijal HB, Tuohy P, Nicol F, Humphreys MA, Samuel A, Clarke J (2008) Development of an adaptive window-opening algorithm to predict the thermal comfort, energy use and overheating in buildings. *Journal of Building Performance Simulation* 1:17–30.

Ronnegard L, Shen X, Alam M (2010) hglm: A package for fitting hierarchical generalized linear models. *The R Journal* 2:20–28.

Schiavon S, Lee KH (2013) Dynamic predictive clothing insulation models based on outdoor air and indoor operative temperatures. *Building and Environment* 59:250–260.

Schweiker M, Ampatzi E, Andargie MS, Andersen RK, Azar E, Barthelmes VM, Berger C, Bourikas L, Carlucci S, Chinazzo G et al. (2020) Review of multi-domain approaches to indoor environmental perception and behaviour. *Building and Environment* 176:106804.

Schweiker M, Brasche S, Hawighorst M, Bischof W, Wagner A (2014) Presenting LOBSTER, an innovative climate chamber, and the analysis of the effect of a ceiling fan on the thermal sensation and performance under summer conditions in an office-like setting In *Proceedings of 8th Windsor Conference: Counting the Cost of Comfort in a changing world Cumberland Lodge, Windsor, UK*, pp. 924 – 937.

Schweiker M, Shukuya M (2009) Comparison of theoretical and statistical models of air-conditioning-unit usage behaviour in a residential setting under Japanese climatic conditions. *Building and Environment* 44:2137–2149.

Schweiker M, Wagner A (2015) A framework for an adaptive thermal heat balance model (ATHB). *Building and Environment* 94:252–262.

Shumway RH (1988) *Applied Statistical Time Series Analysis* Prentice Hall, Englewood Cliffs, NJ.

Simon D (2009) *Optimal State Estimation – Kalman, H_∞, and Non-linear Approaches* John Wiley & Sons, New Jersey.

Spedicato GA (2017) Discrete time markov chains with R. *The R Journal* 9:84–104.

Spliid H (2017) marima: Multivariate arima and arima-x analysis. *R-package* .

Tamási B, Hothorn T (2021) tramME: Mixed-effects transformation models using template model builder. *The R-Journal* 13:398–418.

Therneau T, Lumbrey T, Atkinson E, Crowson C (2022) `survival`: Survival analysis. *R-package version 3-4-0* .

Thygesen UH (2023) *Stochastic Differential Equations for Science and Engineering* CRC Press/Taylor and Francis, Boca Raton.

Venables WN, Ripley BD (2002) *Modern Applied Statistics with S* Springer, New York, fourth edition ISBN 0-387-95457-0.

Visser I, Speekenbrink M (2010) depmixS4: An R package for hidden markov models. *Journal of Statistical Software* 36:1–21.

Wang D, Federspiel CC, Rubinstein F (2005) Modeling occupancy in single person offices. *Energy and buildings* 37:121–126.

Wickham H (2007) Reshaping data with the reshape package. *Journal of Statistical Software* 21:1–20.

Wickham H (2011) The split-apply-combine strategy for data analysis. *Journal of Statistical Software* 40:1–29.

Wickham H (2016) *ggplot2: Elegant Graphics for Data Analysis* Springer-Verlag, New York.

Wolf S, Cali D, Krogstie J, Madsen H (2019a) A markov-switching model for building occupant activity estimation. *Applied Energy* 236:32–41.

Wolf S, Maria Justo A, Cali D, Krogstie J, Mathisen H, Madsen H (2019b) Co2-based grey-box model to estimate airflow rate and room occupancy In *Clima 2019: 13th REHVA World Congress*, Vol. 111. E D P Sciences, Bucharest, Romania.

Wolf S, Møller J, Bitsch M, Krogstie J, Madsen H (2019c) A markov-switching model for building occupant activity estimation. *Energy and Buildings* 183:672–683.

Wolfinger RD, Lin X (1997) Two taylor-series approximation methods for nonlinear mixed models. *Computational Statistics & Data Analysis* 25:465–490.

Wuertz D, Setz T, Chalabi Y, Maechler M, Boshnakov GN (2022) timeSeries: Financial Time Series Objects (Rmetrics) R package version 4021.105 https://CRAN.R-project.org/package=timeSeries.

Xie Y (2014) knitr: A comprehensive tool for reproducible research in R In Stodden V, Leisch F, Peng RD, editors, *Implementing Reproducible Computational Research.* Chapman and Hall/CRC ISBN 978-1466561595.

Xie Y (2015) *Dynamic Documents with R and knitr* Chapman and Hall/CRC, Boca Raton, Florida, 2nd edition ISBN 978-1498716963.

Xie Y (2020) *knitr: A General-Purpose Package for Dynamic Report Generation in R* R package version 1.30.

Yan D, Hong T, Dong B, Mahdavi A, D'Oca S, Gaetani I, Feng X (2017) IEA EBC Annex 66: Definition and simulation of occupant behavior in buildings. *Energy and Buildings* 156:258–270.

Yan D, O'Brien W, Hong T, Feng X, Gunay HB, Tahmasebi F, Mahdavi A (2015) Occupant behavior modeling for building performance simulation: Current state and future challenges. *Energy and Buildings* 107:264–278.

Yoshino H, Hong T, Nord N (2017) IEA EBC Annex 53: Total energy use in buildings-Analysis and evaluation methods. *Energy and Buildings* 152:124–136.

Yun GY, Steemers K (2008) Time-dependent occupant behaviour models of window control in summer. *Building and Environment* 43:1471–1482.

Zucchini W, MacDonald I (2009) *Hidden Markov Models for Time Series, an introduction Using R* Taylor and Francis Group, CRC-Press, Boca Raton.

Index